BIOAVAILABILITY in ENVIRONMENTAL RISK ASSESSMENT

Steve E. Hrudey
Weiping Chen
Colin G. Rousseaux

BIOAVAILABILITY in ENVIRONMENTAL RISK ASSESSMENT

CRC LEWIS PUBLISHERS
Boca Raton New York London Tokyo

Library of Congress Cataloging-in-Publication Data

Hrudey, S. E. (Steve E.)
Bioavailability in environmental risk assessment / S.E. Hrudey, W. Chen, C.G. Rousseaux.
p. cm.
Includes bibliographical references and index.
ISBN 1-56670-186-4
1. Health risk assessment. 2. Bioavailabilty. 3. Hazardous wastes--Risk assessment. 4. Polution--Risk assessment. 5. Pollutants--Health aspects. I. Chen, W. (Weiping) II. Rousseaux, Colin George, 1954- . III. Title.
RA566.27.H77 1995
615.9'0--dc20 95-30402
CIP

Direct all inquiries to CRC Press, Inc., 2000 Corporate Blvd., N.W., Boca Raton, Florida 33431.

Lewis Publishers is an imprint of CRC Press

International Standard Book Number 1-56670-186-4
Library of Congress Card Number 95-30402
Printed in the United States of America 1 2 3 4 5 6 7 8 9 0
Printed on acid-free paper

PREFACE

The legacy of mismanaged hazardous waste became imprinted upon the public conscience with the experience at Love Canal and the subsequent discovery of hundreds of other inadequate waste disposal sites located in proximity to residential dwellings during the late 1970s and early 1980s. These situations often created extreme anxiety among residents who were concerned that they may have been exposed to serious threats to their health. Regulatory agencies reacted to these tragedies with a variety of actions including: programs to evacuate residents believed to be at risk, measures to isolate or remove offending waste and legal actions to recover costs from any parties believed to be responsible for creating the problem. Most parties involved in these actions suffered at least some trauma and substantial costs were incurred in the name of public health protection. As the number of potential problem sites mushroomed and the projected costs of site clean-up have escalated, expectations of environmental restoration to "clean" conditions have confronted several hard economic and technical realities. These circumstances led to the evolution and development of risk assessment as a means to identify needs and options for managing health and environmental risks. The process of risk assessment is briefly outlined in Chapter 1.

Health risk assessment applied to these environmental exposures has suffered from a lack of data which is needed as a foundation for several critical aspects of the evaluation. The basic risk assessment protocols have recognized the importance of exposure assessment in applying knowledge about the dose-response characteristics of toxic substances to the characterization of those risks posed by environmental contaminants. However, that characterization of contaminant exposures has been greatly simplified to overcome deficiencies in data and knowledge. These simplifications have usually favored very cautious assumptions towards upper bound estimates of risk. Uncertainty has generally been resolved by erring on the side of caution. These simplifications have been justified as expedient measures necessary to protect the public's health, particularly when the parties have been involuntarily exposed of contaminants arising from hazardous waste mismanagement by both industry and governments. However, once the most obvious cases of preventable contaminant exposures have been resolved, risk management for the majority of contaminated sites demands more difficult technical, economic and social decisions because lower or less direct contaminant exposures are involved which will demand increasing incremental costs to achieve decreasing incremental risk reduction.

The concept of bioavailability (outlined in Chapter 2) provides some potential for improving risk assessment. Lack of knowledge about bioavailability was resolved easily in the early applications of risk assessment for managing health risks from contaminated sites. The premise of conservatism (erring on the side of caution) was satisfied by assuming that all contaminants from environmental exposures would be completely absorbed into the bloodstream, i.e. 100% bioavailability. For cases where the dose response data used for characterizing a risk was obtained by testing similar chemical forms of a contaminant by similar exposure routes to those being assessed, these simplifying assumptions would not likely cause serious overestimates of risk. In such cases, these similarities would internally compensate for the lack of knowledge about the bioavailability. However, toxicology data is limited for many contaminants which might be assessed at contaminated sites. Risk assessments are likely to encounter situations where assuming 100% bioavailability of contaminants for the locally relevant exposures, which may differ in route and chemical form from those used experimentally, may cause serious overestimates of the relevant health risk. Faced with increasingly difficult choices about the value of investment in remediation of contaminated sites, the negative consequences of overestimating health risks for individual contaminants are becoming increasingly important. The possibility of ruling out viable and effective risk management options is as serious a consequence of unjustifiably pessimistic risk estimates as is the commitment of substantial resources to achieve no substantial reduction of risk. Of course, methodological errors in dose predictions must usually be substantial (e.g. more than a factor of 2) to be consistently important in risk estimates relative to all the uncertainties which are inherent in risk assessment of contaminated sites. This perspective on the relative importance of scale should be borne in mind when considering the variability in reported bioavailability factors which are summarized for various contaminants in Chapter 5.

The foregoing circumstances provide the context for this book. Bioavailability is an important concept which is unavoidable as we seek a more realistic and accurate understanding of the health risks posed by exposures from contaminated sites. The foundations of bioavailability, like those of toxicology, have been initiated and developed by pharmacology because of the importance of bioavailability to the behavior and efficiency of drug delivery. We clearly acknowledge that bioavailability is only one specific element of the more comprehensive framework of physiologically-based pharmacokinetic (PB-PK) modeling, which will yield better understanding and contribute to better prediction of health risks from environmental exposures to contaminants. However, bioavailability warrants full consideration as the health risk assessment process matures from assuming 100% bioavailability of all contaminants by all exposure routes towards fully developed and verified PB-PK models for all relevant contaminants in all relevant chemical forms and exposure circumstances. The latter goal, as desirable as it may appear, will not be achieved for some time. Consequently, a careful consideration of bioavailability offers a realistic approach during the transition while we seek to make health risk predictions more accurate and reliable for justifying increasingly expensive contaminated site remediation measures.

The inspiration for this book can be credited to the HELP contaminated site identification and remediation program of the province of Alberta. This program, a pioneer and leader among risk management programs in Canada, first identified a number of industrial sites which might pose a risk to human health or the environment but which were no longer under the control or management of those parties responsible for the contamination. We undertook a number of research contracts for the Alberta Department of Environmental Protection under the HELP program. These studies were aimed at characterizing aspects of the health risks posed by a number of the HELP sites. The nature of the contaminants commonly encountered at these former industrial sites influenced the selection of contaminants ultimately considered in this book. The foregoing research activities were also directly supported by a University Research Grant from Imperial Oil Ltd. and by the infrastructure support provided by donations from Canadian National Railways, Shell Canada Ltd., Weldwood Canada Ltd. and Gulf Canada Resources Ltd. which funded creation of the Environmental Health Program in the Faculty of Medicine in 1989.

This book project was initiated as a Masters thesis in Environmental Health by one of us (W.C.) and was expanded and developed into this final form as part of the interdisciplinary research program of the *Eco-Research* Chair in Environmental Risk Management currently held by one of us (S.E.H.). The latter is funded by an *Eco-Research* grant from the Tri-Council Secretariat representing the Medical Research Council, the Natural Sciences and Engineering Research Council and the Social Sciences and Humanities Research Council of Canada and by sponsorship from the Alberta Heritage Foundation for Medical Research, Alberta Environmental Protection, Alberta Health, the City of Edmonton, Enviro•Test Laboratories Ltd., Syncrude Canada Ltd., the Wastewater Technology Centre, the Energy Resources Conservation Board (now Alberta Energy and Utilities Board), Acres International Ltd., Bovar Inc., Brownlee-Fryett, CanTox Inc., EBA Engineering Consultants Ltd., Golder Associates Ltd., Gulf Canada Resources Ltd., Agra Earth and Environmental Ltd., Home Oil Company Ltd., Milner Fenerty, O'Connor Associates Environmental Inc., Reid Crowther and Partners Ltd., Stanley Technology Group, the Wyatt Company and UMA Engineering Ltd.

The support and guidance provided toward this work by Dr. Bruce Taylor and Wayne Inkpen of Alberta Environmental Protection, Dr. Warren Kindzierski of Alberta Health and by Alan Hingston and Geraint Edmunds of the HELP project is gratefully acknowledged. Finally, we appreciate the technical assistance with computer graphics and formatting provided by Qishi Chen, the word processing assistance by Nola Low and the careful and thorough proofreading of various drafts of our manuscript by Dawn Torsein and Elizabeth Hrudey.

Steve E. Hrudey *Weiping Chen* *Colin G. Rousseaux*

BIOGRAPHY

Steve E. Hrudey holds a Ph.D. and M.Sc. in Public Health Engineering from the Imperial College of Science, Technology and Medicine, University of London, England and a B.Sc. in Mechanical Engineering from the University of Alberta, Canada. Dr. Hrudey is Professor of Environmental Health and currently holds the Eco-Research Chair in Environmental Risk Management in the Department of Public Health Sciences, Faculty of Medicine, University of Alberta. He teaches courses in risk assessment and risk management of environmental contaminants. Dr. Hrudey has conducted interdisciplinary research on managing health risks from: contaminated industrial sites, hazardous wastes and both natural and industrial hazards to drinking water supplies. He has advised industry and governments on environmental risk management in Canada, the U.S.A., the U.K., Australia and New Zealand.

Weiping Chen holds a M.Sc. in Environmental Health from the University of Alberta and a B.M. (Bachelor of Medicine) from the Zhejiang Medical University, Hangzhou, China. She practiced at the Second Affiliated Hospital of the Zhejiang Medical University with residencies in surgery and internal medicine and practice in gastroenterology. Ms. Chen is a research associate with the Eco-Research Chair in Environmental Risk Management. Her research has focused on the bioavailability of environmental contaminants on human health studies conducted at hazardous waste sites and on environmental risk education for physicians.

Colin Rousseaux holds a B.V.Sc. (Bachelor of Veterinary Science) from the University of Melbourne, Australia and a Ph.D. in toxicologic pathology from the University of Saskatchewan, Canada. Dr. Rousseaux is a Diplomate of the American Board of Toxicologists, a Member of the Royal College of Veterinary Science (London) and the Royal College of Pathologists. He is a principal in **GLOBAL TOX** International Consultants Inc., an adjunct professor in the Department of Public Health Sciences, Faculty of Medicine, University of Alberta and a professeur associé at the Université du Quebec à Montréal. Dr. Rousseaux is the co-editor of the "Handbook of Toxicologic Pathology" and he maintains an active research program in environmental pathology and reproductive toxicologic pathology.

TABLE OF CONTENTS

Chapter 1
INTRODUCTION 1

Chapter 2
FUNDAMENTAL CONCEPT OF BIOAVAILABILITY 7
I. DEFINITION OF BIOAVAILABILITY 7
A. UNALTERED FORM 7
B. SYSTEMIC CIRCULATION 8
C. RELATIONSHIP BETWEEN EXTENT AND RATE OF BIOAVAILABILITY 8
D. EXTENT OF BIOAVAILABILITY 9
1. Absolute bioavailability 10
2. Relative bioavailability 11
II. BIOAVAILABILITY IN HEALTH RISK ASSESSMENT 12
A. BIOLOGICALLY EFFECTIVE DOSE 13
B. INTERNAL DOSE AND AVERAGE DAILY INTERNAL DOSE 13
C. EXTERNAL DOSE AND AVERAGE DAILY EXTERNAL DOSE 16
III. SUMMARY 17

Chapter 3
MECHANISM OF ABSORPTION 19
I. RESPIRATORY SYSTEM 19
A. STRUCTURE AND FUNCTION 19
B. RATE-LIMITING PROCESSES FOR GASEOUS PHASE CONTAMINANTS 22
C. OTHER FACTORS INFLUENCING RESPIRATORY BIOAVAILABILITY FOR GASES AND VAPORS 24
D. RATE-LIMITING PROCESSES FOR PARTICULATE PHASE CONTAMINANTS 26
1. Deposition 28
2. Clearance 33
a. Water Insoluble Aerosols 33
b. Water Soluble Aerosols 34
3. Retention 34
E. OTHER FACTORS INFLUENCING RESPIRATORY BIOAVAILABILITY FOR AEROSOLS 35
II. GASTROINTESTINAL SYSTEM 36
A. STRUCTURE AND FUNCTION 36
B. RATE-LIMITING PROCESSES 38
C. OTHER FACTORS INFLUENCING ORAL BIOAVAILABILITY 41

III. SKIN.. 44
A. STRUCTURE AND FUNCTION........................ 44
B. RATE-LIMITING PROCESSES........................ 45
C. OTHER FACTORS INFLUENCING
DERMAL BIOAVAILABILITY........................ 47
IV. SUMMARY.. 49

Chapter 4
EXPERIMENTAL METHODS FOR
EVALUATING BIOAVAILABILITY........................ 51
I. GENERAL *IN VIVO* METHODS........................ 51
A. MEASUREMENT OF BLOOD LEVELS........................ 51
B. MEASUREMENT OF THE CUMULATIVE
URINARY EXCRETION........................ 53
C. MASS-BALANCE TECHNIQUE........................ 54
D. MEASUREMENT OF TOXIC RESPONSE........................ 56
II. SPECIFIC METHODS FOR THE
INHALATION EXPOSURE ROUTE........................ 56
A. DETERMINATION OF DEPOSITION FRACTION........................ 57
1. Gaseous Phase........................ 57
2. Particulate Phase........................ 59
B. MECHANICAL CLEARANCE FRACTION........................ 59
C. DISSOLUTION RATE........................ 60
D. *IN VITRO* METHODS: ISOLATED PERFUSED
LUNG........................ 61
III. SPECIFIC METHODS FOR THE
INGESTION EXPOSURE ROUTE........................ 62
A. MEASUREMENT OF FECAL EXCRETION........................ 64
B. MEASUREMENT OF LIVER RATIO........................ 65
C. DISSOLUTION TESTING........................ 65
D. METHODS BASED ON THE pH PARTITION
THEORY........................ 66
IV. SPECIFIC METHODS FOR
DERMAL EXPOSURE ROUTE........................ 66
A. SURFACE DISAPPEARANCE........................ 67
B. *IN VITRO* METHODS: BIOLOGICAL
MEMBRANE MODELS........................ 67
C. *IN VITRO* METHODS: ARTIFICIAL MEMBRANE
MODELS........................ 69
V. SUMMARY........................ 70

Chapter 5
EXPOSURE ROUTES AND BIOAVAILABILITY
FACTORS FOR SELECTED CONTAMINANTS........................ 75
I. ARSENIC........................ 75

A. EXPOSURE ROUTES AND RECEPTORS 76
B. TOXICOKINETICS (FOLLOWING ABSORPTION) 77
1. Distribution 77
2. Metabolism and Excretion 77
C. BIOAVAILABILITY 78
1. Inhalation 78
a. Humans 78
b. Animals 78
2. Ingestion 80
a. Humans 80
b. Animals 81
3. Dermal 82
4. Summary 82
D. UNCERTAINTIES 82
E. CONCLUSIONS 89
F. RECOMMENDATIONS 90
II. CADMIUM 91
A. EXPOSURE ROUTES AND RECEPTORS 91
B. TOXICOKINETICS (FOLLOWING ABSORPTION) 92
1. Distribution 92
2. Metabolism and Excretion 92
C. BIOAVAILABILITY 92
1. Inhalation 92
2. Ingestion 94
3. Dermal 95
4. Summary 95
D. UNCERTAINTIES 95
E. CONCLUSIONS 100
F. RECOMMENDATIONS 100
III. CHROMIUM AND CHROMIUM COMPOUNDS 101
A. EXPOSURE ROUTES AND RECEPTORS 102
B. TOXICOKINETICS (FOLLOWING ABSORPTION) 103
1. Distribution 103
2. Metabolism and Excretion 103
C. BIOAVAILABILITY 103
1. Inhalation 103
a. Humans 103
b. Animals 104
2. Ingestion 105
a. Humans 105
b. Animals 106
c. Others 106
3. Dermal 106
4. Summary 107
D. UNCERTAINTIES 107

E. CONCLUSIONS 115
F. RECOMMENDATIONS 116
IV. COPPER 117
A. EXPOSURE ROUTES AND RECEPTORS 117
B. TOXICOKINETICS (FOLLOWING ABSORPTION) 118
1. Distribution 118
2. Metabolism and Excretion 118
C. BIOAVAILABILITY 119
1. Inhalation 119
2. Ingestion 119
3. Dermal 120
D. UNCERTAINTIES 120
E. CONCLUSIONS 121
F. RECOMMENDATIONS 124
V. MERCURY 125
A. EXPOSURE ROUTES AND RECEPTORS 125
B. TOXICOKINETICS (FOLLOWING ABSORPTION) 126
1. Distribution 126
2. Metabolism and Excretion 126
C. BIOAVAILABILITY 127
1. Inhalation 127
2. Ingestion 127
3. Dermal 128
4. Summary 128
D. UNCERTAINTIES 129
E. CONCLUSIONS 129
F. RECOMMENDATIONS 133
VI. LEAD 134
A. EXPOSURE ROUTES AND RECEPTORS 134
B. TOXICOKINETICS (FOLLOWING ABSORPTION) 135
1. Distribution 135
2. Metabolism and Excretion 136
C. BIOAVAILABILITY 136
1. Inhalation 136
2. Ingestion 137
3. Dermal 138
4. Summary 139
D. UNCERTAINTIES 139
E. CONCLUSIONS 144
F. RECOMMENDATIONS 144
VII. 2,3,7,8-TETRACHLORODIBENZO-p-DIOXIN (TCDD) 145
A. EXPOSURE ROUTES AND RECEPTORS 146
B. TOXICOKINETICS (FOLLOWING ABSORPTION) 146
1. Distribution 146
2. Metabolism and Excretion 146

C. BIOAVAILABILITY 147
1. Inhalation 147
2. Ingestion 147
3. Dermal 148
4. Summary 149
D. UNCERTAINTIES 149
E. CONCLUSIONS 154
F. RECOMMENDATIONS 154
VIII. POLYNUCLEAR AROMATIC HYDROCARBONS (PAHs) 155
A. EXPOSURE ROUTES AND RECEPTORS 156
B. TOXICOKINETICS (FOLLOWING ABSORPTION) 157
1. Distribution 157
2. Metabolism and Excretion 157
C. BIOAVAILABILITY 158
1. Inhalation 158
a. Humans 158
b. Animals 158
2. Ingestion 159
a. Humans 159
b. Animals 159
3. Dermal 159
a. Humans 159
b. Animals 160
4. Summary 162
D. UNCERTAINTIES 162
E. CONCLUSIONS 172
F. RECOMMENDATIONS 173
IX. BENZENE 174
A. EXPOSURE ROUTES AND RECEPTORS 174
B. TOXICOKINETICS (FOLLOWING ABSORPTION) 175
1. Distribution 175
2. Metabolism 175
3. Excretion 176
C. BIOAVAILABILITY 176
1. Inhalation 176
a. Humans 176
b. Animals 176
2. Ingestion 177
3. Dermal 177
a. Humans 177
b. Animals 177
4. Summary 178
E. UNCERTAINTIES 178
E. CONCLUSIONS 179
F. RECOMMENDATIONS 183

X. TOLUENE... 184
A. EXPOSURE ROUTES AND RECEPTORS 184
B. TOXICOKINETICS (FOLLOWING ABSORPTION) 184
1. Distribution 184
2. Metabolism and Excretion 185
C. BIOAVAILABILITY 185
1. Inhalation 185
2. Ingestion 186
a. Humans 186
b. Animals 186
3. Dermal 186
a. Humans 186
b. Animals 187
4. Summary 187
D. UNCERTAINTIES 191
E. CONCLUSIONS 191
F. RECOMMENDATIONS 192
XI. POLYCHLORINATED PHENOLS: PENTACHLOROPHENOL, 2,4,5-TRICHLOROPHENOL, 2,4,6-TRICHLOROPHENOL, 2,3,4,5-TETRACHLOROPHENOL... 193
A. EXPOSURE ROUTES AND RECEPTORS 194
B. TOXICOKINETICS (FOLLOWING ABSORPTION) 194
1. Distribution 194
2. Metabolism 194
C. BIOAVAILABILITY 195
1. Inhalation 195
2. Ingestion 195
a. Humans 195
b. Animals 196
3. Dermal 196
4. Summary 196
D. UNCERTAINTIES 197
E. CONCLUSIONS 201
F. RECOMMENDATIONS 202
XII. PHENOL 203
A. EXPOSURE ROUTES AND RECEPTORS 203
B. TOXICOKINETICS (FOLLOWING ABSORPTION) 203
1. Distribution 203
2. Metabolism 203
C. BIOAVAILABILITY 204
1. Inhalation 204
2. Ingestion 204
3. Dermal 204

4. Summary 205
D. UNCERTAINTIES 205
E. CONCLUSIONS 209
F. RECOMMENDATIONS 209
XIII. ANILINE 210
A. EXPOSURE ROUTES AND RECEPTORS 210
B. TOXICOKINETICS (FOLLOWING ABSORPTION) 210
1. Distribution 210
2. Metabolism and Excretion 210
C. BIOAVAILABILITY 211
1. Inhalation 211
2. Ingestion 211
3. Dermal 211
4. Summary 211
D. UNCERTAINTIES 211
E. CONCLUSIONS 213
F. RECOMMENDATIONS 213

Chapter 6
BIOAVAILABILITY OF XENOBIOTICS
FROM CONTAMINATED SITES 215
I. IDENTIFICATION OF EXPOSURE ROUTES 215
II. RELIABILITY AND VALIDITY OF
THE BIOAVAILABILITY FACTORS 216
A. CHOICE OF EXPERIMENTAL METHODS 216
B. ANALYTICAL CAPABILITIES 222
C. SPECIES DIFFERENCES 222
D. MIXTURES AND VEHICLES 222
E. MECHANISM OF TOXICITY 222
F. CHEMICAL FORMS 223
G. EXPOSURE DURATION AND FREQUENCY 223
III. PRIORITY NEEDS FOR FURTHER RESEARCH 224

Chapter 7
CONSIDERATIONS FOR APPLYING
BIOAVAILABILITY TO HEALTH RISK ASSESSMENT 227
I. METALS 227
II. ORGANIC XENOBIOTICS 228
III. OTHER APPLICATION CONSIDERATIONS 228
IV. RISK MANAGEMENT CONSIDERATIONS 230

REFERENCES 237
GLOSSARY OF TERMS 285
INDEX 291

LIST OF TABLES

Table 1-1 Parameters Needed for Calculation of External Dose and Internal Dose 3
Table 3-1 Main Physiological Functions of the Respiratory Tract 23
Table 3-2 Main Physiological Functions of the Gastrointestinal Tract 39
Table 3-3 The Barriers and Rate-Limiting Processes for Absorption of Xenobiotics through Three Exposure Routes 49
Table 3-4 Factors Influencing Bioavailability in Humans 50
Table 4-1 Choice of Experimental Methods for Study of Bioavailability 71
Table 5-1 Respiratory Bioavailability of Arsenic in Humans and Animals 84
Table 5-2 Oral Bioavailability of Arsenic in Humans 85
Table 5-3 Oral Bioavailability of Arsenic in Animals 86
Table 5-4 Dermal Bioavailability of Arsenic in Animals *in vivo* and *in vitro* 88
Table 5-5 Respiratory Bioavailability of Cadmium in Humans and Animals 96
Table 5-6 Oral Bioavailability of Cadmium in Humans 97
Table 5-7 Oral Bioavailability of Cadmium in Animals 98
Table 5-8 Dermal Bioavailability of Cadmium in Humans and Animals 99
Table 5-9 Respiratory Bioavailability of Chromium in Animals 109
Table 5-10 Oral Bioavailability of Chromium in Humans 110
Table 5-11 Oral Bioavailability of Chromium in Animals 111
Table 5-12 Dermal Bioavailability of Chromium in Animals 113
Table 5-13 Experiments of Oral Bioavailability of Copper in Humans 122
Table 5-14 Respiratory Bioavailability of Mercury in Humans and Animals 130
Table 5-15 Oral Bioavailability of Mercury in Humans and Animals 131
Table 5-16 Dermal Bioavailability of Mercury in Humans and Animals 132
Table 5-17 Respiratory Bioavailability of Lead in Humans 140
Table 5-18 Oral Bioavailability of Lead in Humans 141
Table 5-19 Dermal Bioavailability of Lead in Humans and Animals 143
Table 5-20 Respiratory Bioavailability of 2,3,7,8-Tetrachlorodibenzo-p-dioxin in Animals 151

Table 5-21a	Relative Oral Bioavailability of 2,3,7,8-Tetrachlorodibenzo-p-dioxin (TCDD) in Humans and Animals	151
Table 5-21b	Oral Bioavailability of 2,3,7,8-Tetrachlorodibenzo-p-dioxin (TCDD) in Humans and Animals	152
Table 5-22	Dermal Bioavailability of 2,3,7,8-Tetrachlorodibenzo-p-dioxin (TCDD) in Humans and Animals	153
Table 5-23	Respiratory Bioavailability of Some PAHs in Animals	164
Table 5-24	Oral Bioavailability of Some PAHs in Animals	165
Table 5-25	Dermal Bioavailability of Some PAHs in Humans and Animals (*in vivo* and *in vitro*)	166
Table 5-26	Respiratory Bioavailability of Benzene in Humans	180
Table 5-27	Oral Bioavailability of Benzene in Animals	180
Table 5-28	Dermal Bioavailability of Benzene in Humans and Animals (*in vivo*)	181
Table 5-29	Dermal Bioavailability of Benzene in Humans and Animals (*in vitro*)	182
Table 5-30	Respiratory Bioavailability of Toluene in Humans	188
Table 5-31	Oral Bioavailability of Toluene in Humans and Animals	189
Table 5-32	Dermal Bioavailability of Toluene in Humans and Animals	190
Table 5-33	Respiratory Bioavailability of Pentachlorophenol in Humans and Animals	198
Table 5-34	Oral Bioavailability of PCP and 2,4,6-T_3CP in Humans and Animals	199
Table 5-35	Dermal Bioavailability of PCP and 2,4,6-T_3CP in Humans and Animals	200
Table 5-36	Bioavailability of Phenol in Humans and Animals (*in vivo*)	206
Table 5-37	Dermal Bioavailability of Phenol in Humans and Animals (*in vitro*)	208
Table 5-38	Bioavailability of Aniline in Humans and Animals	212
Table 6-1	Potential Exposure Routes to Xenobiotics for the General population	217
Table 6-2	Summary of Bioavailability Factors (%), or Permeability in the Case of Dermal Contact, for Selected Xenobiotics	218
Table 7-1	Available Information on Bioavailability Factors for Selected Xenobiotics	232

LIST OF FIGURES

Figure 2-1 A blood concentration-time curve 9
Figure 2-2 Relationship of various doses, their role and application in quantitative risk assessment 18
Figure 3-1 Proposed model of a cell membrane 20
Figure 3-2 Diagram of lung and bronchial tree 20
Figure 3-3 Diagram of bronchiole and alveoli 21
Figure 3-4 Structure of air-blood barrier 21
Figure 3-5 Gaseous phase absorption of xenobiotics 24
Figure 3-6 Deposition, retention and clearance of aerosols 27
Figure 3-7 Major mechanisms of deposition of inhaled aerosols in the respiratory tract 28
Figure 3-8 Regional deposition fraction for aerosols inhaled by humans 29
Figure 3-9a Total deposition fraction vs. aerosol size distribution via nasal and oral inhalation in humans 31
Figure 3-9b Deposition fraction for the upper respiratory tract vs. aerosol size distribution via nasal and oral inhalation in humans 31
Figure 3-9c Deposition fraction for the tracheobronchial vs. aerosol size distribution via oral inhalation in humans 32
Figure 3-9d Deposition fraction for the pulmonary region vs. aerosol size distribution via nasal and oral inhalation in humans 32
Figure 3-10 Diagram of gastrointestinal tract 37
Figure 3-11 Diagram of the wall of gastrointestinal tract 37
Figure 3-12 Diagram of the intestinal epithelium 38
Figure 3-13 Absorptive steps via ingestion 40
Figure 3-14 Diagram of the skin 44
Figure 3-15 Absorptive steps via dermal route 46
Figure 4-1 Determination of the bioavailability factor based on mass-balance technique 55
Figure 4-2 A procedure for calculation of the respiratory bioavailability factor of aerosols based on deposition-clearance-retention process 57
Figure 4-3 *In vitro* experimental methods for studying oral bioavailability 63
Figure 4-4 Diagram of a typical diffusion cell 69
Figure 5-1 Respiratory bioavailability of inorganic arsenic in human volunteers 79
Figure 5-2 Bioavailability factors of inorganic arsenic in humans and animals by three exposure routes 83

Figure 5-3	Bioavailability factors of cadmium in humans and animals by three exposure routes	95
Figure 5-4	Bioavailability factors of chromium compounds in humans and animals by three exposure routes	108
Figure 5-5	Oral bioavailability factors of copper in humans	120
Figure 5-6	Bioavailability factor vs. daily intake of dietary copper	121
Figure 5-7	Bioavailability factors of mercury in humans and animals by three exposure routes	129
Figure 5-8	Bioavailability factors of lead in humans and animals by three exposure routes	139
Figure 5-9	Bioavailability factors of 2,3,7,8-Tetrachlorodibenzo-p-dioxin in humans and animals by three exposure routes	150
Figure 5-10	Bioavailability factors of benzo[a]pyrene in humans and animals by three exposure routes	163
Figure 5-11	Dermal bioavailability factors for individual compounds after application of the PAHs-containing mixtures to animal skin	163
Figure 5-12	Bioavailability factors of benzene in humans and animals by three exposure routes	179
Figure 5-13	Bioavailability factors of toluene in humans and animals by three exposure routes	187
Figure 5-14	Bioavailability factors for pentachlorophenol in humans and animals by three exposure routes	197
Figure 5-15	Bioavailability factors of phenol (*in vivo*) in humans and animals by three exposure routes	205
Figure 7-1	Framework for application of bioavailability factors in quantitative risk assessment	234

DEDICATION

To the memory of
Steve Hrudey
(1908 - 1994)
craftsman, builder and father

To the memory of
Georges Emile Rousseaux
(1926 - 1993)
teacher, healer and father

Chapter 1
INTRODUCTION

Concern about the possible adverse effects on human health from exposure to xenobiotics, those chemicals largely foreign to the beneficial function of biological systems [Taylor, 1988], has grown steadily over the past three decades. These growing concerns have accelerated the development of toxicology and health risk assessment. Toxicology has been defined as the study of the adverse effects of xenobiotics [Gallo and Doull, 1991]. This field of medical science has a long history with its roots in the earliest uses of animal venoms and plant-derived poisons for hunting and human conflict. As all human knowledge developed in parallel with the scientific method, some fundamental principles of toxicology evolved, with perhaps the best known and most profound being:

> *All substances are poisons; there is none which is not a poison. The right dose differentiates a poison from a remedy.*
>
> *Paracelsus*

While toxicology has clearly developed in concert with the phenomenal developments in medical and pharmaceutical sciences over recent years, environmental concerns and some avoidable health tragedies have spurred the development of a substantial regulatory component to the discipline. In particular, the need to predict and prevent large scale poisoning from new consumer products or environmental contamination has promoted the predictive discipline of health risk assessment.

Toxicology is both a science and an art, the former based upon observational and experimental activities with the latter reflected in the interpretive and predictive activities [Gallo and Doull, 1991]. In comparison, human health risk assessment following exposure to xenobiotics is a process that seeks to predict the adverse health consequences of a specified set of conditions. Because of the emphasis on prediction which is inherent to risk assessment, this activity must be recognized as primarily an application of the "art" of toxicology. This recognition does not imply that the "science" of toxicology is unimportant. On the contrary, the ability of the "art" of toxicology to deliver useful predictions of likely health outcomes is absolutely dependent upon the quality of the "science" of toxicology. However, an adequate practical understanding of health risk assessment requires distinguishing the factual basis (biomedical research findings) for developing health risk predictions (hypotheses or theories) from the predictions themselves being treated as fact. Accurate predictions of health risk require the best possible set of scientifically-derived knowledge about the mechanisms of toxicity and the behavior of xenobiotics. However, possession of high quality findings from toxicological research cannot guarantee accurate health risk predictions unless all major sources of uncertainty and variability are fully resolved. This is an elusive goal. In this context, uncertainty refers to our level of ignorance about mechanisms and nature of behavior while variability refers to measurable differences for the critical parameters governing toxicity. Hence, risk assessment must be recognized for its predictive and correspondingly uncertain nature.

A health risk assessment of environmental exposure to xenobiotics requires a comprehensive evaluation of the major factors governing a prediction of health risk and includes: (1) hazard identification; (2) dose-response assessment; (3) exposure assessment; and (4) risk characterization [USEPA, 1989a].

Hazard identification is the process of identification of xenobiotics that are harmful to human health and a qualitative evaluation of their character and importance.

Dose-response assessment is the process of evaluating and describing the quantitative relationship between the dose of a xenobiotic (external dose, internal dose, or biologically effective dose) and the characterization of an adverse health effect. In this context, external dose is that to which the human receptor or experimental animal is exposed via an environmental medium (air, water, food or direct contact), internal dose is that which reaches the systemic circulation (the blood stream) and biologically effective dose is that which reaches the site of toxic action. These concepts of dose, which underlie the concept of bioavailability, are elaborated in Chapter 2.

The toxicity values are usually obtained from animal experiments by the administration of xenobiotics at external doses (or less commonly with internal doses) involving different environmental media through various exposure routes. The criteria values determined are expressed as a reference dose (RfD) for xenobiotics that produce toxic effects which do not involve initiating cancer (non-carcinogenic) or a potency (slope factor) used to calculate a risk specific dose (RsD) for xenobiotics that initiate cancer (carcinogenic).

Exposure assessment is the process of seeking qualitative insight and/or quantitative data on the magnitude of human exposure to xenobiotics, exposure duration and frequency, sources of exposure, routes of exposure and the characteristics (size, nature and class) of the potentially exposed population. This process evaluates the human exposure pathways (source of contamination, environmental media and transport, point of exposure, and receptor population) and calculates the average daily doses or lifetime average daily doses (external doses and internal doses) for the xenobiotic under assessment.

Parameters needed for calculation of these doses via the main exposure routes are listed in Table 1-1. The environmental parameters (concentrations of xenobiotics in various media) may be directly measured with environmental monitoring or indirectly estimated by mathematical modeling. The required physiological parameters (inhalation rate, ingestion rate, skin surface area and

Table 1-1 Parameters Needed for Calculation of External Dose and Internal Dose

Inhalation and ingestion exposure routes	Dermal exposure route
$ADD_{int} = \frac{\overline{C}\ \overline{IR}\ ED\ OF\ \mathbf{BF}}{BW\ AT}$ $\quad ADD_{ext} = \frac{\overline{C}\ \overline{IR}\ ED\ OF}{BW\ AT}$	*For waterborne contaminants and vapors* $ADD_{int} = \frac{\overline{C}\ P\ \overline{SA}\ ED}{BW\ AT}$
ADD = average daily dose	*For soil-bound contaminants* $ADD_{int} = \frac{\overline{C}\ F_{adh}\ \overline{SA}\ ED\ \mathbf{BF}}{BW\ AT}$ $ADD_{ext} = \frac{\overline{C}\ F_{adh}\ \overline{SA}\ ED}{BW\ AT}$
$\overline{C}$ = average concentrations of a xenobiotic in air (mass/m^3), in drinking water (mass/L), in food (mass/g), and in soil (mass/kg)	
$\overline{IR}$ = average inhalation rate (m^3/day), or ingestion rate for total or tap water (L/day), soil (mass/day), and food (mass/day)	
ED = exposure duration (days or years)	
BW = body weight (kg)	P = dermal permeability (cm/h)
AT = average exposure time period (days or years); For calculating lifetime average daily dose, LA (lifetime average exposure time period) replaces AT	F_{adh} = adherence fraction for soil
BF = inhalation bioavailability factor, or ingestion bioavailability factor (unitless)	$\overline{SA}$ = average exposed skin surface area (cm^2)
OF = other factors, such as *deposition fraction* that is used to adjust to external doses for aerosols, or to adjust internal dose when the respiratory bioavailability factor is based on the deposited dose of an aerosol; *dissolution fraction* that is used to adjust internal doses when soil-bound xenobiotics will dissolve in the body fluid within the respiratory or gastrointestinal tract before absorption	**BF** = dermal bioavailability factor (unitless)

Adapted and modified from USEPA 1992a

body weight) may be obtained by searching in the literature, handbooks and some databases. Bioavailability factors and some other factors (deposition fraction and dissolution fraction) may be obtained from animal experiments (or human experiments from the past).

Risk characterization is a description of the nature and the extent of human risk (i.e. the likelihood of a defined toxic outcome following xenobiotic exposure considering dose duration and frequency) by comparison of the findings from the foregoing steps with a consideration of all the major uncertainties and analytical assumptions. For risk assessment to be accurate and reliable, estimates of dose from an exposure assessment should be expressed in a manner that can be compared with those used to determine toxicity. Within the overall context of risk assessment, a bioavailability factor is primarily used to adjust an external dose from exposure assessment to an internal dose and, occasionally, to adjust toxicity values for dose-response assessment when the exposure media for the environmental conditions under assessment differ from those used in the toxicology studies. For these purposes, bioavailability can be practically defined as the ability of xenobiotics to reach the human systemic circulation (blood stream). Usually, this internal dose will be less than the external dose because of a number of factors involved in the absorption process for various exposure routes. The most common routes of entry for environmental contaminants into humans are inhalation, ingestion, and dermal absorption. These exposure routes and their interactions with bioavailability are elaborated in Chapter 3.

Currently, the validity and reliability of quantitative risk assessments are often restricted by the absence of quantitative data in relation to underlying physicochemical, environmental and biological processes. Determination of bioavailability is one aspect of knowledge that can improve the accuracy of exposure assessment and prediction of adverse health effects. Ultimately, better knowledge of bioavailability of xenobiotics will lead to more accurate evaluation of the potential risk for humans arising from exposure to environmental contaminants, thereby improving the development of environmental policy and regulatory decisions.

In order to study bioavailability, the structure, function, absorption mechanism, and variations in all absorption processes by major exposure routes must be understood. These processes and interactions are described in relation to bioavailability in Chapter 3.

Interpretation of data from bioavailability studies requires an understanding of the design, measurement, strengths and limitations of experimental methods. Experimental methods used in bioavailability studies, including *in vivo* and *in vitro*, are identified and described in Chapter 4.

The roles and applications of bioavailability for selected environmental contaminants (xenobiotics) harmful to human health are explored in Chapter 5. These chemicals were identified as ones which occurred most commonly and which presented the greatest health concerns for orphaned industrial sites in Alberta, Canada. They include some toxic metals (arsenic, cadmium, chromium, copper, lead and mercury), some organic solvents (benzene and toluene), 2,3,7,8-tetrachlorodibenzo-p-dioxin (TCDD), pentachlorophenol-(PCP)-based mixtures and creosote-based mixtures which are used as wood preserving agents. Some of these chemicals are considered as human carcinogens or as possible or probable carcinogens for animals or humans.

Studies on human bioavailability for most environmental contaminants are incomplete, particularly with respect to their presence and bioavailability in complex environmental matrices. Some level of uncertainty will be inherent in the application of bioavailability for risk assessment predictions. However, as a better understanding of the environmental behavior of xenobiotics and of their mechanisms of toxicity develops, a better understanding of how to use the bioavailability concept for quantitative risk assessment will also emerge. Uncertainty related to bioavailability is characterized and information on the needs for further research is provided for each xenobiotic in Chapter 5. More universal questions are considered in Chapter 6.

In summary, current basic knowledge on human bioavailability which is relevant to health risk assessment is briefly reviewed in this book. The potential for wider application of bioavailability data in quantitative risk assessment is illustrated. Important factors which can influence bioavailability are presented. Absorption mechanisms for xenobiotics via inhalation, ingestion, and dermal exposure are discussed and experimental approaches for determining bioavailability are explored. Finally, bioavailability data for xenobiotics commonly identified at contaminated sites which has been extracted from the literature are catalogued in a consistent fashion to facilitate greater access to these data. In presenting this information to the health risk assessment community, the value of a better understanding of bioavailability is underscored. At the same time, the inherent limitations of bioavailability as only one specific aspect in our much broader quest for a fuller understanding of the rates, pathways and mechanisms of toxic action of xenobiotics are fully acknowledged.

Chapter 2
FUNDAMENTAL CONCEPT OF BIOAVAILABILITY

I. DEFINITION OF BIOAVAILABILITY

Toxicology, as it seeks to reveal the adverse effects of xenobiotics (foreign substances), may be subdivided into two phases: toxicity and toxicokinetics. Toxicity is the study of the toxic action of a xenobiotic on the body, including the dose-response relationship, receptor interactions and mechanisms of toxicity. Toxicokinetics is the study of the action of the body on a xenobiotic, including absorption, distribution, metabolism, and excretion (ADME). In this process, there are several modeling and mathematical descriptions used for the time-course of ADME of a xenobiotic. Modeling the complete behavior of xenobiotics in the body is best described using physiologically based-pharmacokinetic (PBPK) modeling. PBPK models allow modeling the time-course of ADME of a xenobiotic in each organ and tissue based on physiological and biochemical parameters (such as blood flow rate, tissue volume and rate of metabolism).

Bioavailability is only one component of the total disposition of the xenobiotic in these models. The term **bioavailability** is generally used to describe the extent and rate of absorption for a xenobiotic which enters the systemic circulation in the unaltered (parent) form from the applied (exposure) site. There are four general aspects in this concept: unaltered (parent) form, systemic circulation, extent of bioavailability and rate of bioavailability.

A. UNALTERED FORM

According to this concept, bioavailability refers to the parent (unaltered) form of a xenobiotic, not its metabolites, in the systemic circulation. Many xenobiotics may be metabolized at the applied sites (skin, lung, or the wall of gastrointestinal tract) or in the liver before they enter the systemic circulation. In some cases, these metabolites are active compounds which exert adverse effects on the body. Benzene is one example which is mainly metabolized in the liver before it reaches the systemic circulation. Only a small fraction of the parent benzene will enter the systemic circulation, but the carcinogenicity of benzene is believed to be related to three products of its metabolism in the liver. Therefore, identification and quantification of its hepatic metabolites is essential for assessing the benzene cancer risk. By extension, in such a case, the internal dose (this chapter, section I. D.1) should include the amount of these metabolites in the systemic circulation. Thus, for some environmental contaminants, limiting bioavailability to the parent form of the xenobiotic will not fully characterize its dose-response relationship. However, in some cases, bioavailability data fails to represent the parent form of a xenobiotic in the systemic circulation because the analytical techniques used have been unable to distinguish the parent form from metabolites in biological fluids.

B. SYSTEMIC CIRCULATION

Systemic circulation, popularly called the bloodstream, refers to both the arterial and venous blood except the portal blood serving the liver [Ritschel, 1972]. A xenobiotic administered via the inhalation or dermal route is directly absorbed into the systemic circulation. However, a xenobiotic in the gastrointestinal tract initially reaches the portal vein and then passes through the liver into the systemic circulation. Systemic circulation is relevant because: (1) a xenobiotic in the systemic circulation is able to reach all target organ sites (with the exception of the liver during ingestion exposure) in the body; (2) in most experiments, the practical site for the measurement of absorbed dose is in the systemic circulation (the amount of a xenobiotic in the portal blood is difficult to measure in humans); (3) the dose-response relationship of a xenobiotic is often closely correlated with the concentration of the xenobiotic in the systemic circulation. Currently, the use of the term bioavailability refers most often to that proportion of a xenobiotic reaching the systemic circulation. This concept has also been termed the "systemic availability".

C. RELATIONSHIP BETWEEN EXTENT AND RATE OF BIOAVAILABILITY

Both extent and rate of absorption are important issues in the bioavailability concept. They are characterized by three basic pharmacokinetic parameters by means of blood measurement *in vivo*. A blood concentration-time curve is obtained by which blood concentrations of a xenobiotic are plotted on graph paper against the time after a single dose exposure (Figure 2-1). From this curve, these three parameters can be determined:

- area under the blood concentration-time curve (AUC) (amount x time/volume);
- peak height (C_{max}) (amount/volume);
- time of peak concentration (t_{max}).

AUC is closely related to the extent of bioavailability because it provides an integrated mass of systemic dose expressed as:

$$AUC = \int_{T^0}^{T\infty} C_i \, dt \qquad (2\text{-}1)$$

AUC alone cannot reflect the nature of bioavailability. Curves A, B, and C in Figure 2-1 have a similar AUC, but the peak concentration of a xenobiotic in the blood (C_{max}) and time needed to reach the peak concentration (t_{max}) differ. Both are critical determinants of toxic effects.

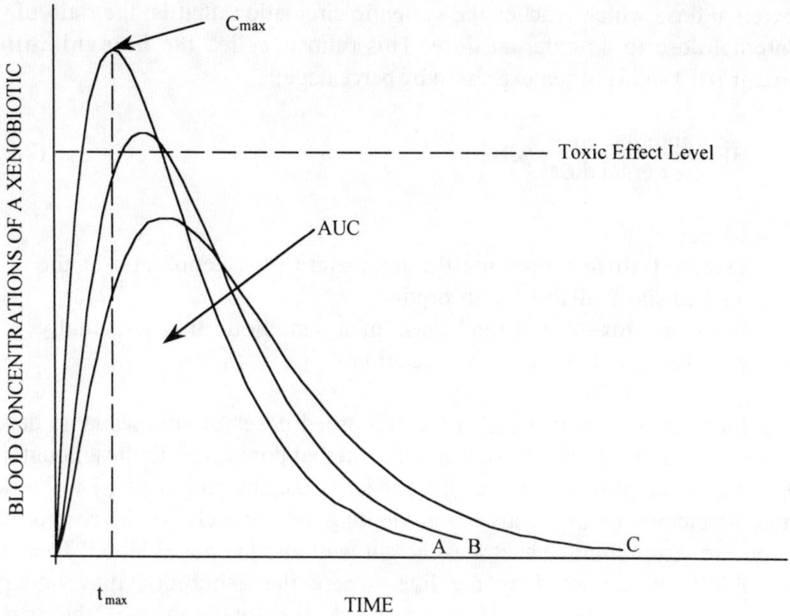

Figure 2-1 A blood concentration-time curve. Curves A, B, and C have a similar area under blood concentration-time curve (AUC), but the peak height of concentration (C_{max}) and time of peak concentration (t_{max}) are different.

C_{max} and t_{max} are closely related to the rate of bioavailability. The value of t_{max} is inversely related to the absorption rate. A slowly absorbed xenobiotic may require a higher external dose to reach the same blood concentration in a given exposure time and to exert a toxic effect equivalent to a more rapidly absorbed xenobiotic. However, when the blood concentration of a xenobiotic reaches its maximum, absorption is not yet completed.

Although there are a number of mathematical models for determining the rate of bioavailability, such as the one-compartment model (Wagner-Nelson method), multicompartment model (Loo-Riegelman method), noncompartment model, statistical moment theory, convolution and deconvolution model [Gibaldi and Perrier, 1982], the measured values of C_{max} and t_{max} are commonly employed to roughly estimate the rate of absorption.

D. EXTENT OF BIOAVAILABILITY

The overall extent of bioavailability for a xenobiotic is determined by either the *absolute bioavailability* or the *relative bioavailability*.

1. Absolute bioavailability

Absolute bioavailability usually refers to the fraction or percentage of an external dose which reaches the systemic circulation, that is, the ratio of an internal dose to an external dose. This ratio is called the **bioavailability factor** (BF) and is (often expressed by percentage):

$$BF = \frac{\text{internal dose}}{\text{external dose}} (x100) \qquad (2\text{-}2)$$

where:

external dose represents the total mass of a xenobiotic at the applied site available for absorption;

internal dose is the total mass of a xenobiotic that is actually absorbed into the systemic circulation.

Internal dose is also called the "absorbed dose" or "bioavailable dose". For dermal and inhalation exposure, the internal dose refers to the amount of a xenobiotic that directly crosses the barrier at the absorption site, i.e. the skin and the membrane of the alveoli in the lung, respectively. In the case of oral exposure, a xenobiotic penetrates the gut wall into the portal blood. Then, the xenobiotic is extracted by the liver where the xenobiotic may undergo presystemic metabolism. If metabolized, the total mass of the parent xenobiotic will decrease before entering the systemic circulation. As a result, the internal dose is less than the dose in the portal blood. This phenomenon is described as the "hepatic first-pass effect" or "hepatic presystemic metabolism". In this case, judgment must be applied to determine whether the metabolite(s) should be considered with the parent compound. If the metabolite(s) is active in exerting toxicity, it should be considered, whereas, if innocuous, the metabolite(s) should not be considered.

External dose in the literature generally implies a variety of doses, such as applied dose, administered dose and exposure dose, depending on the designs and analytical methods involved in the experiments. Theoretically, **applied dose** which refers to the amount of a xenobiotic in contact with the surface of primary absorption sites (e.g. alveoli, gut wall and skin) is an ideal parameter for accurately assessing the bioavailability factor. In practice, applied dose for most xenobiotics is difficult to measure with *in vivo* experiments for inhalation, ingestion and dermal exposure (other than dermal contact with vapors and waterborne contaminants). In most cases, administered dose (labeled dose, potential dose or exposure dose) is employed to describe the bioavailability factor. **Administered dose** is the amount of a xenobiotic that is actually inhaled, ingested or applied on the skin. **Exposure dose** is commonly defined as the total mass of a xenobiotic in the environmental media to which an individual or population is exposed over time [USEPA, 1992a; Lioy, 1990a; NRC, 1991]. Administered dose may be estimated from exposure dose if contact rates (e.g. inhalation rate and ingestion rate) are known.

Additionally, there may exist a great difference between the **administered dose** and the **applied dose** after inhalation of a slightly water soluble aerosol. A xenobiotic administered by inhalation can be initially eliminated via exhalation and/or by mechanical clearance whereby an inhaled xenobiotic is swallowed into the gut or removed to pulmonary lymph nodes. In such cases, only a small fraction of the inhaled xenobiotic which is deposited but not mechanically cleared, i.e. the applied dose, will be available for absorption. Generally, an **external dose** is considered to be an **administered dose** in most bioavailability experiments.

There are two basic *in vivo* methods for evaluating the bioavailability factor in humans and in animals [Wagner, 1969, 1972, 1977]. One is the measurement of the blood levels of a parent xenobiotic over time. The other is the measurement of the total mass of the unaltered form in excreta, such as urine, feces and expired air. In such studies, the blood levels or the cumulative excreta data for a xenobiotic following administration by extravascular routes (e.g. inhalation, ingestion and dermal) are compared to those after intravenous administration. Intravenous dose provides a reference dose with a bioavailability factor equal to one. The other useful *in vivo* method is called the "mass-balance technique", but this is only applicable in experimental animals because it requires post mortem analysis of all major tissues which may act as a sink for the xenobiotic. There are a variety of specific *in vivo* and *in vitro* methods available for determining the extent of bioavailability for each specific exposure route. These will be discussed in Chapter 4.

2. Relative bioavailability

When intravenous administration is not feasible (e.g. because of low solubility), the bioavailability factor may be estimated from relative bioavailability. Relative bioavailability (BF_R) is determined by a comparison of the extent of absorption among several doses of the same xenobiotic, or two or more forms of the same xenobiotic, or different vehicles (food, soil and water) of the same xenobiotic with the same subject [Labaune, 1989]. For instance, when the pure form of a xenobiotic is used as a standard material, any other form of the xenobiotic is used as a test material. The extent of absorption for the test material (f_2) is compared to that for the standard material (f_1), as:

$$BF_R = \frac{f_2}{f_1} \text{ (x100)} \qquad (2\text{-}3)$$

If absolute bioavailability for the standard material is known (BF_1), absolute bioavailability for the test material (BF_2) can be calculated by:

$$BF_2 = BF_R \times BF_1 \tag{2-4}$$

The concept of relative bioavailability is very important for environmental exposure because matrix effects can substantially alter the nature of bioavailability compared with that observed for the pure compound.

II. BIOAVAILABILITY IN HEALTH RISK ASSESSMENT

Environmental health risk assessment has been described as consisting of (1) hazard identification; (2) dose-response assessment; (3) exposure assessment; and (4) risk characterization [NAS, 1983]. Environmental human health risk assessment evaluates the likelihood of adverse human health effects caused by environmental contaminants. The exposure assessment is both an important and complex step. This process involves the characterization of exposure setting, identification of exposure pathways and quantification of exposure [USEPA, 1989a].

With regard to health outcomes, exposure and dose are basic determinants of the toxic responses. The dose-response relationship which is the core of health risk assessment must rely upon an accurate estimation of the exposure and effective dose.

Exposure reflects the environmental contaminants experienced by an individual or a population. Exposure can be readily measured in most cases, but, by itself, offers little insight into health consequences. The extent of exposure (E) is the total concentration of a contaminant at the exposure point over time [Lioy, 1990a; NRC, 1991]:

$$E = \int_{t_1}^{t_2} C(t)\,dt \tag{2-5}$$

where:

$C(t)$ is the concentration at the point of contact as a function of time;

t_2-t_1 represents the exposure duration.

Dose is a concept which seeks to describe the total mass of a contaminant which actually crosses the body boundary, to be absorbed or deposited into the body and which will ultimately be available to reach the target sites over a given period [NRC, 1991]. The general equation of dose (D) is an integration of exposure over time [Lioy, 1990a]:

$$D = \int_{t_1}^{t_2} D(t)\,dt \tag{2-6}$$

Within this boundary, dose has various, more specific definitions for application to risk assessment. It is generally divided into three groups: biologically effective dose, internal dose and external dose.

A. BIOLOGICALLY EFFECTIVE DOSE

The total delivered mass of active toxic species at the sites of toxic action is usually defined as the **biologically effective dose**. Evaluation of this dose (D_{bio}) is based on metabolism, excretion and chemical assimilation [Lioy, 1990a]:

$$\int D_{bio}(t)\, dt = \int TF\ BF\ IR\ C(t)\, dt \tag{2-7}$$

where:
TF represents toxicokinetic fraction (transport, metabolism and excretion);
BF is the bioavailability factor;
IR refers to the contact rate (inhalation rate and ingestion rate).

Theoretically, the biologically effective dose is a critical parameter for evaluating the dose-response relationship. It may be estimated by biological monitoring (measures of DNA and protein adducts in the blood, cells and tissues) and mathematical modeling based on the physicochemical properties of xenobiotics and available physiological and pharmacokinetic parameters for the exposed individual [NRC, 1989, 1991]. However, routine use of this dose for quantitative risk assessment may not be practical because:

- this dose is difficult to measure in the human body and has not been obtained in most animal experiments because of limitations with the experimental designs and analytical techniques;
- the data (key toxicokinetic parameters) for this dose to be estimated by mathematical modeling are rarely available for most environmental contaminants;
- this dose cannot be traced directly back to determine the original source of environmental contaminants or to identify the exposure routes because any xenobiotic-DNA (or protein) adducts are not unique nor specific to any particular exposure source or route.

B. INTERNAL DOSE AND AVERAGE DAILY INTERNAL DOSE

Studies for determining the dose-response relationship for many xenobiotics have shown that the internal dose is closely related to the occurrence of toxic effects. Quantification of internal dose may also involve calculation of the average daily internal dose defined as the internal dose per unit body weight over an average time.

The bioavailability factor has been used to evaluate the internal dose and the average daily internal dose in quantitative risk assessment. The internal dose (D_{int}) is expressed by [Lioy, 1990a; USEPA, 1992a]:

$$\int D_{int}(t)\,dt = \int BF\ IR\ C(t)\,dt \tag{2-8}$$

If C and IR can be approximated by average values, D_{int} can be written by:

$$D_{int} = \overline{C}\ \overline{IR}\ ED\ BF \tag{2-9}$$

where:
$\overline{C}$ is the average concentration of an environmental contaminant;
$\overline{IR}$ is the average value of contact (inhalation or ingestion) rate;
ED is the total exposure duration.

The average daily internal dose (ADD_{int}) is:

$$ADD_{int} = \frac{\overline{C}\ \overline{IR}\ ED\ BF}{BW\ AT} \tag{2-10}$$

where:
BW is the body weight, in units of kg;
AT is the time period over which the dose is to be averaged, in units of days.

For the dermal exposure route, waterborne contaminants or vapors in contact with the skin can directly penetrate the skin barrier into the systemic circulation. The internal dose is dependent on the concentration of contaminants in water, the partial pressure of vapors, the dermal permeability and the surface area exposed. The internal dose is given by [USEPA, 1992a]:

$$\int D_{int}(t)\,dt = \int_{t_1}^{t_2} C(t)\ P\ SA(t)\,dt \tag{2-11}$$

and for average values:

$$D_{int} = \overline{C}\ P\ \overline{SA}\ ED \tag{2-12}$$

where :
$C(t)$ is the concentration of a contaminant in water (mg/L) or air ($\mu g/m^3$) in contact with the skin over time;
$\overline{C}$ is the average concentration value;
P represents the dermal permeability coefficient of chemicals (cm/h);
SA refers to the surface area of skin exposed over time (cm^2);
$\overline{SA}$ is the average value.

The average daily internal dose can be expressed as [USEPA, 1992a]:

$$ADD_{int} = \frac{\overline{C}\ P\ \overline{SA}\ ED}{BW\ AT} \tag{2-13}$$

For soil-bound contaminants, the internal dose can be determined from the contaminant concentration and the amount of carrier medium that is the product of the fraction of soil adhering to the skin, the surface area exposed, exposure duration and bioavailability factor [USEPA, 1992a]:

$$D_{int} = \overline{C}\ M_{medium} = \overline{C}\ F_{adh}\ \overline{SA}\ ED\ BF \tag{2-14}$$

where:
$\overline{C}$ is the average soil-bound contaminant concentration (mg/kg);
M_{medium} refers to the mass of soil applied;
F_{adh} represents the adherence fraction for soil (the proportion of soil contacted that will adhere to the skin and make xenobiotics available for absorption).

This algorithm does not explicitly deal with the reality that contaminants bound to the uneven surface of soil particles will be not uniformly in contact with the skin and will not all be available for transfer across the skin barrier.

ADD_{int} can be calculated by [USEPA, 1992a]:

$$ADD_{int} = \frac{\overline{C}\ M_{medium}\ BF}{BW\ AT} = \frac{\overline{C}\ F_{adh}\ \overline{SA}\ ED\ BF}{BW\ AT} \tag{2-15}$$

A number of factors must be considered in judging the value of estimating bioavailability factors for the purpose of quantitative risk assessment. These include:

- the internal dose will be more specific for quantitative risk assessment than the external dose if toxicity values for a xenobiotic are based on the internal dose. This dose is more likely to be closely related to the toxicity at the target sites;

- the internal dose may be estimated by knowing a few key environmental, physicochemical, physiological and pharmacokinetic parameters;
- the internal dose can reduce uncertainty in the risk assessment provided that the bioavailability factor is reliably estimated from experimental exposures;
- the dose may be traced back to determine the original source of environmental contaminants or to identify the exposure sites of receptors if the quantity of environmental contaminants and exposure pathways are known.

C. EXTERNAL DOSE AND AVERAGE DAILY EXTERNAL DOSE

Quantitative assessment of health risk for groups or subgroups of populations is an important part in health risk assessment. In practice, it has been common to estimate the external dose rather than the internal dose for overall population exposures. The equations for evaluating the external dose (D_{ext}) and the average daily external dose (ADD_{ext}) are [USEPA, 1992a]:

For inhalation and ingestion routes:

$$D_{ext} = \overline{C}\ \overline{IR}\ ED \tag{2-16}$$

and

$$ADD_{ext} = \frac{\overline{C}\ \overline{IR}\ ED}{BW\ AT} \tag{2-17}$$

For dermal contact with soil-bound contaminants:

$$D_{ext} = \overline{C}\ M_{medium} = \overline{C}\ F_{adh}\ \overline{SA}\ ED \tag{2-18}$$

and

$$ADD_{ext} = \frac{\overline{C}\ M_{medium}}{BW\ AT} = \frac{\overline{C}\ F_{adh}\ \overline{SA}\ ED}{BW\ AT} \tag{2-19}$$

These quantitative external dose estimates rarely reflect the most direct input to the dose-response relationship for many environmental contaminants. Using the external dose inevitably contributes to uncertainties in the assessment of health risk. For example, following inhalation of aerosols, the inhaled dose that reaches the surface of alveoli (before entering the systemic circulation) is reduced because of deposition, clearance and retention in other parts of the respiratory system. Consequently, the internal dose in the systemic circulation is considerably less than the external dose. If the external dose is used in quantitative risk assessment, the bioavailability factor is effectively being taken as equal to one.

Currently, many quantitative risk assessment protocols focus on estimation or determination of external dose and the average daily external dose. Reasons for this include:

- a reliable bioavailability factor has not been estimated in a form that is useful for risk assessment for most environmental contaminants;
- the dose-response relationship for many xenobiotics has been estimated by using only the external dose in toxicological experiments;
- use of external dose only requires environmental and intake rate parameters which are easily obtained;
- external dose can be related directly to the original source of environmental contaminants or the exposure sites of receptors.

III. SUMMARY

In order to identify whether the exposure doses from the environment exhibit a potential health risk for humans, the exposure doses can be compared to the doses that cause the toxic effects. Estimation of daily doses (average or lifetime) is commonly used for this purpose. However, various doses may be applied for quantitative risk assessment (Figure 2-2).

The parameter which should most accurately reflect the dose-response relationship is the biologically effective dose. But the use of this dose is generally not feasible because of the lack of biological measurements which are required to provide the basis for the mathematical modeling.

The most common dose parameter is the external dose because it is readily determined through relatively simple approaches (direct measurement, biological monitoring and mathematical modeling). Use of this dose involves a weaker level of understanding of toxic effects than would be the case if the knowledge was available to support the use of a biologically effective dose.

Between the ideal of the biologically effective dose and the more commonly used external dose is the internal dose which is often well correlated with toxic responses. Bioavailability, as reflected in the extent and rate of xenobiotic absorption, will govern the internal dose.

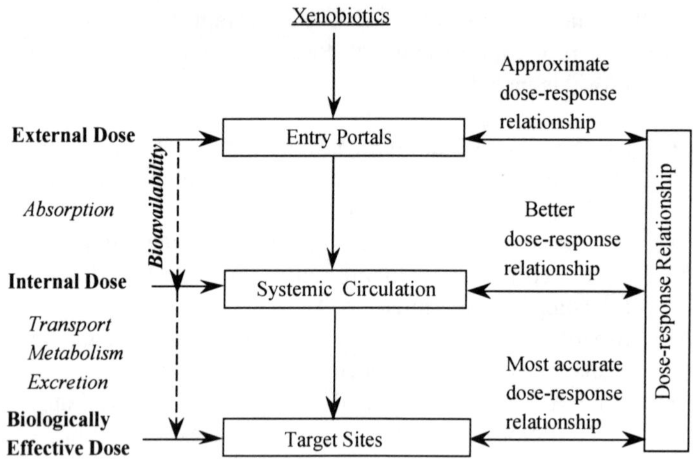

Figure 2-2 Relationship of various doses, their role and application in quantitative risk assessment

Chapter 3

MECHANISM OF ABSORPTION

Xenobiotics must pass through one or more biological membranes during absorption. The biological barriers may consist of a single layer of the cell membrane (such as alveolar wall) or multiple layers of those membranes (such as skin). Cell membranes have a similar structure and composition, which has been described as a "mayonnaise sandwich" (Figure 3-1). Within the cell membrane, proteins are inserted in the lipid bilayer. The lipid composition varies from cell to cell, including various phospholipids, fatty acid, cholesterol and other steroids.

Transport of the xenobiotics within an organism can be classified into: (1) passive transport, including simple diffusion and filtration; and (2) special transport, including active transport, facilitated diffusion, phagocytosis, pinocytosis and persorption. The mechanism of absorption of the xenobiotics via inhalation, ingestion and dermal exposure is explored with relation to the concept of bioavailability in the following sections.

I. RESPIRATORY SYSTEM

A. STRUCTURE AND FUNCTION

The respiratory tract is a very complex system (Figure 3-2, 3-3). Based on anatomic structure, particle deposition and clearance features, the respiratory tract is divided into three regions [Morrow *et al.*, 1966]:

- the nasopharyngeal region, including the nose, nasopharynx and larynx;
- the tracheobronchial region, containing trachea, bronchi, and bronchioles (diameter of 1 mm or less);
- the pulmonary region, comprising the respiratory bronchioles (diameter of about 0.06 mm), alveolar ducts, alveolar sacs, and alveoli surrounded by blood capillaries and lymphatic vessels.

In a newly revised lung model proposed by the International Commission on Radiological Protection (ICRP) Task Group [Bair, 1989; James *et al.*, 1989], the thoracic lymph nodes are included in the pulmonary region, the so-called parenchymal-nodular region.

Absorption of the xenobiotics may occur mainly in the alveolar epithelium because of the large surface area (140 to 150 m^2) available [Gordon and Amdur, 1991]. The air-blood barrier (0.36 to 2.25 μm thick) consists of the aqueous layers on the surface of the alveoli, the epithelium cells (type I), the thin interstitial space containing fibroblasts and the capillary endothelial cells (Figure 3-4) [Phalen, 1984].

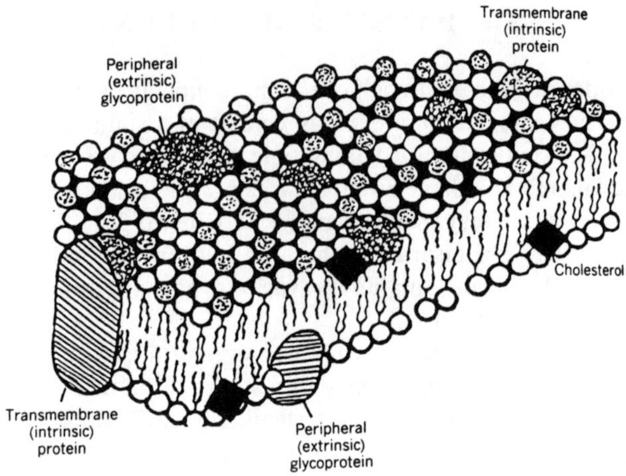

Figure 3-1 Proposed model of a cell membrane. Integral proteins are inserted in the lipid bilayer. The lipid composition consists of various phospholipids, fatty acid, cholesterol and other steroids. The carbohydrate chains are attached to proteins. This structure has been called a "mayonnaise sandwich". (From author, Dinman, B.D., The mode of absorption, distribution and elimination of toxic materials, in *Patty's Industrial Hygiene and Toxicology,* 4th ed., Vol. 1, Part A, Clayton, G.D. and Clayton, F.E., Eds., John Wiley & Sons, New York, 1991, 214. With permission.)

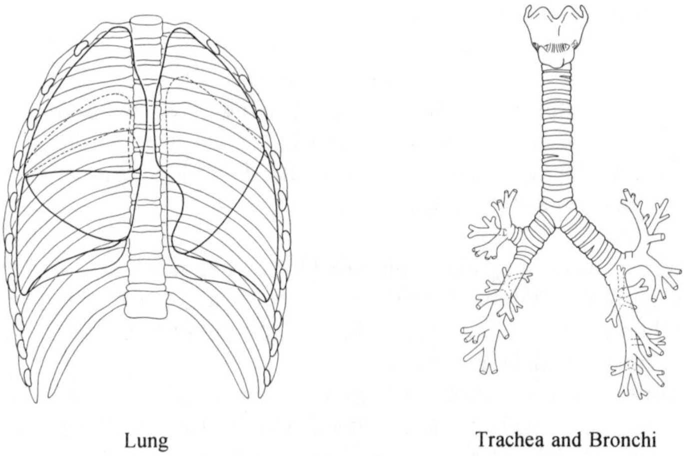

Figure 3-2 Diagram of lung and bronchial tree (Modified from author, Cull, P., Ed., *The Sourcebook of Medical Illustration,* The Parthenon Publishing Group Limited, Lancs, England, 1989, 71, 83. With permission.)

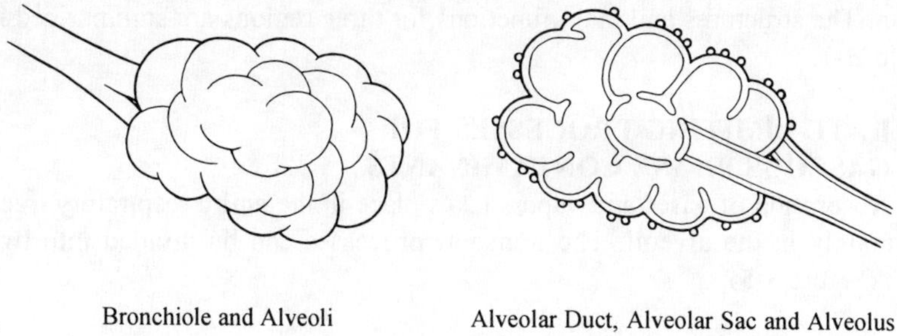

Figure 3-3 Diagram of bronchiole and alveoli (Modified from author, Cull, P., Ed., *The Sourcebook of Medical Illustration,* The Parthenon Publishing Group Limited, Lancs, England, 1989, 83. With permission.)

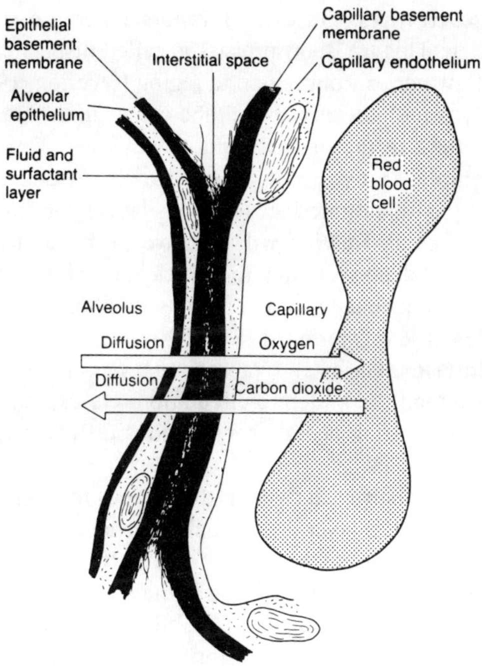

Figure 3-4 Structure of air-blood barrier. This barrier consists of the aqueous and surfactant layers on the surface of the alveoli, the alveolar epithelium cells and their basement membranes, the thin interstitial space containing fibroblasts and the capillary endothelial cells and their basement membranes (From author, Guyton, A.C., *Textbook of Medical Physiology*, 8th ed., W.B. Saunders Company, Philadelphia, PA, 1991, 422-432. With permission.)

The primary function of the respiratory system is gaseous exchange for the uptake of oxygen and excretion of carbon dioxide. However, absorption, deposition, metabolism and clearance of the xenobiotics also occurs in this organ. The structures and main functions for three regions are summarized in Table 3-1.

B. RATE-LIMITING PROCESSES FOR GASEOUS PHASE CONTAMINANTS

Absorption of gases and vapors takes place in the entire respiratory tract, but mainly in the alveoli. The transport processes can be divided into two steps (Figure 3-5):

- transfer of the gas and vapor from the ambient environment into the alveoli, occurring in the *conducting airways*;
- penetration of the *air-blood barrier* and dissolution in the blood.

Conducting Airways. Gases and vapors enter the respiratory tract following inhalation. This transport process is called dispersion. The velocity of convective airflow varies from region to region [Overton, 1984, 1990]. The maximum velocity usually occurs at the tube center in the trachea and at the inner wall of the tube in the bronchi and bronchioles. The airflow, in the respiratory bronchioles, moves axially and slowly. Along the wall of the airways, there is a gas-liquid interface (aqueous layer). During the passage of gases and vapors, a small fraction will dissolve in this aqueous layer. This process is primarily dependent on the thickness of the aqueous layer, concentrations in the aqueous layer and the gas-liquid partition coefficient of xenobiotics [Phalen, 1984; Overton, 1984, 1990].

Air-Blood Barrier. Gases and vapors must pass through the air-blood barrier prior to reaching the systemic circulation. Gases and vapors exist as molecules, so that absorption occurs from gaseous diffusion following Fick's law [Guyton, 1991].

The total resistance barrier to gaseous diffusion is the sum of resistance of the aqueous layer, epithelial layer, interstitial space and plasma fluid [Weibel, 1970]. With exposure to a gas or vapor for a sufficient time, tissues will effectively reach equilibrium. For diffusion in the liquid phase (aqueous layer, blood and other tissue fluids), the water solubility of a gas or vapor is a key determinant for absorption. At equilibrium, the solubility ratio of a xenobiotic between the blood and air is constant [Klaassen and Rozman, 1991]. This constant is termed the "blood/air partition coefficient" ($P_{B/A}$):

$$P_{B/A} = \frac{C_B}{C_A} \tag{3-1}$$

Table 3-1 Main Physiological Functions of the Respiratory Tract

Region	Structure	Main function
<u>Nasopharyngeal</u>		
Nose Nasopharynx Larynx	Mucous membrane Epithelium Glands	Warming and humidification of inhaled air Filtration of large inhaled particles Absorption of water-soluble compounds Metabolism Olfaction
<u>Tracheobronchial</u>		
Trachea Bronchi Bronchioles	Mucous membrane Epithelium Cartilage Smooth muscle Glands	Conducting air and aerosols Deposition of medial particles Metabolism Fast particle clearance Uptake of gases and vapors
<u>Pulmonary</u>		
Respiratory bronchioles Alveolar ducts Alveolar sacs Alveoli	Single-layer epithelium with surfactant and capillaries	Gas exchange Deposition of small particles Absorption of gases, vapors and aerosols Slow or very slow particle clearance Synthesis and secretion of surfactant Metabolism

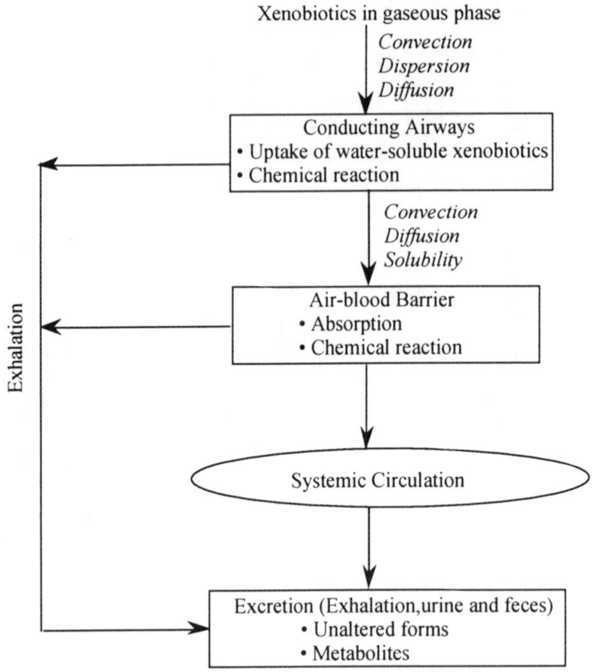

Figure 3-5 Gaseous phase absorption of xenobiotics

where:
C_B is the concentration of a xenobiotic in the blood;
C_A is the concentration of a xenobiotic in the airspace.

At equilibrium, the rate of xenobiotic diffusion through the air-blood barrier is equal to the rate of xenobiotic removal by blood flow from the airspace. Obviously, for a xenobiotic with a low $P_{B/A}$, the concentration of the xenobiotic that dissolves in the blood will be low. In such a case, a small amount of the xenobiotic will be removed by blood from the airspace. When the rate of blood flow increases, a larger amount of the xenobiotic will be removed by blood from the airspace and the rate of xenobiotic diffusion will be enhanced. This is a *perfusion limited step*. In contrast, a xenobiotic with a high $P_{B/A}$ is more readily dissolved in the blood. For this case, an increase of ventilation (i.e. an increase of xenobiotic concentration available in the airspace) will accelerate the rate of absorption of a xenobiotic with high $P_{B/A}$, thereby corresponding to a *ventilation limited step*.

C. OTHER FACTORS INFLUENCING RESPIRATORY BIOAVAILABILITY FOR GASES AND VAPORS

In the case of gas or vapor uptake by simple diffusion, the factors influencing diffusion will affect respiratory bioavailability. Although the gas or vapor may diffuse in both directions across the barrier, the net diffusion is attributed to a variety of factors. These factors are pressure or concentration gradients, thickness of the air-blood barrier, surface area, water solubility, molecular weight, diffusion coefficient, blood/air partition coefficient, permeability of the membrane, ventilation, blood flow, chemical reactivity, and disease states. Of all the factors, the water solubility, blood/air partition coefficient and chemical reactivity generally play the most important roles in determining the respiratory bioavailability for xenobiotics.

Water Solubility. There is a linear relationship between the solubility and the extent of deposition of a xenobiotic in the aqueous layer and tissue fluids [Overton and Miller, 1988]. The highly water soluble and chemically reactive compounds tend to deposit to a high degree in the aqueous layer or tissue fluids of the upper respiratory tract unless the concentrations are extremely high [Cuddihy and McClellan, 1989]. Thus, such compounds are effectively removed to the aqueous layer from the conducting airway surface. These include compounds like sulfur dioxide, ammonia and iodine vapor [Pattle, 1961; Speizer and Frank, 1966; Black and Hounam, 1968; Frank *et al.,* 1969; Frank, 1970; Webber and Harvey, 1976; Kleinman, 1984]. For lower solubility but reactive compounds (e.g. ozone), more than 50% of the inhaled gas or vapor may reach the pulmonary region [Miller *et al.,* 1978; Cuddihy and McCellan, 1989]. Lipid-soluble and inert compounds (e.g. anesthetic) may primarily reach the alveolar space, independent of their concentration in the air [Saidman, 1983; USEPA, 1989b]. For lipid-soluble compounds, the aqueous layer provides a major barrier for absorption.

Pressure or Concentration Gradient. The pressure or concentration gradient refers to the difference between the partial pressure or concentration of a xenobiotic in the airspace of the alveoli and in the blood. When the pressure or concentration of the compound is greater on one side of the barrier than the other, the net diffusion will occur from the region of high pressure or concentration toward the region of the low pressure or concentration [Dinman, 1991; Guyton, 1991].

Thickness of the Air-Blood Barrier. In terms of Fick's law, the rate of diffusion across the barrier is inversely proportional to the thickness of the barrier. The greater the thickness, the lower the rate of diffusion [Dinman, 1991; Guyton, 1991].

Surface Area. The rate of diffusion is a function of the alveolar surface area and the blood capillaries. Under resting conditions, gas exchange is only slightly affected by the loss of up to two-thirds to three-fourths of the normal surface area of the alveoli, whereas the same loss of surface area will cause a serious problem to gas exchange during physical exercise [Guyton, 1991].

Molecular Weight. The velocity of thermal motion of a dissolved gas is related to the square root of its molecular weight. Small molecules will be more permeable to the air-blood barrier [Bend *et al.* ,1985; Guyton, 1991].

Diffusion Coefficient. The diffusion coefficient (D) of a gas is determined by the water solubility of a xenobiotic (S) and its molecular weight (MW) according to [Phalen, 1984; Guyton, 1991]:

$$D = \frac{S}{\sqrt{MW}} \tag{3-2}$$

Permeability of the Air-Blood Barrier. Permeability of the barrier is defined as the rate of diffusion of a xenobiotic through each unit area of the barrier for a unit concentration difference [Guyton, 1991]:

$$P = \frac{D}{A} \tag{3-3}$$

where:
D is the diffusion coefficient;
A is the total alveolar surface area.

Ventilation and Blood Flow. Ventilation and blood flow vary both at rest and during physical exercise. Normally, the ratio of ventilation (V_A) to blood flow (Q) remains constant although the ventilation or blood flow changes with time [Astrand, 1983]. This is termed the "ventilation-perfusion ratio", given by V_A/Q. However, at the beginning of physical exercise, the uptake of gas or vapor is enhanced until the blood becomes saturated with the gas. When exercising at a constant work load or resting, the rate of absorption is constant.

Chemical Reactivity. Chemical reactions will be important for highly reactive gases and vapors. Using a mathematical model, Overton *et al.* [1987] predicted that 60% to 92% of ozone uptake, at different human-activity levels, occurred in the pulmonary tissues and only 2% to 3% reached the systemic circulation. The low systemic values are caused by rapid chemical reactions between ozone and the lung tissues, which also produces pulmonary injury.

Disease State. Some pulmonary diseases cause edema and fibrosis in the interstitial space which increases the thickness of the respiratory membrane, and emphysema which decreases the effective surface area of the alveoli. As a result, these disease states reduce the rate of diffusion of the gas and vapor into the bloodstream [Bend *et al.,* 1985; Guyton, 1991].

D. RATE-LIMITING PROCESSES FOR PARTICULATE PHASE CONTAMINANTS

Bioavailability for aerosols involves a series of processes: deposition, clearance and retention (Figure 3-6).

Once inhaled, aerosols may directly deposit on the surface of various regions of the airways. However, not all airborne aerosols may be respired. Currently, inhalable particulate matter (IPM) often refers to the size of

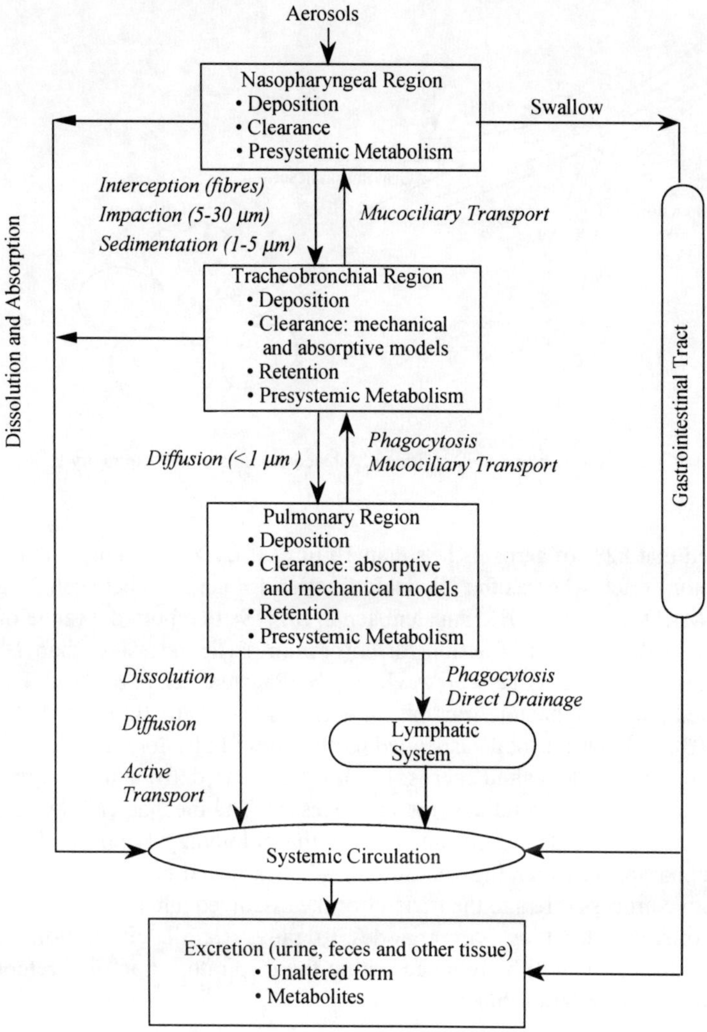

Figure 3-6 Deposition, retention and clearance of aerosols

aerosols less than 10-15 μm mass median aerodynamic diameter (MMAD). This measure reflects the reality that sampling methods used to characterize particle size distribution depend upon the aerodynamic properties rather than physical size of particles. Particulate matter, described as IPM_{10} or IPM_{15}, has the properties to reach the tracheobronchial and pulmonary regions [Lioy, 1990b; USEPA, 1990]. The inhalable fraction (intake efficiency or inhalability) is defined as the portion of the total mass of airborne aerosols that actually crosses the nose and mouth. Vincent and Armbruster [1981]

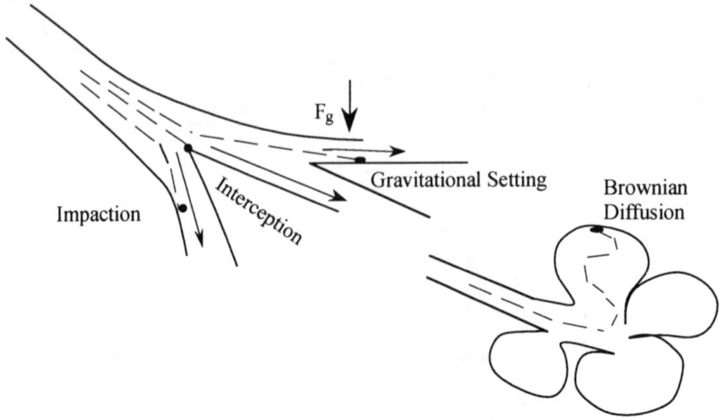

Figure 3-7 Major mechanisms of deposition of inhaled aerosols in the respiratory tract

reported that 82% of aerosols less than 10 μm MMAD could enter the human respiratory tract, whereas the inhalable fraction for aerosols between 30 to 80 μm MMAD was only 50%. Paustenbach *et al.* [1991] reported a range of 6% to 30% of the inhalable fraction for chromium particulates less than 10 μm MMAD. USEPA data that were reviewed by Paustenbach *et al.* [1986, 1987] indicated that the inhalable fraction for aerosols less than 10 μm MMAD was only 30% and that for total suspended particulates (TSP) not more than 50%.

After being deposited, aerosols may be cleared from the respiratory system by mechanical and absorptive processes. The mechanical clearance is the removal of the deposited materials to the pulmonary lymph nodes or to the oropharynx in which the xenobiotics are swallowed or expectorated. The absorptive process refers to the transfer of the dissolved inhaled materials from the respiratory tract or lymph nodes to the systemic circulation. The uncleared aerosols will be retained within the respiratory tract, i.e. retention, according to the relationship:

Retention = Deposition - Clearance.

1. Deposition

Deposition is defined as the process whereby the mass of the inhaled xenobiotic is deposited in various regions of the respiratory tract during inhalation exposure. The principal mechanisms of deposition include interception, impaction, sedimentation and diffusion (Figure 3-7) [Raabe, 1979, 1982]. (Note that, in this section, the aerosol size parameters are aerodynamic diameters, MMAD, for aerosols ≥ 0.5 μm and diffusion diameters, reflecting the diffusion properties of aerosols for those < 0.5 μm):

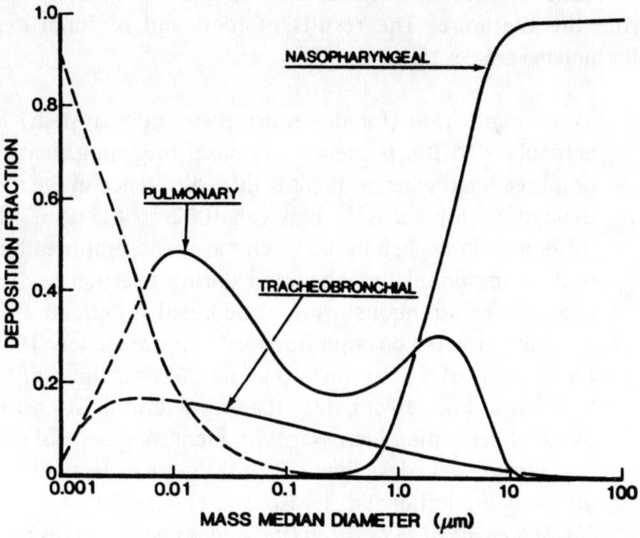

Figure 3-8 Regional deposition fraction for aerosols inhaled by humans (From author, Snipes, M.B., Long-term retention and clearance of particles inhaled by mammalians, *Crit. Rev. Toxicol.*, 20(3):175-211, 1989. With permission.)

- *Interception* at bifurcations of bronchial trees occurs only for long fibrous particulates (e.g. asbestos). In this case, the length and shape of particles are more important determinants of deposition than settling velocity.
- *Inertial impaction* means that particulates (ranging from 5 to 30 μm) continue to travel along their original direction because of their inertia despite changes in the airstream direction. Those larger particles can be impacted on the surface of the large airways.
- *Sedimentation* is the process whereby the small particles (1 to 5 μm) settle in the small bronchi, bronchioles and the alveolar space by gravitational force.
- *Diffusion* means that aerosols less than 1 μm (particularly <0.5 μm) will reach the surface of the small airways and alveoli primarily by diffusion.

Deposition location and degree is highly variable with the aerosol size distribution (Figure 3-8) and breathing patterns (nasal inhalation and oral inhalation). In principle, in humans, inhaled aerosols between 5 to 30 μm may be trapped in or pass the nasopharyngeal region. Inhaled aerosols of 1 to

5 μm may be deposited in the tracheobronchial region. Aerosols less than 1 μm tend to reach the alveolar surface. Schlesinger [1989] analyzed several studies from the literature. The results of total and regional deposition fractions for humans are described as:

- *Total deposition* (for the entire respiratory system) for aerosols >0.5 μm is greater via nasal breathing than via oral breathing, whereas there is little difference in the total deposition for aerosols between 0.02 to 0.5 μm. The difference in deposition between nasal and oral breathing will be important for inhalation during exertion such as exercise or strenuous work. The results indicated the maximum total deposition fraction is approximately 100% for aerosols of 8 to10 μm and about 20% for those of 0.3 to 0.5 μm. For aerosol sizes ranging within 0.5 to 8 μm, the total deposition increases with increasing aerosol size. For aerosol sizes less than 0.2 μm, the total deposition is inversely proportional to aerosol size (Figure 3-9a).
- *Nasopharyngeal deposition* (the upper respiratory tract) is still greater by nasal inhalation than by oral inhalation. For instance, the fraction for 2 to 10 μm of aerosols is 40% to 90% via nasal inhalation and only 2 to 35% via oral inhalation. The predominant aerosol size deposited in this region ranges from 1 to 10 μm with most larger than 5 μm. Within this range, the deposition fraction increases with increasing aerosol size (Figure 3-9b).
- *Tracheobronchial deposition* has been found to be relatively constant, ranging from 20% to 40% for aerosol sizes between 2 to 10 μm. As the aerosol size decreases to less than 1 μm, the deposition fraction markedly declines (Figure 3-9c).
- *Pulmonary deposition* via oral breathing is proportional to the aerosol size larger than 0.5 μm. The deposition fraction for aerosols between 2 to 4 μm ranges from 30% to 60%, whereas aerosols between 0.1 to 1 μm have a deposition fraction between 15% to 20%. The deposition fraction via nasal breathing is relatively constant (20% to 22%) for aerosols of 0.7 to 3 μm. Conversely, when the aerosol size is larger than 3 μm, the deposition fraction reduces with an increase of aerosol size. Clearly, nasal breathing gives rise to lower deposition fractions than oral breathing in the pulmonary region of the respiratory tract for larger aerosols (Figure 3-9d).

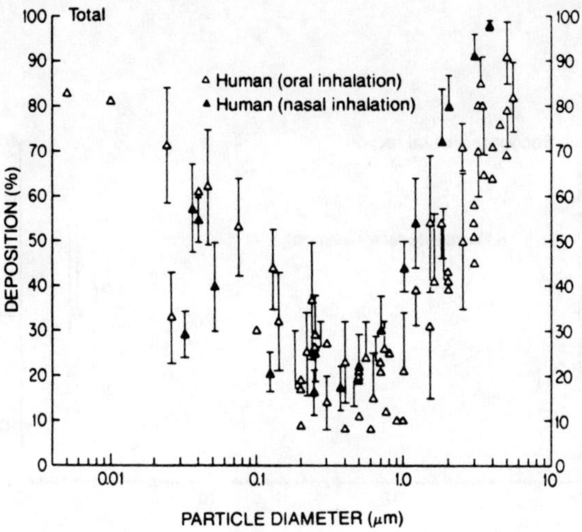

Figure 3-9a Total deposition fraction vs. aerosol size distribution via nasal and oral inhalation in humans (From author, Schlesinger, R.B., Biological disposition of airborn particles: basic principles and application to vehicular emissions, in *Air Pollution, the Automobile, and Public Health,* Watson, A.Y., Bates, R.R., and Kennedy, D., Eds., National Academy Press, 1988, 256. With permission.)

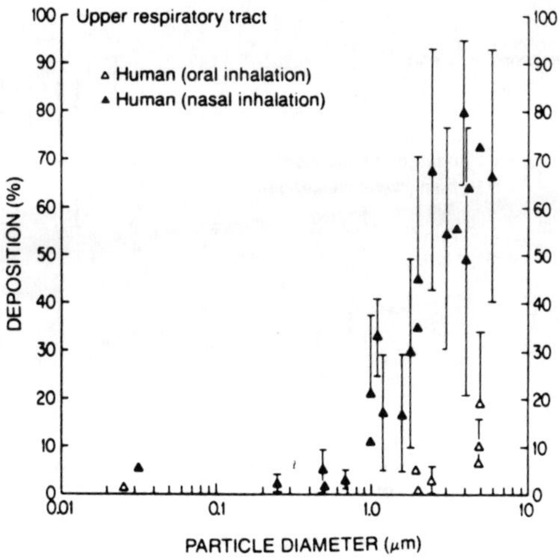

Figure 3-9b Deposition fraction for the upper respiratory tract vs. aerosol size distribution via nasal and oral inhalation in humans (From author, Schlesinger, R.B., Biological disposition of airborn particles: basic principles and application to vehicular emissions, in *Air Pollution, the Automobile, and Public Health,* Watson, A.Y., Bates, R.R., and Kennedy, D., Eds., National Academy Press, 1988, 256. With permission.).

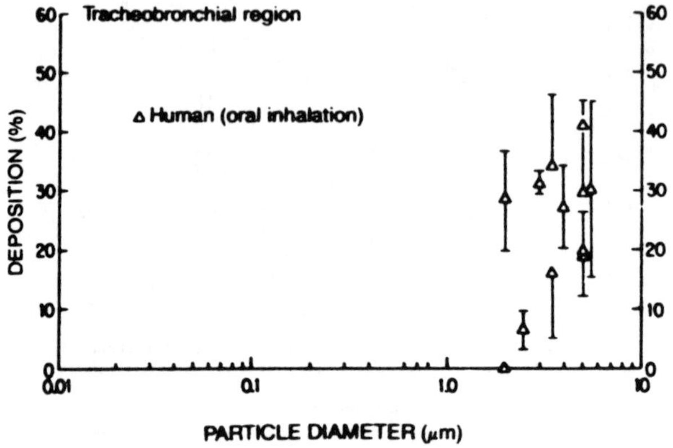

Figure 3-9c Deposition fraction for the tracheobronchial vs. aerosol size distribution via oral inhalation in humans (From author, Schlesinger, R.B., Biological disposition of airborn particles: basic principles and application to vehicular emissions, in *Air Pollution, the Automobile, and Public Health,* Watson, A.Y., Bates, R.R., and Kennedy, D., Eds., National Academy Press, 1988, 257. With permission.)

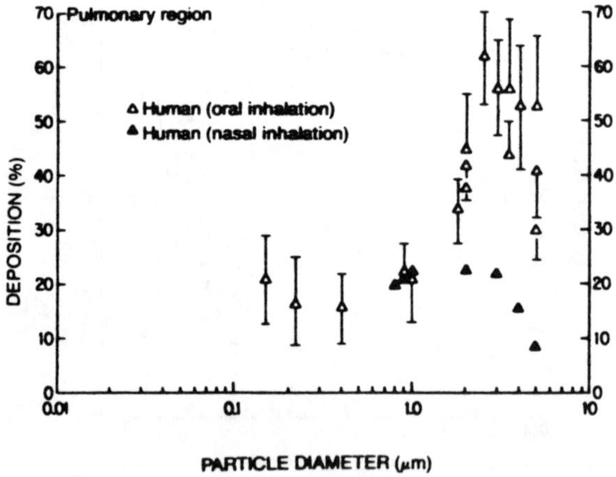

Figure 3-9d Deposition fraction for the pulmonary region vs. aerosol size distribution via nasal and oral inhalation in humans (From author, Schlesinger, R.B., Biological disposition of airborn particles: basic principles and application to vehicular emissions, in *Air Pollution, the Automobile, and Public Health,* Watson, A.Y., Bates, R.R., and Kennedy, D., Eds., National Academy Press, 1988, 257. With permission.)

When data on the regional deposition fraction and particle size distribution are not known, ICRP [1968] assumes that 25% of total inhaled aerosols are exhaled, 50% are deposited in the nasopharyngeal and tracheobronchial regions and 25% reach the pulmonary region. About 60% of inhaled aerosols appear to be swallowed into the gastrointestinal tract [Kimbrough *et al.*,1984; Paustenbach *et al., 1986*; Sheehan *et al.*, 1991].

2. Clearance

Clearance of aerosols can be divided into two processes:

- *Non-absorption processes* (mechanical translocation). These processes usually occur for soluble aerosols and relatively insoluble aerosols that very slowly dissolve in the fluid of the respiratory tract.
- *Dissolution-absorption processes.* Water soluble aerosols undergo these processes. The solubility of the deposited aerosols is a dominant factor for absorption.

a. Water Insoluble Aerosols

Generally, insoluble aerosols tend to be retained within the respiratory tract to cause pulmonary injury and diseases by chemical reactions and metabolism. These deposited aerosols are removed by various processes, depending on the deposition regions.

In the nasopharyngeal region, the major clearance processes are mucociliary transport, sneezing, wiping and blowing. Aerosols deposited in the oral passage may be swallowed into the gastrointestinal tract. The clearance half-time (time for 50% removal) by sneezing, wiping and blowing is estimated as less than 30 minutes and mucociliary transport is about 12 hours [Fry and Black, 1973; Morrow, 1977; Schlesinger, 1989; USEPA, 1989c].

In the tracheobronchial region, insoluble aerosols are cleared by mucociliary transport and coughing. Clearance rates vary considerably with the depth of the airways. The average clearance half-time is 48 hours (ranging from 25 minutes to 5 months) [Bailey *et al.,* 1985]. About 1% of the deposited aerosols in this region are likely to be retained long-term and ultimately penetrate into the deep tissues (epithelium, gland and connective tissue) [Brain and Mosier, 1977; Watson and Brain, 1979; Gore and Patrick, 1982; Schlesinger, 1989].

Insoluble aerosols from the pulmonary region undergo alveolar-macrophage associated particle clearance and lymphatic system clearance.

Alveolar-Macrophage Associated Particle Clearance. After ingestion by alveolar macrophages, aerosols may be moved upward to the oropharynx, followed by swallowing into the gastrointestinal tract. Clearance rates in this step are independent of aerosol size, but depend on the rate of phagocytosis.

The clearance half-time is estimated as weeks corresponding to slow particle clearance [Schlesinger, 1988, 1989].

Lymphatic System Clearance. Either the ingested or free insoluble aerosols may be cleared from lymphatic drainage via direct penetration or phagocytosis within a few hours or days [Schlesinger, 1988, 1989]. Clearance rates for free aerosols in this step are size-dependent [Raabe *et al.,* 1978; Stradling *et al.,* 1978]. These aerosols may be immobilized and accumulated in the lymph nodes which serve as reservoirs for the systemic circulation with the clearance half-time of months [Schlesinger, 1988, 1989].

b. Water Soluble Aerosols

In addition to mechanical clearance, the dominant clearance of soluble aerosols in each region takes place via the dissolution-absorption processes.

Dissolution-Absorption Clearance. Water soluble constituents eluted from the matrix of particles may be initially dissolved within the alveolar aqueous layer and directly pass the air-blood barrier, by simple diffusion, into the systemic circulation. Alternatively, the aerosols may be engulfed by alveolar macrophages and dissolved within the macrophages. Then, the dissolved constituents are released out of the macrophages and become available for absorption. In specific cases (e.g. the pesticide paraquat), the absorption occurs by an active transport process. The clearance rates in this step are very slow, taking months or years corresponding to long-term clearance or very slow clearance [Cuddihy and Yeh, 1988].

Typically, the absorption rate is dissolution-dependent and aerosol-size-dependent. Highly water soluble aerosols appear to be rapidly passed through aqueous paths or pores in the air-blood barrier following deposition, particularly in the nasopharyngeal region where there is an abundant vascular network. The rate of absorption is inversely related to the molecular size [Mercer, 1967].

Conversely, lipophilic compounds can readily penetrate the air-blood barrier. The rate of absorption varies linearly depending on the octanol/water partition coefficient. The absorption rate of weakly acidic or basic compounds is determined by their degree of ionization in the aerosol fluid or alveolar aqueous layer. The non-ionized form will be more lipophilic.

In some cases, small fractions of dissolved constituents from aerosols may remain in the specific local regions for a long period by binding the pulmonary tissue [Mewhinney and Griffith, 1982; Cuddihy, 1984]. This binding process can reduce the absorption of the deposited substances, but may promote cytotoxicity to the pulmonary tissue.

3. Retention

Following repeated inhalation exposure to aerosols for a sufficient time, the rate of deposition and clearance will reach a balance. At equilibrium, the rate of retention of aerosols in the respiratory tract remains constant. The total mass of the retained aerosols is defined as the "equilibrium lung burden" which may indicate the relative concentrations in exposure air. On the other hand, if the rate of clearance is less than that of deposition, the lung burden increases, which is termed as "lung overloading". Because of these interactions, the retention pattern can be used to predict the equivalent concentration of exposure aerosols, that is, the concentrations of aerosols that would produce equivalent deposition of inhaled aerosols per unit mass of lung [Snipes, 1989].

E. OTHER FACTORS INFLUENCING RESPIRATORY BIOAVAILABILITY FOR AEROSOLS

Respiratory bioavailability for aerosols is dependent on both deposition and clearance. These processes are affected by a number of variables. In general, the variables can be categorized into five groups of characteristics: (1) environmental; (2) physicochemical; (3) respiratory tract; (4) breathing pattern; and (5) general.

Physicochemical factors involve particle size, shape, density, surface area, hygroscopicity, electrical charge and conductivity, dissolution rate, solubility and ionization. As mentioned previously, the structure of the various parts of the respiratory tract contributes to determining the site of deposition and the rate of clearance. Also, breathing patterns, such as respiratory rate, volume of inhaled air, air velocity and breath-holding, dramatically influence the deposition and clearance of aerosols. Moreover, general factors (age, gender, and disease states) have to be considered for respiratory bioavailability studies. Of all these factors, the properties of aerosols are the most critical for determining the deposition and clearance behavior. Environmental factors include temperature, humidity, composition of air and wind velocity. These will not be discussed further.

Particle Size. As discussed in Section D, particle size is a key determinant for deposition and clearance processes.

Shape of Particle. The shape of particles will influence deposition pattern and phagocytic efficiencies [Mercer, 1967; Raabe, 1971; Phalen, 1984; Cuddihy and Yeh, 1988]. Particles with an irregular shape have a larger surface and lower density. As a result, they may deliver a greater proportion of xenobiotics into the alveoli.

Density. If the geometric diameters of two aerosols are the same, the settling velocities are directly proportional to density [Mercer, 1967; Raabe, 1971; Phalen, 1984; Cuddihy and Yeh, 1988].

Surface Area. As aerosols become smaller, there is a marked increase in surface area as well as in physicochemical activities [Mercer, 1967; Raabe, 1971; Phalen, 1984; Cuddihy and Yeh, 1988]. The surface area of aerosols will affect the ability of aerosols to adsorb gases and the rate of dissolution of particles in the respiratory tract [Phalen, 1984]. An increase in surface area of particles appears to enhance their toxicity because there is a larger exposed surface for reaction with cells or tissues [Hatch and Gross, 1964].

Hygroscopicity. Some water soluble aerosols will grow in moist air until they reach equilibrium, e.g. sea salt, tobacco smoke and sulfuric acid droplets. Typically, the smaller particles grow more rapidly than the larger ones. At equilibrium, the size of hygroscopic particles can be utilized to estimate inhalation dose and deposition pattern [Ferron, 1977; Phalen, 1984].

Dissolution Rate. In principle, the dissolution rate is proportional to the surface area of aerosols [Cuddihy and Yeh, 1988]. The smaller particles may dissolve and be cleared faster than the larger ones. Also, dissolution makes constituents of the particle available for metabolism within the respiratory tract.

Solubility and Extent of Ionization. As discussed in Section D, these two factors will influence the penetration of the air-blood barrier for aerosols.

Anatomy. The diameter of the airway influences the displacement required by the aerosols [Snipes, 1989; Schlesinger, 1985, 1990]. The airflow velocity is attributed to the cross section of the airways.

Breathing Pattern. During exercise, an increase of respiratory rate and inhaled volume or breath-holding may accelerate the settling velocity and rate of deposition [Palmes *et al.,* 1973]. Oral breathing may deliver the larger aerosols deeper into the respiratory tract.

Metabolism. The respiratory system also contains a group of biotransformation enzymes. Detoxication and toxication of aerosols may occur before a xenobiotic reaches the systemic circulation, termed "respiratory presystemic metabolism" [Benford and Bridges, 1986]. However, this effect is usually small and generally insignificant for assessing the bioavailability for most xenobiotics.

Age. Mucociliary transport rate appears to become slower with age [Clarke and Pavia, 1980].

Disease States. Pulmonary diseases may decrease the rate of aerosol clearance by various mechanisms [Brain, 1980].

II. GASTROINTESTINAL SYSTEM

A. STRUCTURE AND FUNCTION

The gastrointestinal (GI) tract comprises the esophagus, stomach, small intestine (the duodenum, jejunum and ileum) and colon (Figure 3-10). In general, the typical structure of the wall of the gastrointestinal tract consists of the four main components (Figure 3-11) [Smith, 1986; Kutchai, 1988; Selen, 1991]:

- *Mucosa* that contain an epithelial cell layer with goblet cells and endocrine cells, the lamina propria in which there are lymph nodules and capillaries, and the muscularis mucosae;

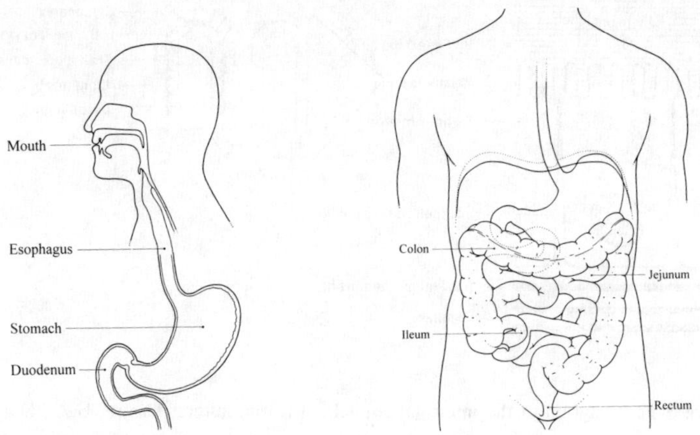

Figure 3-10 Diagram of gastrointestinal tract (Modified from author, Cull, P., Ed., *The Sourcebook of Medical Illustration,* The Parthenon Publishing Group Limited, Lancs, England, 1989, 96, 98. With permission.)

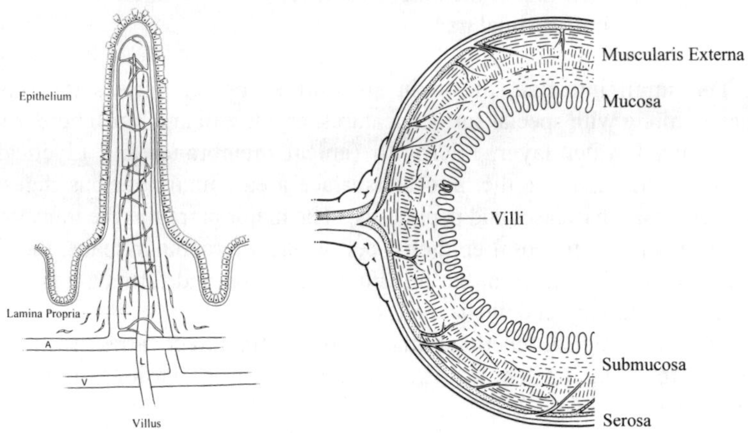

Figure 3-11 Diagram of the wall of gastrointestinal tract (Modified from author, Cull, P., Ed., *The Sourcebook of Medical Illustration,* The Parthenon Publishing Group Limited, Lancs, England, 1989, 108, 109. With permission.)

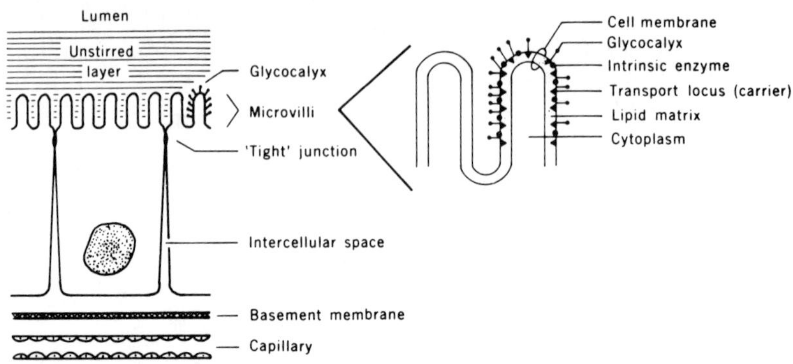

Figure 3-12 Diagram of the intestinal epithelium (From author, Castro, G.A., Digestion and absorption, in *Gastrointestinal Physiology*, 4th ed., Johnson, L.R., Ed., Mosby-Year Book, Inc., St. Louis, Missouri, 1991, 110. With permission.)

- *Submucosa* that is constituted by loose connective tissue, with glands and the larger blood vessels of the gastrointestinal wall;
- *Muscularis externa* that consists of two layers of smooth muscle cells;
- *Serosa* that is the outermost layer from the interior of the gastrointestinal tract.

The small intestine, the main absorptive site, comprises the typical structure along with special surface features: crypt, villi and brush border with the unstirred water layer, microvilli (apical membrane) and glycocalyx. Microvilli that increase the mucosal surface area contain various digestive enzymes (disaccharidases and peptidases). The major barrier to the transport of xenobiotics is the mucosal epithelial cell layer, intercellular space, the tight junction which is located at the juxtaluminal region and forms a continuous barrier, and capillary endothelial cells (Figure 3-12).

The physiological function of the gastrointestinal tract varies greatly from site to site along its length. The main functions of various regions are summarized in Table 3-2.

B. RATE-LIMITING PROCESSES

After entering the gut lumen and, if possible, dissolving in the gut fluid, a xenobiotic may cross the gut/blood barrier (mucosal barrier), via transcellular or intercellular route, by one of the two major mechanisms of gastrointestinal absorption: passive (simple diffusion and filtration) and active

Table 3-2 Main Physiological Functions of the Gastrointestinal Tract

Region	Structure and conditions	Main function
Esophagus	*Typical structure with squamous epithelium	Delivery of food
Stomach	*Typical structure with more circular muscle layer (pH 1 - 3.5)	Mixing Emulsion formation Protein digestion Absorption of certain lipid-soluble compounds (e.g. ethanol)
Small Intestine	*Typical structure with certain surface feature (crypt, villi, microvilli)	Absorption
Duodenum	Presence of Brunner's gland in the submucosa (pH 5 -6)	Influx of biliary and pancreatic secretions Iron and calcium absorption
Jejunum	(pH 6 - 7)	Hydrolysis of disaccharides Absorption of nutrients, water, electrolytes and certain chemicals
Ileum	(pH 7 -8)	Absorption of bile salt, sodium, vitamin B_{12} and certain chemicals
Colon	Absence of well-defined villi	Absorption of water and sodium Secretion of potassium

*Typical structure: mucosa, submucosa, muscularis externa, and serosa.

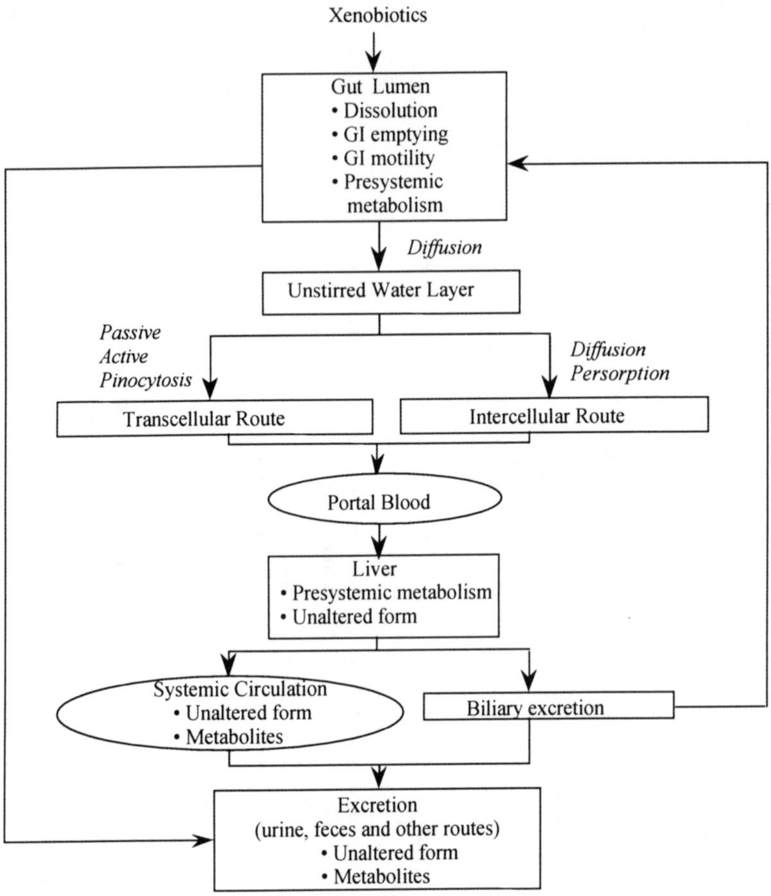

Figure 3-13 Absorptive steps via ingestion

transport. Absorption of macromolecules may be involved in phagocytosis, pinocytosis and persorption [Houston and Wood, 1980; Csáky, 1984; Aungst and Shen, 1986; Abdou, 1989; Klaassen and Rozman, 1991; Dawson, 1991]. A xenobiotic that crosses the gut/blood barrier appears in the portal blood, passes through the liver and then enters the systemic circulation. A xenobiotic that is not absorbed is eliminated with the feces. The absorptive processes are depicted in Figure 3-13.

Unstirred Water Layer. This mucous covering (300 to 400 nm thick) is the initial diffusion barrier for xenobiotics [Houston and Wood, 1980; Csáky, 1984; Thomson and Dietschy, 1984; Hänninen, 1986]. Water soluble xenobiotics pass through this layer by simple diffusion following Fick's law. For hydrophobic xenobiotics, this layer will represent the rate-limiting barrier.

Transcellular Route. The absorptive cell layer (approximately 1 μm thick) is constituted by a series of membranes: the apical membrane, the basolateral and basement membranes of the epithelial cell, and the wall membrane of the lymphatic or blood capillary [Esposito 1984a]. Hydrophobic substances readily permeate this layer by diffusion [Capraro 1984].

A variety of nutrients and endogenous compounds, for instance, sugar, amino acid, bile salt, minerals, water soluble vitamin and certain xenobiotics (lead and aluminum), are known to be absorbed from the small intestine by the active transport process [Stein, 1984; Foulkes, 1984; Esposito, 1984b]. Macromolecules may be absorbed via pinocytosis.

Intercellular Route. There is an intercellular route between the two epithelial cells, including the tight junction, the gap conjunction, the pores, the intercellular space, the basement membrane of the epithelial cell, and the wall membrane of the lymphatic or blood capillary [Esposito, 1984a]. Fluid absorption is believed to occur through aqueous pores (0.3 to 0.88 nm) by diffusion, filtration, osmosis and active transport. Small molecules may penetrate the pores of the membrane or the tight junction with this water flux [Turnheim, 1984]. Large molecules, e.g. starch granules and diatomaceous earth particles, may enter the intercellular space from a gap at the tip of the villi between two cells which are sloughed off. This process is termed persorption [Csáky, 1984; Aungst and Shen, 1986].

First Pass Effect. This effect has been described as presystemic xenobiotic metabolism. Some xenobiotics will be metabolized at the applied sites (the airway wall, the gut wall and the skin) or during passage through the liver from the portal vein prior to the systemic circulation [Abdou, 1989; Labaune, 1989; Dinman, 1991; Selen, 1991]. The first-pass effect reduces the total mass of xenobiotics that can reach the systemic circulation. Usually, the hepatic first-pass effect will dramatically influence the oral bioavailability for most xenobiotics in humans because the liver is the primary organ for metabolic activity. For exposure routes other than ingestion, primary metabolism is less significant at reducing systemic concentration.

C. OTHER FACTORS INFLUENCING ORAL BIOAVAILABILITY

The bioavailability of a xenobiotic via the ingestion route varies greatly, depending on a variety of factors. These factors include physicochemical (water solubility, octanol/water partition coefficient, dissolution rate, molecular size, dosing forms, vehicles and complexation), physiological (gastrointestinal emptying and motility, blood flow, first-pass effect and enterohepatic circulation) and general factors (age, gender and diseases). Of all the factors, the dissolution rate, gastrointestinal emptying and motility, blood flow and first-pass effect are the major determinants for the oral bioavailability.

Dissolution Rate. The dissolution rate of solid xenobiotics in the gastrointestinal tract has been found to be one of the rate-limiting factors in gastrointestinal absorption [Houston and Wood, 1980; Smolen and Ball,

1984; Welling, 1984, 1988; Abdou, 1989; Labaune, 1989; Dinman, 1991; Selen, 1991]. The dissolution rate (DR) is described by the Noyes-Whitney law:

$$DR = K\,(C_s - C) = \frac{DS}{h}\,(C_s - C) \tag{3-4}$$

where:
K is a dissolution rate constant or partition coefficient;
C_s is the solubility of the xenobiotic in solvent;
C is the concentration of dissolved substance in the bulk medium;
D is the diffusion coefficient of the solute molecules in the medium;
h is the thickness of the diffusion layer;
S is the solid surface area.

The dissolution rate may be altered by the diffusion coefficient, surface area of particles, thickness of aqueous film, dissolution constant and solubility. However, a key determinant of dissolution is the solubility of xenobiotics. Because a low water-solubility solid particulate or nonpolar compound dissolves poorly in the aqueous gastrointestinal fluids, it may not be effectively absorbed regardless of the ease of penetration of the cell membranes.

The solubility can be altered by the entire dissolution medium in the gastrointestinal fluid or the diffusion layer surrounding a particle. Thus, pH range in the gastrointestinal fluid may affect the dissolution rate. Also, the dissolution rate is a function of the surface area of a particle.

Furthermore, the dosing form and vehicle are important considerations in the dissolution process. For example, some salts of weakly acidic or basic compounds (with large molecular weights) are more easily dissolved and absorbed more quickly than are the free acids or bases that are less dissolved. Generally, these water soluble compounds with large molecular weights (>100) pass through the gut/blood barrier via the pores of the cell membranes.

Gastrointestinal pH. The absorption of weak acids or bases (with small molecular weights) is influenced by gastrointestinal pH [Labaune, 1989; Dinman, 1991; Klaassen and Rozman, 1991; Selen, 1991]. In most cases, these compounds pass through the gut/blood barrier (with lipid characteristics) via the passive diffusion of the nonionized form (more lipid-soluble). Based on the Henderson-Hasselbalch equations, the bioavailability for an organic acid or base is related to the degree of its nonionized form as determining by the dissociation constant:

For weak acids

$$\log \frac{[\text{nonionized}]}{[\text{ionized}]} = pK_a - pH \tag{3-5}$$

For weak bases

$$\log \frac{[\text{ionized}]}{[\text{nonionized}]} = pK_a - pH \tag{3-6}$$

In an acidic environment (i.e. stomach), the nonionized form for a weak acid is predominant, and is more readily absorbed. In an alkaline environment (i.e. the small intestine), a weak base is more easily absorbed. Conversely the rate of absorption will be slower for a weak acid. However, because of the huge surface area of the small intestine, the overall capacity of absorption for weak acids is still greater than in the stomach. Generally, the presence of food or other compounds like antacids could alter the pH range of the gastrointestinal tract.

Gastrointestinal Emptying and Motility. Gastrointestinal emptying and motility have been recognized as controlling steps in gastrointestinal absorption [Johnson, 1992]. To some extent, the degree of absorption is proportional to the xenobiotic residence time. The food types may delay, reduce or increase the xenobiotic absorption because of the changes of the gastrointestinal emptying and motility. For instance, liquid meals have shorter residence times than solid meals. In the fasting state, the motility rates may decrease, resulting in an increased absorption of xenobiotics.

Blood Flow. Blood flow has been considered to be a rate-limiting step for the absorption of highly lipophilic xenobiotics [Labaune, 1989; Dinman, 1991; Selen, 1991]. After crossing the intestinal epithelium, the xenobiotic is rapidly removed by blood flow which maintains a large concentration gradient (i.e. sink condition). Thus, for lipophilic compounds, this is a *perfusion rate-limited process*. As a result, bioavailability is proportional to blood flow. In contrast, the absorption of hydrophilic xenobiotics is usually independent of blood flow and is controlled by diffusion, that is, a *diffusion rate-limited process*.

Interaction and Complexation. A xenobiotic may interact with the intestinal mucin, bile acid, inorganic salts and other chemicals in the gastrointestinal tract and form a complex which is incapable of penetrating the mucosal barrier [Gregus and Klaassen, 1986]. Thus, the complex may inhibit the absorption of xenobiotics. For instance, cadmium decreases zinc and copper absorption. Phosphate and calcium may inhibit absorption of aluminum. The complexation may also increase the extent of absorption. Lipophilic xenobiotics or xenobiotics with oily vehicles in the small intestine may form micelles with bile salts. Those mixed lipid-bile salt micelles are more easily absorbed. Absorption of some heavy metals (e.g. lead) may be enhanced by oral administration of a chelating agent like EDTA because the lipid solubility of the EDTA-complexed ions is increased [Klaassen and Rozman, 1991]. Consequently, the treatment of lead intoxication to extract previously absorbed lead involves intravenous rather than oral delivery of EDTA.

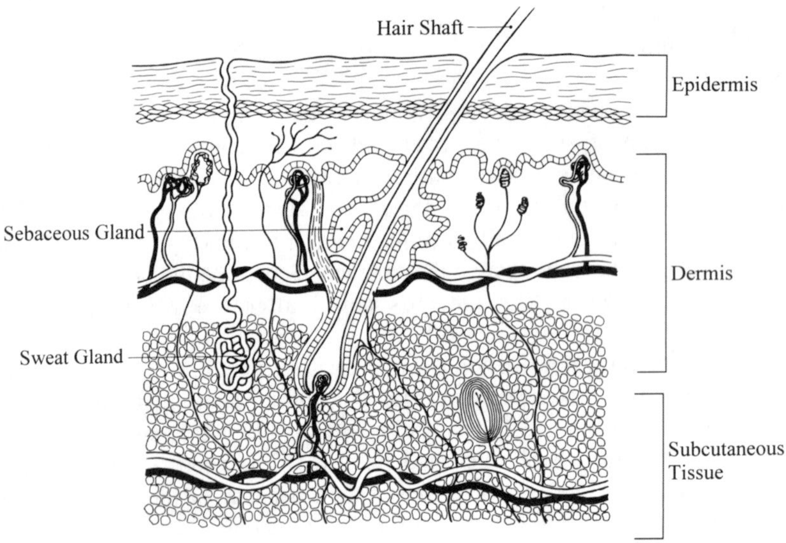

Figure 3-14 Diagram of the skin (Modified from author, Cull, P., Ed., *The Sourcebook of Medical Illustration,* The Parthenon Publishing Group Limited, Lancs, England, 1989, 270. With permission.)

Enterohepatic Circulation. This is a process whereby xenobiotics and their metabolites entering the intestine with biliary excretion may be reabsorbed into the portal blood and returned to the liver [Gregus and Klaassen, 1986]. This effect may influence both the bioavailability and disposition of the xenobiotic by prolonging the residence time of the xenobiotic in the body.

Age. In humans, infants and young children may absorb some xenobiotics faster than adults (e.g. lead absorption via ingestion). On the other hand, the elderly appear to absorb some xenobiotics more slowly [Greenblatt *et al.*, 1982; Eller, 1992].

Disease State. Diseases or gastrointestinal surgery could markedly alter the bioavailability of xenobiotics [Smolen and Ball, 1984].

III. SKIN

A. STRUCTURE AND FUNCTION

The human skin comprises two distinct tissues: dermis and epidermis, the latter with appendages (hair follicles, sweat and sebaceous glands) (Figure 3-14) [Barry, 1983; Klaassen and Rozman, 1991]:

- *Dermis* (3 to 5 mm thick, the deepest layer) is a matrix of connective tissue containing blood vessels, nerves and lymphatics, in which any xenobiotic can be removed into the systemic circulation.
- *Epidermis* (50 to 100 μm thick) consists of proliferating cells and serves as the primary barrier to the absorption of substances through the skin. The outermost layer of the epidermis known as the *stratum corneum* (10 to 15 μm thick) is composed of 10 to 15 layers of cornified cells with cross-linked keratin filaments, i.e. a nonpolar lipid matrix surrounding a large number of thin parallel protein plates (the so-called brick-and-mortar model). This lipid matrix contains much more neutral lipids and less phospholipids than the living epithelial tissue [Lampe *et al.,* 1983].

The skin has many varied functions. The most important one is the protective function of acting as the barrier to xenobiotics, microorganisms, radiation, heat and other foreign agents. The stratum corneum is considered to be the major rate-limiting external barrier in dermal absorption of xenobiotics. The skin also serves as an important metabolizing organ for xenobiotics.

B. RATE-LIMITING PROCESSES

Percutaneous absorption is defined as the xenobiotic transport through the epidermis into the dermis to reach the systemic circulation *in vivo*. For *in vitro* study (particularly using non-viable skin), xenobiotics penetrate the skin barrier by diffusion and accumulate on the other side of the skin because they are not removed by systemic circulation. This process is described as skin permeation.

The mechanism of percutaneous absorption for xenobiotics is likely to be passive diffusion rather than an active transport process [Klaassen and Rozman, 1991]. The steady-state transport of lipophilic chemicals through biological membranes is a solubility-diffusion process [Aguiar *et al.,* 1969; Barry, 1983; Smith, 1990; Potts and Guy, 1992].

Xenobiotics initially dissolve into the surface layer of the stratum corneum; then diffuse into the stratum corneum, the rest of epidermis and dermis, respectively. Thereafter, they partition into body fluids at the stratum corneum-epidemis interface, epidermis and dermis, respectively, before ultimately entering the systemic circulation (Figure 3-15).

Based on experimental data, there are two possible transport routes for percutaneous absorption [Barry, 1983; Idson and Behl, 1987; Flynn, 1990; Behl *et al.*, 1990; Illel *et al.,* 1991]:

Transappendageal Route. The appendages provide an aqueous pore pathway through the skin. For very hydrophilic xenobiotics, this route may play an important role in dermal absorption.

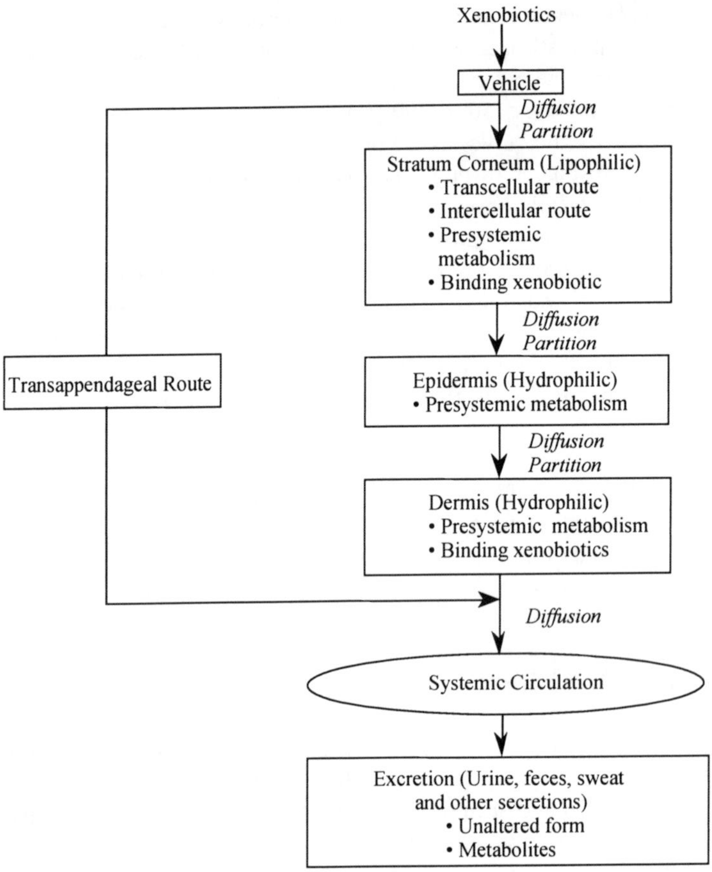

Figure 3-15 Absorptive steps via dermal route

Transepidermal Route. This includes intercellular and transcellular pathways. The intercellular (lipid partitioning) pathway for penetration is through the lipid of the stratum corneum. The transcellular pathway for permeation is through the aqueous tissue of the skin.

For non-polar xenobiotics, there is a linear relationship between skin permeability and the physicochemical properties of their xenobiotics in accordance with solution-diffusion theory and experimental data [Blank *et al.,* 1967; Blank and Scheuplein, 1969; Ackermann *et al.,* 1987; Hatanaka *et al.,* 1990; Potts and Guy, 1992]. The rate of permeability of these compounds is mainly related to their solubility and octanol/water partition coefficients, and is inversely proportional to the molecular weight [Marzulli *et al.,* 1965; Lieb and Stein, 1971; Potts and Guy, 1992]. The diffusion coefficients within the lipid phase are significantly lower than within the porous phase [Barry, 1983].

Thus, this kind of partition-limiting behavior may play an important role in skin permeation although the space of the lipid phase contributes only 10% to 30% of the total stratum corneum volume [Grayson and Elias, 1982]. The existence of the lipid pathway is useful for developing simple lipid membranes as a model skin barrier [Houk and Guy, 1988].

For polar xenobiotics, permeability can be better described by pore theory. Based on studies of hydrophilic drugs [Ackermann *et al.,* 1987; Hatanaka *et al.,* 1990], the permeability coefficients of these compounds through hairless mouse skin or artificial porous membranes depend on their molecular size, rather than their octanol/water partition coefficients.

The stratum corneum provides a two phase lipid-protein heterogeneous membrane with parallel lipophilic and hydrophilic pathways. Once a xenobiotic crosses the stratum corneum, it partitions into the underlying layers of epidermis, dermis and systemic circulation, which are more hydrophilic than the stratum corneum. These underlying layers provide more mass transfer resistance for lipophilic compounds.

C. OTHER FACTORS INFLUENCING DERMAL BIOAVAILABILITY

The factors influencing rate of skin absorption include the physicochemical factors of the xenobiotic (such as structure, molecular weight, pKa, ionization, stability, volatility, solubility, partition coefficients and diffusivity) and physiological factors (such as skin hydration, metabolism, age, skin condition and site of application). These factors can be examined by reference to published values, theoretical calculations or experimental measurements.

The most important parameters affecting dermal bioavailability for a xenobiotic are its water solubility, permeability coefficient, octanol/water partition coefficient and diffusivity.

Water Solubility. Water solubility will directly limit the rate of permeability for lipophilic xenobiotics [Scheuplein, 1965; Aguiar *et al.,* 1969; Ravis, 1990; Flynn, 1974, 1990].

Octanol/Water Partition Coefficient (k_{ow}). The k_{ow} is used to predict the relative rate of skin absorption from partitioning experiments between two immiscible liquid phases, typically octanol/water or mineral oil/water [Ravis, 1990; Flynn, 1974, 1990]. The k_{ow} and water solubility measurement can provide a good indicator of solubility in the membrane, but cannot predict diffusivity.

Diffusivity. Based on Fick's law of diffusion, the permeability rate of a xenobiotic is proportional to diffusivity [Ravis, 1990; Flynn, 1974, 1990]. Diffusivity is a kinetic property that depends on the state of substance (e.g. gas, liquid, or solid), temperature, viscosity, polarizability, electronic effects, hydrophobicity and molecular size. Nevertheless, the difference between diffusion coefficients is relatively small compared to difference in the k_{ow}.

Permeability Coefficient. Dermal permeability is experimentally measurable. The permeability coefficients of some organic compounds from water or vapor for humans are available in the literature [Dutkiewicz and Piotrowski, 1961; Hanke *et al.,* 1961; Piotrowski, 1967, 1971; Blank *et al.*, 1967; Dutkiewicz and Tyras, 1967, 1968; Scheuplein and Blank, 1971, 1973; Roberts *et al.*, 1977; Albery and Hadgraft, 1979; Baranowska-Dutkiewicz, 1982; Southwell *et al.,* 1984; Blank and McAuliffe, 1985; Wieczeorek, 1985; Bronaugh *et al.*, 1986; Hadgraft and Ridout, 1987]. These values, however, can only be gathered under steady-state conditions. Permeability is commonly applied for evaluating human exposure to vapors or to waterborne contaminants during swimming and bathing. Although the permeability may be approximated as a constant over time, it is highly variable according to the nature of xenobiotics (molecular size and shape, hydrophobicity and solubility), concentration of xenobiotics applied to the skin and thickness of the skin barrier.

Ionization. The surface of human skin is weakly acidic (pH 4.2 to 5.6), while the lower layers have a more neutral pH of 7.4. Penetration of acidic and basic organic xenobiotics is influenced by the pH of the skin or a carrier vehicle [Smith, 1990]. Generally, a weak acid placed on the relatively acidic skin will exist mainly in the un-ionized form that will more readily penetrate the skin. If the pH of the skin is changed by other media, the properties of the skin may also be altered, resulting in changes of the dermal absorption. Extremes of pH can damage the structure of the skin and dramatically alter dermal bioavailability.

Hydration. Skin hydration (wet skin) may cause the tissue to soften, swell or wrinkle. This state can increase the permeability of the skin [Barry, 1983].

Metabolism. Some xenobiotics (e.g. some carcinogenic PAHs) are sensitive to enzymatic metabolism within the skin [Scheuplein and Bronaugh, 1983; Barry, 1983; Ravis, 1990]. The resulting reactive metabolites may covalently bind with DNA or protein in the skin tissue to exert local toxicity or carcinogenicity. In this manner, the dermal bioavailability (access to systemic circulation) will be reduced.

Age. Infants or young children demonstrate much greater bioavailability than adults because of their under-developed skin barrier [Fisher, 1989]. The skin absorption of certain compounds may be lower in the elderly [Roskos *et al.*, 1989; Hall *et al.,*1992].

Skin Condition. A diseased or damaged skin usually increases permeability and the resultant rate of skin absorption [Scheuplein and Bronaugh, 1983; Barry, 1983; Ravis, 1990].

Site of Application. The skin absorption in humans or animals varies with the site of chemical application, which is termed regional variation [Wester and Maibach, 1989a; Gorsline *et al.*, 1992]. Permeability in the genital area or scrotum and around the head and face is higher than that in the limbs. The foot or palm areas exhibit much lower permeability.

Table 3-3 The Barriers and Rate-Limiting Processes for Absorption of Xenobiotics through Three Exposure Routes

Exposure route	Barrier	Rate-limiting determinant
Inhalation	Air/blood barrier: (1) aqueous layer; (2) epithelial cell; (3) interstitial space (4) capillary endothelial cell	Gas/vapor Water solubility Blood/air partition coefficient Chemical reactivity Aerosol Water solubility of a xenobiotic in the aerosol Dissolution rate in the respiratory tract Particle size distribution
Ingestion	Gut/blood barrier: (1) mucosal epithelial cell; (2) tight conjunction; (3) intercellular space; (4) capillary endothelial cell	Dissolution rate Blood flow Gut motility First-pass effect
Dermal	Stratum corneum	Lipid solubility Permeability Octanol/water partition coefficient (k_{ow}) Diffusivity

Physical Form. The xenobiotic may be exposed to the skin with a vehicle which may be solid, aqueous, aqueous paste, suspension, volatile, solvent or gas. The properties of the vehicle can be one of the major factors influencing dermal bioavailability [Meyer, 1991; Hugest *et al.,* 1992].

Dermal Decontamination. In most cases, a soap-and-water wash will remove contaminants from the skin. However, there is also evidence that dermal washing may enhance the skin absorption for certain xenobiotics [Wester and Maibach, 1989b].

IV. SUMMARY

Humans may be exposed to environmental contaminants via inhalation, ingestion and dermal contact. The mechanisms of absorption for each exposure route were summarized from the literature and discussed. The absorptive barriers and rate-limiting determinants are unique for each exposure route and are summarized in Table 3-3.

Table 3-4 Factors Influencing Bioavailability in Humans

Exposure factors	Physicochemical factors	Physiological (host) factors
Concentration	Solubility	Age
Duration	Partition coefficients	Race
Dosing vehicle	pK_a	Gender
Dosing form	Dissociation rate	pH of site application
Frequency of exposure	Molecular size	Surface area
Species difference	Molecular weight	Permeability of barrier
Environmental conditions (temperature, humidity, and wind velocity)	Reactivity	Specialized transport
	Polarity or solubility	Interaction with mucous
	Diffusion coefficient	Complexing components
	Volatility	Local blood flow
	Ionization	Surface activity
	Partial pressure	Bile interaction
	Vapor pressure	Food effects
	Properties of particle (shape, density, surface area, electrical charge and hygro-socopicity)	Gastric emptying
		Hydration of skin
		First-pass effect
		Motility of GI
		Absorption sites
		Plasma protein levels
		Metabolism
		Recirculation
		Clearance
		Ventilation
		Breathing pattern
		Disease states

The exposure state, physicochemical or physiological properties of biological membranes and xenobiotics affect their bioavailability. Various factors influencing bioavailability in humans are summarized in Table 3-4.

The values for bioavailability which are reported in Chapter 5 are subject to all the variability and uncertainty associated with the issues discussed in this chapter. Users of the data reported in Chapter 5 are cautioned to seek the original publications for data they may wish to use, in order to be sure that adequate attention has been given to these factors for the application being considered.

Chapter 4

EXPERIMENTAL METHODS FOR EVALUATING BIOAVAILABILITY

Both *in vivo* and *in vitro* methods have been used for bioavailability studies in humans and in animals. Because *in vivo* methodology is inherently more representative, this approach is greatly preferred for assessing bioavailability. However, the direct administration of xenobiotics to humans may pose a hazard. In a few cases, such studies have been conducted with young healthy male adult volunteers [Capel *et al.,* 1972], but ethics approval for non-medical trials are very unlikely to be approved in the future. More realistically, laboratory animals are utilized for these studies. Because indirect methods are most commonly used, there will be substantial uncertainties involved in interpreting the results. Additionally, uncontrollable biological variables, individual organism variability, animal care ethics and high cost must be resolved.

There are a variety of *in vivo* methods for investigating bioavailability via the three main exposure routes. Indirect methods (e.g. measurement of xenobiotics in biological samples) are conducted in living animal models. Direct methods (e.g. mass-balance technique) use live animals, but require the animals to be sacrificed to confirm the extent of absorption.

Attempts have been made to develop *in vitro* methodology for bioavailability studies. Compared to *in vivo* methods, *in vitro* methods are less expensive, less time-consuming, and involve fewer ethical issues. Good agreement of *in vivo* and *in vitro* for some xenobiotics has been observed. Currently, most *in vitro* methods are not directly applicable for quantitative determination of bioavailability. Nevertheless, *in vitro* methods are widely used for the determination of some important physicochemical factors, such as solubility, water/air or lipid/water partition coefficient, dissolution rate and particle size distribution.

I. GENERAL *IN VIVO* METHODS

A. MEASUREMENT OF BLOOD LEVELS

Measurement of blood levels of xenobiotics is employed to quantitatively evaluate the bioavailability for most xenobiotics through all three main exposure routes [Barr, 1972; Wester and Noonan, 1978; Nugent and Wood, 1980; Hwang and Danti, 1983; Wester *et al.,* 1983a; McDougal *et al.,* 1985; Aungst and Shen, 1986; Ogiso *et al.,* 1987; Abdou, 1989; Hall and Shah, 1990]. The procedure is described as:

- subjects are prepared (test animals);

- a single dose or multiple doses of a tested xenobiotic, often radioactively labeled, is administered via (1) the extravascular (ev) routes, including inhalation, ingestion and dermal, and (2) the intravenous (iv) route. The intravenous dose is used as a reference dose because it is directly introduced into the systemic circulation and may be taken to have a bioavailability factor equal to one;
- serial blood samples are collected at desired intervals until absorption and distribution are completed;
- blood levels of the xenobiotic are measured by using appropriate analytical techniques. Chromatographic techniques and liquid scintillation spectrometry are commonly employed for the qualitative and quantitative analysis of xenobiotics in biological samples;
- blood levels are plotted vs. time to determine the area under the plasma concentration-time curve (AUC), in units of mass x time/volume (see equation 2-1) [Abdous, 1989];
- the bioavailability factor (BF) is calculated from:

For a single exposure [Abdous, 1989]:

$$BF = \frac{AUC_{ev}/D_{ev}}{AUC_{iv}/D_{iv}} \ (x\ 100) \qquad (4\text{-}1)$$

where:
D_{ev} refers to extravascular dose;
D_{iv} is intravenous dose, assuming that the excretion rate constants of both administration routes are consistent.

For multiple exposures [Abdous, 1989]:

$$BF = \frac{(AUC_{n(ev)}/D_{ev})\tau_{ev}}{(AUC_{n(iv)}/D_{iv})\tau_{iv}} \ (x\ 100) \qquad (4\text{-}2)$$

in which τ is the dosing time.

For relative bioavailability [Labaune, 1989]:

$$BF_R = \frac{AUC_2}{AUC_1} \ (x\ 100) \qquad (4\text{-}3)$$

where:
AUC_1 is the blood concentration for a standard reference material;
AUC_2 is the blood concentration for the material under study.

Strengths:

- This approach is the most sensitive and accurate way for the assessment of both the extent and rate of bioavailability.

Limitations:

- This method is inconvenient for small animal studies because of the small total volume of blood in these animals.
- This method can be technically demanding.

B. MEASUREMENT OF THE CUMULATIVE URINARY EXCRETION

Because the blood concentration and the urine excretion concentration will be proportional, measurement of the cumulative concentration of xenobiotics in urine is used to roughly estimate bioavailability. Typically, urinary data represents the lower bound of bioavailability factor for most xenobiotics [Abdou, 1989]. The procedure is similar to the blood level method and needs:

- subjects are prepared (test animals);
- a xenobiotic under study is administered via the extravascular routes and the intravenous route;
- serial urine samples are collected at desired intervals until excretion is completed;
- urinary levels of the xenobiotic are measured by using appropriate analytical techniques;
- the bioavailability factor is calculated from [Abdou, 1989]:

$$BF = \frac{A_{e\,(ev)}^{\infty} / D_{ev}}{A_{e\,(iv)}^{\infty} / D_{iv}} \ (\text{x } 100) \tag{4-4}$$

where:

A_e^{∞} is the total mass of a xenobiotic excreted in urine from dosing time to infinity.

For relative bioavailability [Labaune, 1989]:

$$BF_R = \frac{A_{e2}^{\infty}}{A_{e1}^{\infty}} \ (\text{x } 100) \tag{4-5}$$

where:

A_{e1}^{∞} is the total mass of a xenobiotic in urine for a standard material;

A_{e2}^{∞} is the total mass of a xenobiotic in urine for a material under study.

In many cases, the bioavailability factors are directly estimated from the fraction of a xenobiotic excreted in urine without using a reference dose (i.e. the intravenous dose) for a correction of metabolism and distribution of the xenobiotic. This measured urinary level of the xenobiotic may include the parent compound and its metabolites. Use of this urinary fraction to estimate the bioavailability factor of a xenobiotic is only valid when the xenobiotic is poorly metabolized and distributed in the body but is mainly excreted in urine. This urinary fraction will represent a lower bound of the bioavailability factor.

Strengths:

- Urine samples are simple and convenient to use for animal studies.
- This method is both inexpensive and rapid.

Limitations:

- This method may not be reliable. The approach underestimates the bioavailability factor in most cases because only a small fraction of the parent compound is found in urine, and because the measured values are confounded by other factors, such as the urinary pH being different from blood pH, variable amount of free form in plasma, variable urinary flow rate, inconsistent collection method and the lack of sensitivity of analytical methods.
- This method is valid only when at least 20% of the total dose is eliminated in urine following intravenous administration.

C. MASS-BALANCE TECHNIQUE

The mass-balance technique has been extensively employed for bioavailability studies in animals [Pegg *et al.,* 1979; Schumann *et al.,* 1980; Shah *et al.,* 1981, 1987; Frantz and Watanabe, 1983; Shah and Guthrie, 1983; Reifenrath *et al.,* 1984a, 1984b; Bucks *et al.,* 1988; Hall and Shah, 1990; Schlesinger, 1990; Banks and Birnbaum, 1991; Hughes *et al.,* 1992; Tsai *et al.,* 1992]. This approach is based on overall conservation of xenobiotic mass in relation to the known administered dose. Because xenobiotics present in the systemic circulation will be distributed to various body fluids, organs and tissues, the total mass of a xenobiotic at these sites other than the applied sites is part of the internal dose. Thus, the internal dose is estimated as the total recovery from all body fluids, organs and tissues minus the mass recovered from the applied site (Figure 4-1).The procedure is:

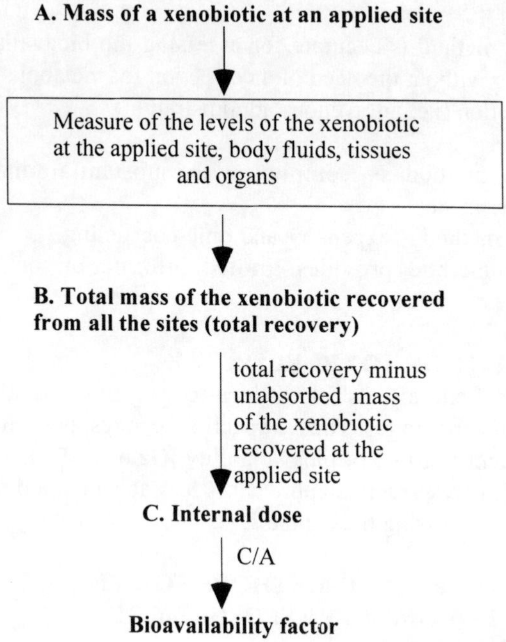

Figure 4-1 Determination of the bioavailability factor based on mass-balance technique

- subjects are prepared (test animals);
- a xenobiotic is administered by inhalation, ingestion or dermal contact;
- blood and urine samples are collected at desired intervals;
- animals are sacrificed at selected times;
- the xenobiotic is analyzed in the body fluids, tissues and organs to determine the total mass recovery (note that the measured compounds may include the metabolites as well as the parent compounds. These metabolites are biotransformed from the parent compounds, and then distributed to various tissues and organs);
- the ratio of internal dose (D_{in}) to external dose (D_{ex}) is calculated to determine the bioavailability factor, as:

$$BF = \frac{D_{in}}{D_{ex}} (x100) \tag{4-6}$$

where:

D_{in} includes the mass of xenobiotics in the body fluids, tissues and organs.

Strengths:

- This method is accurate for assessing the bioavailability factor without the need of a correction for metabolism and excretion (i.e. intravenous administration).

Limitation:

- This method is complex with substantial analytical requirements.
- This method is expensive and time-consuming.
- This method provides limited information on kinetic factors.

D. MEASUREMENT OF TOXIC RESPONSE

Occurrence of toxic effects, either local or systemic, usually indicates some absorption of xenobiotics. Thus, a specific toxic response might be used to estimate the extent and rate of bioavailability [Gaines, 1960, 1969; Gaines and Linder, 1986]. However, this approach is less accurate and substantially less quantitative for assessing bioavailability.

II. SPECIFIC METHODS FOR THE INHALATION EXPOSURE ROUTE

Bioavailability within the respiratory tract is a very complex process. For evaluating gas and vapor uptake, the water solubility of gas and vapor and the total intake are often examined. For aerosols, assessment of bioavailability is associated with a series of interactive events: deposition-clearance-retention. Bioavailability may be estimated by calculation of the difference between the deposition fraction from a combination of the retention fraction and the mechanical clearance fraction with a correction for the amount of xenobiotic cleared to the gastrointestinal tract (Figure 4-2).

In principle, there are both *in vivo* and *in vitro* methods available for the study of respiratory bioavailability. With *in vivo* methods, the exposure approaches include: (1) *Whole Body Exposure.* Humans and animals are placed in a room or in a chamber (only for animals) supplied by clear air dosed with the test substance [Phalen *et al.,* 1984; MacFarland, 1987; Drew, 1988; Dorato, 1990; Snelling and Dodd, 1990; Karg *et al.,* 1992]; (2) *Nose-only Exposure.* Animals are placed in stocks or whole-body tubes with only the nose protruding into an exposure chamber [MacFarland, 1987; Henry and Kouri, 1987; Holländer 1988]; (3) *Head-only Exposure.* This is similar to nose-only exposure except that the animals are restrained in the chamber by a collar around the neck, so that the mouth, eyes and head are exposed to the test substance; and (4) *Intratracheal Instillation.* In this method, a blunt needle is inserted to 1 or 2 mm above the bifurcation of the airway and the test substances are directly introduced into the respiratory tract [Bend *et al.,* 1985; Bevan and Ruggio, 1991]. One *in vitro* method, the isolated perfused lung, has been used for respiratory bioavailability. In this method, the

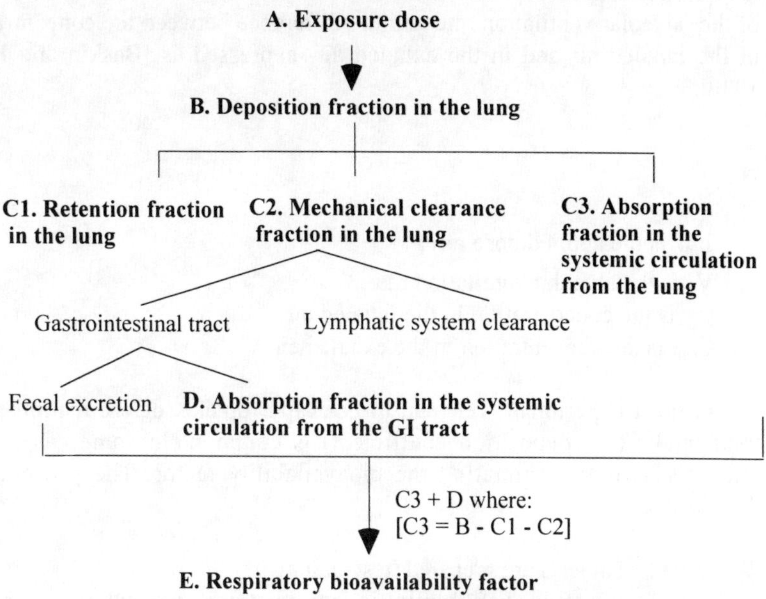

Figure 4-2 A procedure for calculation of respiratory bioavailability factor of aerosols based on the deposition-clearance-retention process

isolated perfused lung still retains the pulmonary structure and functions during a short experimental period. The parameters of metabolism as well as bioavailability for a xenobiotic can be obtained. Selection among these *in vivo* and *in vitro* methods must be justified to animal care ethics committees for approval of experimental protocols.

A. DETERMINATION OF DEPOSITION FRACTION

After inhalation exposure, only the fraction of a xenobiotic deposited within the respiratory tract is available for further mechanical clearance and absorption. Consequently, the deposition fraction is related to the applied dose.

1. Gaseous Phase

Typically, characterization of an inert insoluble gas or vapor (such as CO), including mixing and distribution, can be determined by lung function tests [Ultman, 1988]. For toxic and reactive gases or vapors, the deposited pattern in various regions is derived from mathematical models rather than from experiments [Miller *et al.,* 1985, 1988; Overton, 1984, 1990; Overton *et al.,* 1987; Cuddihy and McClellan, 1989].

Deposition of gases and vapors onto the walls of airways accounts for reduction of concentrations reaching the pulmonary region. Thus, the applied dose is less than the administered dose. The applied dose rate is the product of the alveolar ventilation rate and the difference between the concentrations in the inhaled air and in the exhaled air, expressed as [Baskin and Falco, 1989]:

$$J_{app} = \dot{V}_A (C_{in} - C_{out}) \tag{4-7}$$

where:
J_{app} is the applied dose rate;
$\dot{V}_A$ is the alveolar ventilation rate;
C_{in} is the concentration in the inhaled air;
C_{out} is the concentration in the exhaled air.

In most experiments relevant to gas/vapor uptake, determination of the total intake (i.e. deposition coefficient) is common. In some cases, total intake is used for estimating the bioavailability factor. The procedure is described as:

- subjects are prepared (test animals);
- a tested xenobiotic is administered via whole body exposure, nose-only exposure or head-only exposure;
- the xenobiotic concentrations in the inhaled air and in the exhaled air are measured;
- deposition fraction (total intake) is calculated, as expressed by:

$$I = \frac{C_{in} - C_{out}}{C_{in}} \ (\text{x } 100) \tag{4-8}$$

Strengths:

- This approach is simple and convenient.

Limitations:

- The bioavailability factor is overestimated for most xenobiotics because the total intake reflects the fraction of the total inhaled xenobiotic deposited within the respiratory tract rather than those which actually cross the air-blood barrier in the air sac of the lung.
- The bioavailability factor is applicable only for relatively low water-solubility xenobiotics because the absorption of gases and vapors is dependent on water-solubility. Highly water soluble xenobiotics tend to deposit on the aqueous layer of the upper airways. Lipophilic xenobiotics primarily reach the alveoli where there is a large surface available for absorption and the compounds readily penetrate the air-blood barrier. Therefore, the deposition

fraction for highly lipophilic xenobiotics is more representative of the bioavailability factor than for hydrophilic xenobiotics.

- The bioavailability factor is variable with breathing rate, blood/air partition coefficient, and blood flow rate.

Moreover, determinations of physicochemical factors, such as blood/air partition coefficient and water solubility of xenobiotics, are required for fully interpreting a bioavailability study.

2. Particulate Phase

Total deposition fraction can be determined by comparing the difference in concentrations between the administered test aerosols of the inhaled air and the exhaled air. Regional deposition fractions may be estimated by measurement of aerosol concentrations in different volume fractions of the exhaled air by making assumptions about mixing and dead space [Schlesinger, 1985, 1988, 1989]. Alternatively, deposition fractions may be estimated from the inhalable fraction and particle size distribution.

A variety of analytical techniques are applied for assessing the aerosol size distribution, such as optical or electron microscopy, cascade impactor, nuclei counters, charge spectrometer, light extinction device, diffusion battery, surface area measurement device and elutriators [Macias and Husar, 1976; Phalen, 1984; Holländer, 1988; Cai *et al.,* 1992; Cheng *et al.,* 1992; Hunt *et al.,* 1992].

B. MECHANICAL CLEARANCE FRACTION

The majority of aerosols deposited within the respiratory tract will be removed by mechanical clearance processes. Clearance studies generally focus on examining the rate for removal of inhaled xenobiotics from the whole lungs (including a portion of the tracheobronchial region and the entire pulmonary region) and from individual airways, alveolar macrophage function and accumulation in the lymph nodes. Measurement of the long-term clearance rate to lymph nodes is rarely available in human studies. Such data may be derived from animal studies. The common methods for assessing mechanical clearance are:

- *Mucociliary Transport Velocity.* Mucociliary transport rate which refers to the time required for aerosols to move between two defined areas can be identified by the monitoring of markers (e.g. colloid albumin, India ink droplet, color solution and dye) on the epithelium or measuring the movement of radioactive aerosols within the airways [Schlesinger, 1990]. This technique is simple but invasive.

- *Whole Lung Clearance.* There are two methods for measurement of whole lung clearance [Lippmann, 1981; Schlesinger, 1990]. First, the clearance rate can be determined by using a scintillation detection technique. The second technique involves the use of magnetopneumography. Aerosols which produce magnetization can be detected by magnetopneumography [Cohen, 1973; Cohen *et al.,* 1979; Halpern *et al.,* 1981].
- *Fecal Analysis.* Following inhalation of radioactive aerosols, fecal samples collected at selected times are analyzed by liquid scintillation detector [Kenoyer *et al.,* 1981; Mannix *et al.,* 1983]. This measurement reflects the xenobiotic clearance by the mucociliary transport into the gastrointestinal tract.
- *Macrophage Function.* Classic approaches for studying the pulmonary macrophage function consist of ultrastructure, biochemistry, *in vitro* culture and immunologic bioassays. The *in vivo* procedure for assessing phagocytosis is a comparison of the difference between free gold and cell-associated gold washout pattern after administration of labeled colloidal gold [Beck *et al.,* 1981]. Magnetic particles and sensitive magnetometers have been applied for studying the pulmonary macrophage motility *in vivo* [Brain, 1992].

C. DISSOLUTION RATE

Dissolution testing is used to study the behavior of aerosols when the dissolution is a determinant for rate of absorption. Methods are:

- *In vivo methods.* Radioactive particles are injected into the muscle of the experimental animals, where the test substances are dissolved in the biological fluids. Clearance profiles can be obtained by analyzing the retention of the test substances at the injected site. Clearance parameters from the intramuscular injection were found to correspond to data from the lung [Morrow *et al.,* 1968, 1979; Thomas *et al.,* 1973]. Intramuscular injection procedures provide information about the dissolution-absorption clearance.
- *In vitro methods.* Constituents of the aerosols may be eluted by using ultrafiltration with buffered and unbuffered biological methods and other techniques [Bevan and Worrell, 1985; Bevan and Yonda, 1985; Törnquist *et al.,* 1988; Bevan and Ruggio, 1991].

D. *IN VITRO* METHODS: ISOLATED PERFUSED LUNG

The isolated perfused lung preparation is well suited to the investigation of xenobiotic bioavailability and metabolism in the lung [Mehendale *et al.*, 1981; Niemeier, 1984; Rohades, 1984; Rao and Mehendale, 1987; Niven and Byron, 1988; Roth and Bassett, 1989]. The procedure involves:

- an isolated lung is prepared by removal of a lung from an experimental animal;
- a perfusion system is prepared, including a negative pressure ventilation-perfusion system [Niemeier and Binghham, 1972], positive pressure ventilation-perfusion system [O'Neil and Tierney, 1974] and a split lung perfusion system [Tucker and Shertzer, 1980];
- the isolated lung is placed in the perfusion apparatus where it is suspended by the pulmonary arterial cannula;
- the medium is perfused into the pulmonary arterial cannula at a constant flow and pressure;
- serial samples are collected;
- concentrations of collected samples are measured by appropriate analytical techniques.

Strengths:

- Pulmonary structure and function integrity are maintained.
- The role of the lung, compared to the whole animal or lung slice preparation, is evaluated.
- Mass balance of xenobiotics may be determined by measuring the mass recovery in the perfusate, the tissues and the exhaled air via the trachea.
- Exposure routes are similar to those under physiological conditions. Test xenobiotics may be administered by inhalation, pulmonary artery or trachea.
- A variety of compositions of perfused medium may be used.
- Experimental parameters, such as volume of perfusion medium, concentration of test substances, perfusion pressure, ventilation rate and blood flow, may be easily controlled.

Limitation:

- The perfused lung is only viable for a short time.
- This method is technically demanding to set up and perform.

III. SPECIFIC METHODS FOR THE INGESTION EXPOSURE ROUTE

In principle, there are three general approaches for determining gastrointestinal bioavailability:

- *In vivo* methods, using unanesthetized animals, provide direct assessment of the bioavailability factor of a xenobiotic via ingestion. In addition to the basic methods mentioned in Section I, the measurement of cumulative fecal excretion of the xenobiotic is frequently used. Other methods consist of the measurement of liver ratio of a xenobiotic discussed later and the chronic isolated loop model. Chronic isolated loop technique is performed with unanesthetized large animals (e.g. dogs), as well as small animals [Ings, 1984; Csáky, 1984; Poelma and Tukker, 1987; Lu *et al.,* 1992]. A segment of the intestine with its blood supply and lymphatic connection intact is isolated and attached to the abdominal wall. The test substance is introduced into the loop and samples from the loop content are collected at selected times.
- *In situ* methods, using anesthetized animals, involve taking the intestine outside of animal while keeping the blood supply intact. These methods provide experimental conditions to mimic true *in vivo* conditions. This approach may be used to study the mechanism of transport, site of absorption and gastrointestinal metabolism. Techniques for this method include using perfused loop (Figure 4-3a), closed loop (Figure 4-3b) and mesenteric collection (Figure 4-3c).
- *In vitro* methods, using isolated gastrointestinal tissue, dissolution and partition coefficient tests, involve using model membranes selected to reflect the functionality of biological membranes. The isolated gut wall method is used to study the transport characteristics of a xenobiotic substance whose absorption is passive. These techniques include: everted intestinal sac (intestine turned inside out) (Figure 4-3d), isolated perfused intestinal segment, or biological and artificial membrane models. Dissolution and partition coefficient tests are useful for determining physicochemical factors relevant to bioavailability.

The latter two categories, *in situ* and *in vitro* methods, only provide insight into the intestinal membrane-level behavior of xenobiotic absorption.

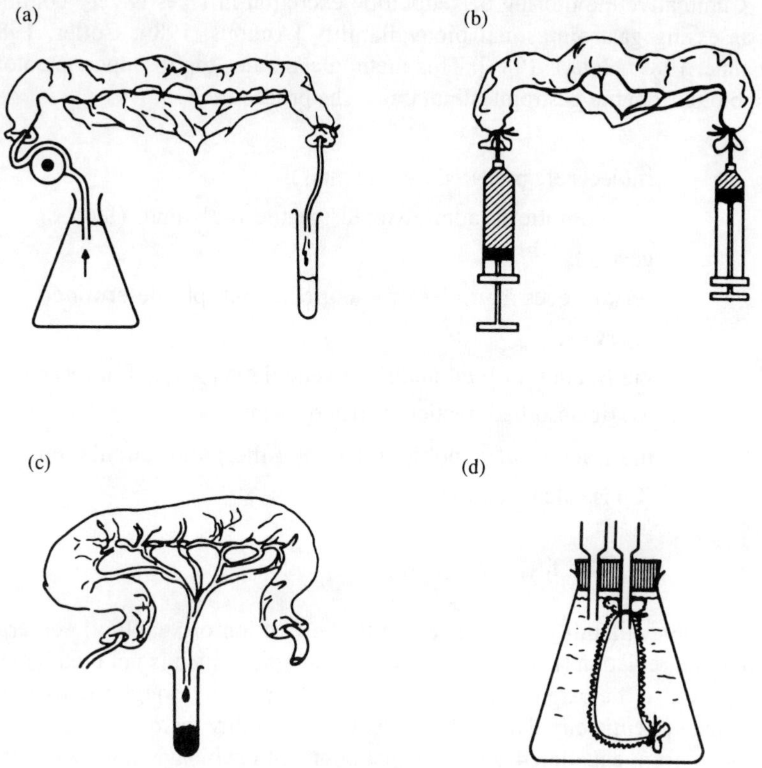

Figure 4-3 *In vitro* experimental methods for studying oral bioavailability. (a) Perfused loop preparation (*in situ*). The small intestinal loop is isolated by ligation from an anesthetized small animal and cannulated at the duodenal and ileal ends with polythene cannula having an infusion pump. The isolated loop is perfused with the perfusion and tested solution. (b) Closed loop preparation (*in situ*). The small intestinal loop is isolated by ligation from an anesthetized small animal. The tested solution is introduced into the loop by a syringe with a narrow-gauge needle. At a selected time, the animal is sacrificed and the concentration of the compound in the loop content or in the blood is assayed. (c) Isolated loop with complete collection of mesenteric venous (*in situ*). The isolated loop from a small animal is cannulated and perfused with the tested solution. The hepatic portal vein or mesenteric vein from the loop is cannulated and the complete mesenteric blood is collected, while the donor blood infuses to the animal via the renal vein. (d) Everted intestinal sac (*in vitro*). A small segment of the intestine is cut from laboratory animals, everting the segment, filling the sac with a small volume (0.5-1 ml) of physiological buffer, tieing off both ends and incubating the sac in buffer solution containing the test substrate at 37 ^{o}C for 15-60 min. After incubation, the serosal fluid is assayed. (From author, Barr, W.H., The use of physical and animal models to assess bioavailability, *Pharmacol.* 8, 55, 1972. With permission.)

A. MEASUREMENT OF FECAL EXCRETION

Cumulative monitoring of xenobiotic excretion in feces is very common for assessing gastrointestinal bioavailability [Abdous, 1989; Cutler, 1986; Labaune, 1989; Notari, 1987]. This method estimates the disappearance of a xenobiotic from the gastrointestinal tract. The procedure is:

- subjects are prepared (test animals);
- a xenobiotic is administered via the oral route (feed or gavage);
- serial feces samples are collected at pre-determined intervals;
- the fraction of total unaltered xenobiotic excreted in feces (f_f) (unabsorbed fraction) is determined;
- the fraction of xenobiotic entering the portal circulation (F') is calculated by:

$$F' = 1 - f_f \tag{4-9}$$

The bioavailability factor is equal to the fraction of xenobiotic entering the portal circulation (F') for the case of a xenobiotic which is not metabolized in the liver, so the parent form of the xenobiotic passes through the liver and enters the systemic circulation. Thus, the bioavailability factor of a xenobiotic calculated from equation 4-9 is the upper bound of the bioavailability factor.

However, estimation of the fraction of presystemic metabolism which occurs in the liver (f_h) for most xenobiotics is needed for an assessment of systemic bioavailability. Estimation of f_h is relatively simple in animals, but more difficult in humans. The equation for f_h is:

$$f_h = \frac{CL}{Q} \tag{4-10}$$

where:

CL is the hepatic clearance;

Q represents the hepatic blood flow.

Thus, the overall bioavailability factor is:

$$BF = F' - f_h = 1 - f_f - f_h \tag{4-11}$$

For instance, when a xenobiotic completely crosses the gut wall into the portal blood (that is, f_f=0 and F'=1) and the fraction of hepatic presystemic metabolism is 0.7, the bioavailability factor, BF, will be 0.3.

Hepatic presystemic metabolism is readily saturated. Consequently, the bioavailability factor is dependent not only on the concentration of a xenobiotic in the portal blood, but also on the capacity of hepatic metabolism for the xenobiotic. Under this circumstance, the extent of absorption in the systemic circulation is non-linearly related to an increase of the administered dose.

Strengths:

- This approach is relatively simple and convenient.

Limitations:

- The bioavailability factor will be overestimated unless a correction for hepatic presystemic metabolism is used for those xenobiotics which are readily metabolized in the liver, a substantial correction in many cases.

B. MEASUREMENT OF LIVER RATIO

Oral bioavailability may be estimated by comparing the ratio of the levels of a test form to those of a standard form (e.g. pure form) for the same xenobiotic in the liver following oral administration in animals [McConnell *et al.,* 1984; Umbreit *et al.,* 1986; Shu *et al.,* 1988]. This approach is based on the premise that the level of the xenobiotic in the liver may represent the systemic level of the xenobiotic in the body and that the bioavailability of the standard material is assumed to be 100%. Some studies indicated that this approach is applicable for 2,3,7,8-tetrachlorodibenzo-p-dioxin (TCDD) [Shu *et al.,* 1988]. Two reasons can be presented. Firstly, the liver is a major organ which uptakes and retains TCDD, so that the liver levels are likely to reflect the systemic levels of TCDD. Secondly, the levels of TCDD in the liver slowly decrease after oral dosing, so that significant quantities of TCDD may be measured.

C. DISSOLUTION TESTING

Dissolution testing for determination of physicochemical factors is commonly employed to predict the oral bioavailability of solid xenobiotics. Most experimental techniques for estimating water solubility of xenobiotics are derived from bioavailability study methods developed for drugs, such as USP rotating basket, rotating paddle, column flow-through apparatus, sartorius apparatus (*in vivo* simulative method), BioPredictor (*in vivo* predictive methods) and equilibrium geochemical speciation model MINTEQA 2 [Smolen and Ball, 1984; Ruby *et al.,* 1992; Davis *et al.,* 1992].

Scientists have attempted to establish *in vitro-in vivo* correlation of the compounds by using *in vitro* bioavailability studies because the primary absorption parameters of a solid xenobiotic for gastrointestinal bioavailability depend on its solubility and dissolution characteristics [Abdou, 1989]. One approach is to conduct *in vivo* experiments and *in vitro* experiments (e.g. dissolution testing) simultaneously. *In vivo* and *in vitro* correlation is examined. Some scientists also collected *in vivo* and *in vitro* data for a similar group of xenobiotics from the literature, and examined *in vivo* and *in vitro* correlation. In some selected cases, dissolution tests may predict *in vivo* behavior. Most of those have been unsuccessful. However, the *in vivo* environment is more complex than the *in vitro* test environment. *In vitro* dissolution tests cannot reliably predict differences among xenobiotics with low *in vivo* bioavailability, very high bioavailability and those subject to changes in bioavailability by physiological factors.

D. METHODS BASED ON THE pH PARTITION THEORY

Based on pH partition theory, the relative oral bioavailability of a xenobiotic can be estimated by comparing its partition coefficients between two immiscible solvents. A number of methods are reviewed by Abdou [1989]. These methods include inverted Y-shaped tube, circular cell, aqueous-octanol models, finite-difference methods and a two-phase model with dissolution layer. The inverted Y-shaped tube and circular cell methods are restricted to the study of xenobiotics in solution.

IV. SPECIFIC METHODS FOR DERMAL EXPOSURE ROUTE

In vivo methods for dermal bioavailability involve the surface disappearance of a xenobiotic at an applied area of skin. *In vitro* methodology is also a powerful tool for evaluating the dermal bioavailability. Because the stratum corneum is non-living and is the rate-limiting membrane of the skin, *in vitro* approaches using this layer are likely to accurately measure the skin bioavailability and reflect *in vivo* skin absorption profiles while providing the ability to control experimental conditions.

Waterborne contaminants or vapors in contact with skin can directly penetrate the skin barrier to reach the systemic circulation. Because there are difficulties with determining the contact rate for waterborne contaminants or vapors, it is very difficult to directly apply a bioavailability factor. Rather, the extent of absorption can be estimated by using dermal permeability. In most cases, the permeability coefficients are derived from *in vitro* tests. For some hydrophilic compounds, good agreement between *in vivo* and *in vitro* dermal absorption has been found. However, correlations are poor for lipophilic and extremely hydrophilic compounds [USEPA, 1992b].

A. SURFACE DISAPPEARANCE

Because a xenobiotic directly penetrates the skin into the systemic circulation, dermal bioavailability can be estimated from examination of surface disappearance [Hall and Shah, 1990]. The bioavailability factor is the difference between the applied dose and the fraction of a xenobiotic that disappears from the applied site. Several appropriate techniques for measuring the surface disappearance have been reported in the literature. In principle, these procedures investigate the loss and retention of a xenobiotic within the skin in humans and animals and provide the dermal absorption profiles. The techniques include:

- *Tape Stripping.* The stratum corneum is stripped off by successively sticking it to cellophane tape (20 to 30 times) following a short application periods (30 minutes). These tape strippings are assayed by using appropriate methods [Baker and Kligman, 1967].
- *External Counting Technique.* A radioactively labeled xenobiotic is applied on the skin surface. A radiation detector is placed over the application site and the radiation for the xenobiotic is measured at pre-determined intervals by the radiation detector [Marty *et al.,* 1989; Hall and Shah, 1990]. This technique is easy to use, non-invasive, objective, precise and inexpensive if the xenobiotic is available in labeled form.
- *Histological Method.* A biopsy of the site of application is taken at a desirable time and the location of the compounds within the membrane structure is determined by light microscopy [Harris *et al.,* 1974; Hall and Shah, 1990; Srikrishna *et al.,* 1992]. This technique is strictly qualitative.

B. *IN VITRO* METHODS: BIOLOGICAL MEMBRANE MODELS

Biological membranes involve living or non-living skin in humans or animals [Behl *et al.,* 1990; Bronaugh, 1990]. Human skin is preferred for most skin permeation studies. Human skin is obtained from autopsies (fresh cadaver skin) or surgical specimens (face, breast, and abdominal). Depending on the need of the study, whole skin, stripped skin, epidermis and dermis could be chosen. Human skin can not be readily obtained for experimental purposes. A number of animal skins, including hairless mouse, rat and guinea pig, Swiss mouse, Athymic nude mouse, furry rat, nude rat, cat, dog, rhesus monkey and rabbit, have been commonly used for skin permeation studies.

Some researchers have been exploring new approaches for skin permeation studies [Itoh, 1990a, 1990b; Rigg and Barry, 1990; Harada *et al.,* 1992]. Currently, shed snake skin, which is a non-living pure stratum corneum with no follicles, has been investigated as a model membrane. Compared to rat or mouse skin through which the permeability coefficients are usually 100-1000 fold greater than that through human skin, the data from shed snake skin are only several times different from human skin. Overall, shed snake membrane may prove to be a good model membrane system for the study of human skin permeability because of its similarities to human skin, low cost, and ease of storage and handling.

In model membrane studies, the thickness of the membranes will play a critical role in influencing the overall skin permeability. There are several ways to prepare the membranes:

- *Full-thickness Skin Resection.* This method uses the epidermis and the full-thickness dermis (even subcutaneous tissue) and is suitable for animals with a thin skin (e.g. mouse and rabbit) [Bronaugh and Stewart, 1984].
- *Split Resection.* The thickness of human and some animal skins (guinea pig, pig, and monkey) is almost 1 mm or more than 1 mm. These need to be split into thinner sections for experiments. These sections of skin contain the viable epidermis and some portion of the capillary dermis with approximately 200 μm thickness needed to include the microcirculation [Barry, 1983].
- *Isolated Perfused Porcine Skin Flap (IPPSF).* This is a living *in vitro* method which uses the intact skin of a pig with a viable and defined arterial and venous supply [Riviere and Monteiro-Riviere, 1991]. The preparation, a single-pedicle axial pattern tube, may be perfused under ambient environmental conditions. The tested xenobiotic can be placed on the surface of IPPSF and assayed from the venous effluent over time.
- *Congenitally Athymic (Nude) Animal Models.* This skin model consists of a split thickness human or pig skin that is grafted to the subcutaneous surface of the epigastric skin of the nude mouse or rat [Klain and Black, 1990]. Following the application of a tested xenobiotic, animals are sacrificed and a variety of tissue samples are analyzed to determine the fate of the xenobiotic.

Currently, there are two basic designs in model membrane studies: static cells and flow-through cells [Frantz, 1990; Gummer and Maibach, 1991]. A typical diffusion cell experiment involves three compartments (Figure 4-4): the donor, the barrier and the receptor. Two chamber systems can be divided into either an air/fluid phase chamber in which the surface of the epidermis is exposed to room air and the dermis is bathed in the aqueous medium, or a

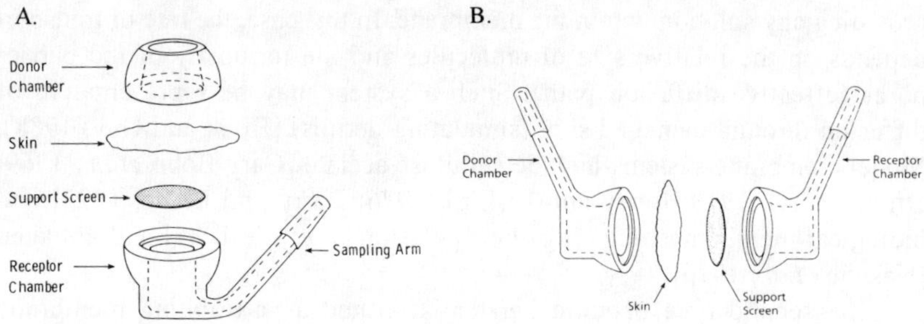

Figure 4-4 Diagram of a typical diffusion cell. A. the air/fluid type of the diffusion cell. B. the fluid/fluid of the diffusion cell. (From author, Dugard et al., Absorption of some glycol ethers through human skin *in vitro, Environ. Health Persp.*, 57, 193, 1984. With permission.)

fluid/fluid phase chamber in which the aqueous medium is in contact with both sides of the skin preparation. The receptor fluid should be chosen to allow free diffusion of the xenobiotic into it. Normal saline and aqueous buffer solutions are commonly used. The xenobiotic is delivered to the membrane from the donor compartment, where it penetrates the barrier and appears in the receptor compartment. Diffusion parameters are determined by sampling and analyzing the solute in either the donor or receptor phase as a function of time. The results can be expressed as either absorptive rate vs. time or cumulative absorption vs. time. In the aqueous/aqueous phase static diffusion cells, the parameters are evaluated by serial sampling of the receptor compartment. With flow-through diffusion cells, automatic replenishment of the receptor liquid can be provided to keep concentrations in the receptor liquid at a constant level.

Xenobiotic samples are quantified by various analytical methods, including radioisotope detectors, various chromatographic detectors, spectrophotometry, fluorometry and specific assays [Loden, 1990].

Permeability, diffusion coefficient and partition coefficient can be determined from these experiments.

C. *IN VITRO* METHODS: ARTIFICIAL MEMBRANE MODELS

Artificial membrane systems are generally divided into two categories: a liquid barrier and a solid barrier. In the liquid membranes which mimic the lipid nature of biomembranes, organic liquids (isopropyl myristate and tetradecane) are soaked on filter membranes [Albery *et al.,* 1976; Albery and Hadgraft, 1979; Guy and Fleming, 1979; Hadgraft and Ridout, 1987, 1988; Houk and Guy, 1988; Ridout and Guy, 1988; Tanaka *et al.,* 1978]. In contrast, solid membranes are involved in employing particular polymers or other solid materials as a barrier between two liquid phases. Based on the mechanism of transport, two types of solid barriers can be distinguished.

The first type is termed a dialysis membrane (heterogeneous barriers) in which the substances can penetrate through pores in the membrane without providing any solution within the membrane. In this case, the rate of transport depends on the relative size of molecules and the tortuosity of the barrier pores (effective diffusion path). Such a system may be representative of diffusion through damaged skin (simulating dermis) [Houk and Guy, 1988]. Model membrane systems include cellulose acetate [Gary-Bobo *et al.,* 1969; DiPolo *et al.,* 1970; Barry and El Eini, 1976; Barry and Brace, 1977] and biological membranes, e.g. egg shell [Washitake *et al.,* 1980] and collagen [Nakano *et al.,* 1976].

The second type of solid barrier is termed a partitioning membrane (homogeneous or nonporous polymers) which acts as a non-porous barrier to diffusion, requiring the diffusing substance to dissolve within the membrane. The rate of permeability will be determined by the absolute water solubility of the xenobiotic, its relative affinity for the membrane and the xenobiotic's molecular structure [Garrett and Chemburkar, 1968a]. Thus, such a system is used as a lipoidal barrier simulating the stratum corneum. The most common membranes for this purpose are Silastic® (polydimethyl siloxane) and polyurethane [Behl *et al.,* 1984, 1985, 1986; Bottari *et al.,* 1977; Flynn and Smith, 1972; Flynn *et al.,* 1979; Garrett and Chemburkar, 1968a, 1968b, 1968c; Hunke and Matheson, Jr., 1981; Kincl *et al.,* 1968; Lovering *et al.,* 1974a, 1974b; Most, 1970].

Multimembrane systems refer to a combination of two or more different membranes within a composite, such as cellulose acetate: silastic:cellulose acetate [Nacht and Yeung, 1985] and silicone:poly (2-hydroxylethyl methacrylate) [Hatanaka *et al.,* 1992]. Other less extensively studied membranes include zeolites [Dyer *et al.,* 1979], diaflo-ultrafiltration membrane [O'Neill, 1980] and Supor® [Behl *et al.,* 1990]. Diffusion cell design, detection methods and data interpretation are similar to those for the biological membranes.

V. SUMMARY

Both *in vivo* and *in vitro* methods are applicable for the study of bioavailability. The experimental methods for measuring bioavailability are summarized in Table 4-1. The bioavailability factor measured for a xenobiotic may vary according to the different experimental methods used to measure it and the subjects involved. The use of *in vivo* methods in humans is not feasible because of ethical considerations. Consequently, most experimental values are obtained from animal experiments but these are also being subjected to increasing scrutiny by animal care ethics committees.

The values for bioavailability which are reported in Chapter 5 are subject to all the variability and uncertainty associated with the issues discussed in this chapter. Users of the data reported in Chapter 5 are cautioned to seek the original publications for data they may wish to use, in order to be sure that adequate attention has been given to the influence of experimental methods on the results for the application being considered.

Table 4-1 Choice of Experimental Methods for Study of Bioavailability

General method	Objective	Technique	Administered route	Subject	Strength	Limitation
In vivo	Bioavailability factor	Measurement of blood level	Inhalation Ingestion Dermal	Humans* Animals	Accurate Reliable	Inconvenient
In vivo	Bioavailability factor (lower bound)	Measurement of urinary level	Inhalation Ingestion Dermal	Humans* Animals	Simple Convenient Inexpensive Rapid	Underestimation Unreliable
In vivo	Bioavailability factor	Mass-balance	Inhalation Ingestion Dermal	Animals	Accurate Reliable	Technically demanding Expensive Time-consuming
In vivo	Dissolution rate (Prediction of bioavailability)	Various techniques	Inhalation Ingestion Dermal	Animals	Simple Inexpensive Rapid	Not accurate Unreliable
In vitro	Dissolution rate (Prediction of bioavailability)	Various techniques	-	-	Simple Inexpensive Rapid	Not accurate Unreliable
In vivo	Deposition fraction (gas/vapor and aerosol)	Measurement of exhaled air level	Inhalation	Humans* Animals	Simple Inexpensive Rapid	Not accurate (Overestimation of bioavailability factor in some cases)

* In the past, *in vivo* bioavailability studies of some xenobiotics were conducted in human volunteers but human study is no longer viable for xenobiotics.

(Continued)

General method	Objective	Technique	Administered route	Subject	Strength	Limitation
In vivo	Mechanical clearance fraction and retention fraction (Prediction of bioavailability factor)	Various techniques	Inhalation	Humans* Animals	Simple Inexpensive Noninvasive (In some cases)	Not accurate
In vitro	Inhalable fraction Particle size Water solubility Blood/air or octanol/water partition coefficients (Prediction of deposition fraction and bioavailability factor)	Various techniques	-	-	Simple Inexpensive	Not accurate
In vitro	Bioavailability factor	Isolated perfused lung	Inhalation	Animals	Accurate Remaining pulmonary function and structure integrity	Expensive Viable in short period of time Technically demanding
In vivo	Bioavailability factor	Measurement of fecal level	Ingestion	Humans* Animals	Simple Convenient	Overestimation Unreliable
In vivo	Bioavailability factor	Chronic isolated loop	Ingestion	Animals	Controlling variables Remaining physiological function	Overestimation Unreliable

(Continued)

General method	Objective	Technique	Administered route	Subject	Strength	Limitation
In vivo	Bioavailability factor	Measurement of liver ratio	Ingestion	Animals	Direct measurement of bioavailability factor	Only available by making an assumption that the levels of xenobiotics primarily reaching the liver represent the systemic levels
In vitro	Lipid/water or water/ lipid partition coefficient in the gastrointestinal tract (various pH environment) (Prediction of bioavailability factor)	Various techniques based on pH partition theory	-	-	Simple Inexpensive Rapid	Not accurate
In vivo	Bioavailability factor	Surface disappearance	Dermal	Humans* Animals	Simple Inexpensive Noninvasive (in some cases)	Not accurate (Overestimation)
In vitro	Bioavailability factor Permeability Diffusivity Partition coefficient	Diffusion cell	-	Skin from humans * and animals	Simple Inexpensive Controlling variables	Unreliable Unknown relationship between *in vivo* and *in vitro* for most xenobiotics
In vitro	Bioavailability factor Permeability Diffusivity Partition coefficient	Diffusion cell	-	Artificial membrane	Simple Inexpensive	Unreliable Not accurate

Chapter 5

EXPOSURE ROUTES AND BIOAVAILABILITY FACTORS FOR SELECTED CONTAMINANTS*

I. ARSENIC

Arsenic is an element commonly found on earth. Common oxidation states include 0, -3, +3 and +5 for inorganic and organic forms [NAS, 1977a; USEPA, 1984b; Lide, 1993].

Arsenic is frequently released to the air, soil, water and biota. Arsenic in particulate matter may occur in inorganic and organic forms [WHO, 1981]. Arsenic (III) in the air is readily oxidized to arsenic (V). The oxidized arsenic is relatively stable in air [NAS, 1977a]. Both inorganic and organic compounds are also present in surface water. Arsenic (V) usually predominates in surface water. Arsenic apparently accumulates in soils, with the inorganic form, complexed and chelated by organic material, iron, aluminum and calcium [ATSDR, 1989b, 1993a]. Waterborne and soil-bound arsenic forms readily undergo a series of transformations, such as oxidation/reduction, methylation by microorganisms, volatilization, leaching, adsorption and precipitation [NRCC, 1978; Woolson, 1983; ATSDR, 1989b, 1993a]. Plants rarely accumulate high concentrations of arsenic because this element is phytotoxic. Plants tolerant to high levels of arsenic in soil include potatoes, tomatoes, cabbage, carrots, grapes, Sudan grass, and corns [Walsh *et al.*, 1977]. Roots usually concentrate higher levels of arsenic than leaves. Bioaccumulation of arsenic takes place in food chains, particularly in aquatic organisms. Data from food monitoring programs in the U.S. indicate that most arsenic was present in the meat-fish-poultry class of the total diet [Jelinek and Corneliussen, 1977].

All soluble inorganic arsenic compounds appear to have toxic effects on humans. There is sufficient evidence to indicate an association between lung or skin cancer and respiratory or oral exposure to inorganic arsenic. In humans, occupational exposure to inorganic arsenic, particularly in metal smelting operations and pesticide production, is primarily associated with an increase of the incidence of lung cancer [Ott *et al.*, 1974; Mabuchi *et al.*,

---

* **CAUTION.** The data presented in this chapter have been summarized from the scientific literature and all reasonable care has been taken to insure the accuracy of the data reported here. However, there are many factors described in Chapter 3 and Chapter 4 which will affect the applicability of the data for specific applications. Consequently, readers are cautioned to seek out the original publications for any data they wish to rely upon, in order to insure that adequate consideration has been given to the manner in which data have been obtained relative to how it may be used.

1979; IARC, 1980, 1987; WHO, 1981; USEPA, 1984b; Enterline *et al.*, 1987, 1995; Mass, 1992; Hertz-Picciotto and Smith, 1993; Tsuda *et al.*, 1995]. The active arsenic species linked to the lung cancer is not known, but it is possibly related to the refining of As_2O_3 [Tinwell *et al.*, 1991]. The biochemical mechanism by which arsenic causes cancer has not been found [ATSDR, 1993a].

There is also sufficient evidence of a relationship between the development of skin cancer and chronic oral exposure to inorganic arsenic (potassium arsenite) in drinking water [Tseng *et al.*, 1968; Tseng, 1977; IARC, 1980, 1987; USEPA, 1984b; Mass, 1992]. Some studies suggested that a variety of internal tumors (bladder, kidney, liver and leukemia) may also be related to oral exposure to inorganic arsenic [ATSDR, 1989b, 1993a; Bates *et al.*, 1992; Mass, 1992; Smith *et al.*, 1992]. To date, carcinogenesis via dermal exposure has not been observed [ATSDR, 1989b, 1993a].

In the respiratory tract and gastrointestinal tract, water-soluble inorganic and some organic arsenic compounds appear to be highly bioavailable.

A. EXPOSURE ROUTES AND RECEPTORS

Humans may be routinely exposed to arsenic at low levels by inhalation, ingestion and dermal route because of the extensive distribution of arsenic in environmental media. In an occupational setting, especially in metal smelters and arsenic pesticide manufacture, there are high levels of airborne arsenic (as much as 20 $\mu g/m^3$) [Vahter, 1988]. Inhalation of these compounds is a primary exposure route for workers. Also, workers may accidentally have dermal contact through aqueous solutions of inorganic arsenic.

General population exposure to arsenic mainly results from ingestion of contaminated drinking water, food and drugs. Residents living near industrial areas may be subject to arsenic exposure by inhalation, ingestion and dermal contact. Arsenic concentrations in drinking water vary extensively with geographic regions. In some areas, high levels of arsenic in drinking water have been observed [WHO, 1981]. Oral exposure to contaminated drinking water with inorganic arsenic may pose a significant health risk to humans [Abernathy and Ohanian, 1992].

A major arsenic source for the general population via ingestion comes from food, in particular, seafood that contains some organic arsenic compounds. With the exception of seafood, arsenic concentrations in most foods are very low [Jelinek and Corneliussen, 1977]. Moreover, organic arsenic forms present in most foods (e.g. arsenobetaine-rich seafood) are readily eliminated in an unchanged form from the human body and seldom accumulate to toxic levels while organic arsenic species (arsenosugars) in seaweed are metabolized [Foa *et al.*, 1984; Le *et al.*, 1993, 1994]. Organic arsenic in seafood (i.e. fish and seaweed) generally has a low toxicity to humans. As a result, the potential hazard by consumption of food may be insignificant [Abernathy and Ohanian, 1992]. The subgroup of the population who consumes excessive seafood or locally grown vegetables containing a large amount of inorganic arsenic has been of concern [WHO, 1981; Vahter 1988]. In some cases, residents living near contaminated areas may be exposed to arsenic by ingestion of soil or dust. Typically, soil-bound arsenic is unlikely to accumulate to a high level. Exposure by ingestion of soil and dust is rarely significant. Dermal exposure through contact with soil is likely to be minor because of the very low bioavailability [USEPA, 1984b].

Generally, the inhalation exposure route is insignificant for the general population unless people inhale excessively high levels of airborne arsenic from contaminated areas [Murphy *et al.,* 1989].

Additionally, the determination of exposure routes depends on the population sensitivity [ATSDR, 1989b, 1993a]. According to the metabolic process by which inorganic arsenic is detoxified in the liver by methylation, different degrees of enzymatic activity in individuals strongly influences the toxic effects of arsenic to humans.

B. TOXICOKINETICS (FOLLOWING ABSORPTION)

1. Distribution

Once absorbed, inorganic arsenic is initially bound to the globin portion of hemoglobin in the red blood cells [Malachowski, 1990]. Then, the majority of arsenic in the blood is redistributed to different organs within 24 h. The transfer process from the blood is likely to follow a 2 or 3 phase course [Vahter, 1983]. Fast clearance half-life from the blood was estimated to be 1 to 2 h, while the other two phases were 30 and 200 h, respectively. Another study indicated that the clearance half-life of arsenic from blood in humans was 1, 5, and 35 h in a 3 phase exponential model [Tam *et al.,* 1979].

Redistribution of the compounds occurs rapidly in the liver, kidney, spleen, lung and gastrointestinal tract during the first day, incorporating arsenic into hair, nail, skin and bone within 2 to 4 wk [Wester *et al.,* 1981; Vahter, 1983; USEPA, 1985a; Malachowski, 1990]. Bone, muscle and skin are the major depots of the arsenic body burden. Furthermore, arsenic absorbed via ingestion has a shorter clearance half-life than that via inhalation [Magos, 1991]. Organic arsenic that is absorbed will also be distributed throughout the body, reaching blood, kidney, liver, lung, muscle and skin.

2. Metabolism and Excretion

Metabolism of inorganic arsenic in the human body is not completely understood. In general, some of the dissolved arsenic (V) can be rapidly reduced to arsenic (III) in blood, while some arsenic (III) is oxidized to arsenic (V) [Vahter, 1983; Malachowski, 1990]. Both arsenic (III) and (V) are metabolized in the liver to methylated compounds: methylarsonic acid, dimethylarsinic acid, and trimethylarsenic compounds. Methylation detoxifies inorganic compounds and enhances arsenic excretion. One form of organoarsenicals, arsenosugars, is metabolized by the human body [Le *et al.,* 1994].

The major excretion route for all arsenic species is in urine. A small fraction of arsenic compounds appear in the feces, bile, sweat and breast milk [Malachowski, 1990]. The urinary excretion rate in humans rises sharply within the first day and then declines rapidly in the following days [Tam *et al.,* 1979; Farmer and Johnson, 1990; Malachowski, 1990]. About 80 to 90% of the total inorganic arsenic excreted in urine occurs as methylated compounds [Hopenhayn-Rich *et al.,* 1993]. On the other hand, some organic arsenic species (arsenobetaine) are absorbed but are directly eliminated in urine without metabolism [ATSDR, 1989b, 1993a; Le *et al.,* 1993].

C. BIOAVAILABILITY

1. Inhalation

a. Humans

Inhaled airborne arsenic has been found to be rapidly deposited in the respiratory tract and absorbed in the systemic circulation. There is a relationship between inhalation of inorganic arsenic and an increase of the urinary arsenic levels [Pinto *et al.,* 1977; Smith *et al.,* 1977; Vahter *et al.,* 1986]. The extent of deposition, retention and clearance is controlled by the size of aerosols and solubility.

Bioavailability of inorganic arsenic in human volunteers had been studied by Holland *et al.* [1959]. Two volunteers with terminal lung cancer inhaled ^{74}As labeled sodium arsenite ($NaAsO_2$) by smoking a cigarette and being nebulized with arsenic aerosols. Using external chest counting, researchers found that 5% and 8% of arsenic contained in the cigarette and 32% and 62% of a nebulized arsenic solution were deposited within the lung. After 14 d, 85% to 90% of the deposited arsenic was cleared from the lungs. This clearance included the fraction of deposited arsenic directly entering the systemic circulation from the lungs and the fraction entering the gastrointestinal (GI) tract. Part of the arsenic cleared to the GI tract was absorbed into the systemic circulation and the unabsorbed fraction was excreted in the feces. However, the fecal excretion following 10 d dosing was 2.5% with 45% of the deposited arsenic being excreted in urine after 10 d. Overall, 85% to 90% of the deposited water-soluble arsenic (III) in the lungs can be bioavailable to humans. From this study, we calculated that about 4% to 7% of the administered arsenic in the cigarette and 27% to 55% of the nebulized arsenic solution were absorbed, as shown in Figure 5-1.

These values are derived from the deposited dose of the compounds. Consequently, it is important to recognize the need to take into account the deposition fraction.

b. Animals

In animal experiments, there are differences of respiratory bioavailability among species, chemical forms and water solubility. Water soluble inorganic and organic arsenic can be more rapidly absorbed than slightly water soluble

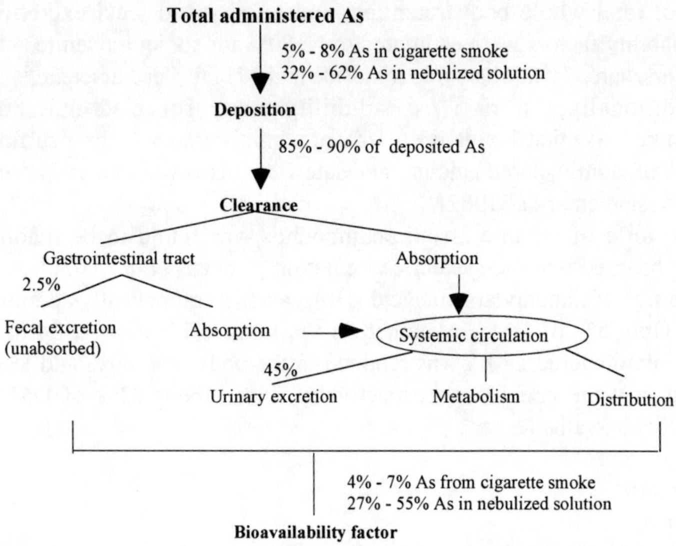

Figure 5-1 Respiratory bioavailability of inorganic arsenic in human volunteers [Adapted from source: Holland *et al.*, 1959]

inorganic and organic arsenic [Dutkiewicz, 1977; Stevens *et al.,* 1977; Inamasu *et al.,* 1982; Marafante and Vahter, 1987].

Some studies only reported the fraction of arsenic which was retained in the lung. When other parameters (deposition fraction, fractions of fecal and urinary excretion) are not provided, the bioavailability factors cannot be calculated. However, the information on retention in the lungs will be useful for evaluating the risk arising from exposure to some arsenic species which directly exert toxic or carcinogenic effects in the lungs. Inamasu *et al.* [1982] investigated the degree of retention of arsenic trioxide (III) and calcium arsenate (V) in rats. One day after intratracheal administration, only 50% of arsenic trioxide was retained in the respiratory tract. After 7 d, 1% remained in the lung. Compared to arsenic trioxide, the retention fraction of calcium arsenate at 1 and 7 d was 50%. In this study, no data were provided to indicate urinary or fecal excretion for these two compounds. Accordingly, it is difficult to determine the bioavailability factor. In another study, high retention levels of arsenic in the lungs were also found in hamsters following intratracheal administration of calcium arsenate [Pershagen *et al.,* 1982]. However, the high retention of calcium arsenate in the lung corresponds to an observed increased lung cancer incidence following intratracheal administration of calcium arsenate in animals.

Marafante and Vahter [1987] reported the solubility of inorganic arsenic trisulfide (As_2S_3) and lead arsenate ($Pb_3(AsO_4)_2$) *in vivo* was 80% and 10%, respectively. Following intratracheal dosing, the lung retention fractions for sodium arsenite ($NaAsO_2$), sodium arsenate (Na_3AsO_4), As_2S_3 and

$Pb_3(AsO_4)_2$ were 0.06%, 0.2%, 1.3%, and 45.5%, respectively. Based on the results of total whole body retention, and urinary and fecal excretion, the bioavailability factors were estimated to be 80% for sodium arsenite, 71% for sodium arsenate, 75% for arsenic trisulfide and 47% for lead arsenate.

Additionally, an *in vitro* solubility study revealed that 80% of administered arsenic trioxide dust, 16% of administered arsenic trisulfide dust and 23% of administered calcium arsenate were dissolved in a saline solution at 6 d [Pershagen *et al.,* 1982].

One form of organic arsenic compounds was found to be readily and rapidly absorbed into the systemic circulation [Stevens *et al.,* 1977]. A single dose (33 μg) of dimethylarsenic acid (DMA) was intratracheally administered in rats. Only 5% of DMA remained in the lung at 15 min. At 24 h, about 24% of administered DMA was retained in the body, and 60% and 8% were excreted in urine and feces, respectively. Thus, about 92% of DMA was apparently bioavailable.

2. Ingestion

a. Humans

Water soluble inorganic arsenic is rapidly and substantially absorbed by humans. Bettley and O'Shea [1975] treated seven subjects orally with 8.5 mg of Liquor arsenicalis (arsenic III). The total recovery of arsenic in feces was only 3.5% at 10 d. Accordingly, about 96% of administered arsenic (maximum value of bioavailability) was apparently absorbed into the portal blood. 52% of this administered arsenic was eliminated in urine (minimum value of bioavailability).

There was a linear relationship between the total daily amount of arsenic excreted in urine and the absorbed daily amount after equilibrium was reached [Mappes, 1977; Buchet *et al.,* 1981a]. Under this circumstance, the amount of daily urinary excretion reflects a minimum estimate of daily arsenic absorption. Buchet *et al.* [1981a] investigated the absorption of sodium arsenite ($NaAsO_2$) in four volunteers. The subjects ingested the compound at doses of 125 μg, 250 μg, 500 μg and 1,000 μg once a day for five consecutive days. After 14 days dosing, the fractions of urinary excretion were 54%, 73%, 74% and 64% for these four doses, respectively. Only 9%, 5%, 14% and 17% of parent form of administered arsenic at these four dose levels were detected in urine.

In another study, Mappes [1977] reported 62% to 72% of daily urinary excretion when one volunteer took 0.76 mg arsenic (As_2O_3) daily by oral ingestion for five days (with equilibrium). Conversely, no increase of urinary excretion took place following oral administration of 12 mg of water insoluble arsenic selenide (As_2Se_3), implying that this dose was entirely eliminated by fecal excretion without any absorption.

Organic arsenic is bioavailable, depending on water solubility. Tam and Lacroix [1980] found only 0.21% and 0.44% of arsenic in feces from two volunteers who ingested 10 mg of organically bound arsenic in fish.

Therefore, this seafood arsenic was completely absorbed although this form of arsenic is not metabolized and will be excreted in unaltered form in urine. In contrast, arsanilic acid, which is only slightly water-soluble, was less bioavailable [Calesnick *et al.*, 1966]. The oral bioavailability factor was 26% for arsanilic acid solution, and 36% for tissue bound arsenic as measured in chicks.

b. Animals

Animals studies show that oral bioavailability of arsenic varies with species, chemical forms and water solubility. Either arsenic (III) or arsenic (V) was completely absorbed in mice [Vahter and Norin, 1980]. Sodium arsenite ($NaAsO_2$) (III) and sodium arsenate (Na_2HAsO_4) (V) were administered to mice by gavage and by subcutaneous injection at doses of 0.01 mg and 0.1 mg. Two days later, the fractions of total dose by oral and subcutaneous injection excreted by the fecal route were 8% and 6% for arsenic (V), and 7% and 4% for arsenic (III), respectively. For subcutaneous injection, the fecal fraction represented biliary excretion. With a correction of biliary excretion (8% minus 6% for arsenic (V), and 7% minus 4% for arsenic (III), the bioavailability factors were 98% and 97% for these two compounds. In another study, the bioavailability factor of sodium arsenate (Na_2HAsO_4) in rabbits was 50% after administration of 1.95 mg arsenic solution by gavage [Freeman *et al.*, 1993].

The absorption fraction of arsenic trioxide in rats and rabbits was similar, being 31% at 4 d and 37% at 7 d (based on fecal excretion data), respectively [Ariyoshi and Ikeada, 1974]. But the absorption of arsenic trioxide in monkeys was extremely high during 14 d following oral administration of the compound (1 μg/kg) [Charbonneau *et al.*, 1978]. The bioavailability factors ranged from 73% (based on urinary excretion data) to 98% (based on fecal excretion data). Arsenic pentoxide (As_2O_5) showed a relatively low bioavailability in hamsters [Charbonneau *et al.*, 1980]. Arsenic pentoxide solution was given to hamsters by gavage or intravenously at doses of 0.01 μg arsenic per animal. During 24 h, 70% of the administered compound given by gavage and 6% given by intravenous were eliminated in feces. The fraction by intravenous route excreted by the fecal route represented biliary excretion. With a correction of biliary excretion (70% minus 6%), about 64% of the administered compound was not absorbed. Thus, the oral bioavailability factor was 36%. However, arsenic pentoxide was found to be highly absorbed in dogs (95%) after the first day administration [Hollins *et al.*, 1979].

Bioavailability of inorganic arsenic compounds by ingestion depends directly on water solubility. An experiment by Marafante and Vahter [1987] showed bioavailability factors (urinary excretion data plus whole body retention) in hamsters to be 43% for $NaAsO_2$ (water-soluble), 81% for Na_3AsO_4 (water-soluble), 13% for As_2S, (slightly water-soluble) and 24% for $Pb(AsO_4)_2$ (slightly water-soluble). Compared to mice [Vahter and Norin,

1980], the degree of absorption of $NaAsO_2$ was lower in hamsters. Bioavailability of soil-bound arsenic compounds is very low. Freeman *et al.* [1993] orally administered soil-bound sodium arsenate (Na_2HAsO_4) (in capsule) to rabbits at three different doses: 0.2 g (0.78 mg As), 0.5 g (1.95 mg As) and 1.0 g (3.9 mg As) soil/kg body weight. 24% of the administered arsenic compound was absorbed. The bioavailability factor of arsenic pentoxide (As_2O_5) from ore-containing soil was 8% in dogs after oral administration of a single dose of 0.6 mg As/kg [Groen *et al.*, 1994].

The bioavailability factor for an organic arsenic compound, methylarsine sulfide (MAS), was reported to be 64% in rabbits [Ariyoshi and Ikeda, 1974]. About 66% of DMA was bioavailable in rats [Stevens *et al.,* 1977]. About 90% of seafood arsenic (arsenobetaine) in monkeys was absorbed (based on 10 % of administered seafood arsenic excreted in feces because arsenobetaine is not metabolized) [Charbonneau *et al.,* 1978].

One study reported that about 60% of arsenic bound to soil was dissolved in a pH 2 aqueous solution at 6 h [Murphy and Toole, 1989].

3. Dermal

A few studies have shown that the absorption of inorganic arsenic (e.g. arsenic acid and arsenic trichloride) through the skin occurred in humans and animals [WHO, 1981; Vahter, 1983]. Wester *et al.* [1993] studied the skin absorption of arsenic from water and soil *in vivo* and *in vitro.* About 2 to 6% of arsenic-73 labeled solution (H_3AsO_4) was absorbed through monkey abdominal skin at a low dose of 0.00024 $\mu g/cm^2$ of skin and a high dose of 2.1 $\mu g/cm^2$ of skin. Only 0.8% of H_3AsO_4 solution penetrated the human skin *in vitro.* The dermal bioavailability factors of soil-bound arsenic was 3% to 4.5% in monkey at a low dose of 0.00004 $\mu g/cm^2$ of skin and a high dose of 0.6 $\mu g/cm^2$ of skin. 1.9% of soil-bound arsenic penetrated the human skin *in vitro.* Dutkiewicz [1977] reported a dermal permeability of 2.7×10^{-6} cm/h in rats after applying 0.1M sodium arsenate solution to the rat skin.

4. Summary

Overall, both inorganic and organic arsenic compounds are bioavailable in humans and animals by inhalation, ingestion and dermal exposure routes (Table 5-1, 5-2, 5-3, and 5-4). Quantitative data are available for inhalation, ingestion and dermal exposure routes (Figure 5-2). Water soluble arsenic compounds are more readily absorbed and therefore more bioavailable.

D. UNCERTAINTIES

At former industrial and waste disposal sites, arsenic pentoxide (water-soluble) has contaminated surface soils. Airborne arsenic aerosols may be transported to residential areas by physical processes. Airborne arsenic which is water soluble is a potential health hazard for residents.

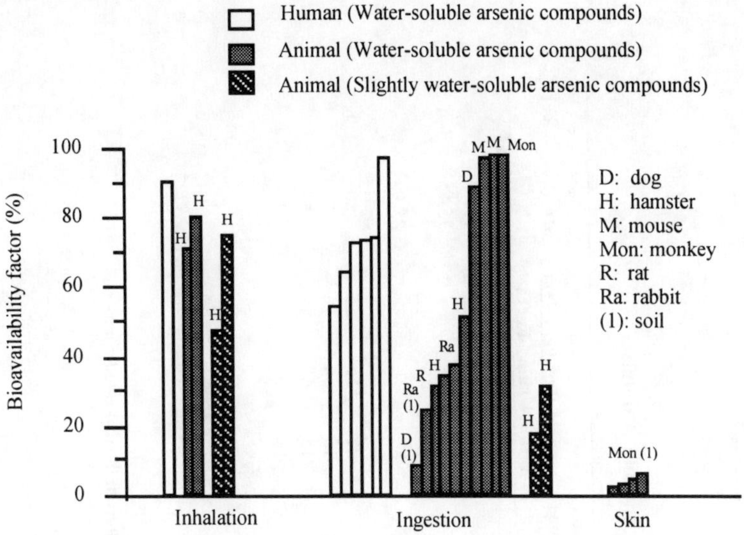

Figure 5-2 Bioavailability factors of inorganic arsenic in humans and animals by three exposure routes [Adapted from sources in Table 5-1: a, b (inhalation); Table 5-2: a, b, c (ingestion); Table 5-3: a, b, c, d, e, f, g, h, i (ingestion), and Table 5-4: a (skin)].

Water soluble arsenic is highly bioavailable in the respiratory tract in humans. The reported bioavailability factors are derived from experiments with terminal lung cancer patients. These values may differ from normal healthy persons. On the other hand, the inhalable fraction and arsenic particle size in the air, as well as deposition fraction in the respiratory tract, are important considerations in the assessment of respiratory bioavailability.

However, the inhalable fraction of arsenic aerosols distributed at sites in the respiratory tract and arsenic particle size are not well documented. The deposition fraction of arsenic inhaled in the lungs has been estimated from smoking or experimental exposures. These processes do not reflect normal inhalation patterns. Net absorption of inhaled arsenic is extremely dependent on the deposition fraction of arsenic-bearing particles.

Arsenic pentoxide in the soil and lagoons at contaminated sites may leach into groundwater or be removed to surface water by runoff and fallout. Reliable quantitative data on water solubility for soil-bound arsenic pentoxide under natural or physiological conditions are lacking.

One major concern is whether potable groundwater near these sites has been contaminated because of the apparent association between an increased incidence of skin cancer, and perhaps some internal tumors, resulting from exposure to inorganic arsenic in drinking water. Chromated copper arsenate is found to have a relatively low solubility at physiological pH, but no quantitative data are available [Vahter, 1988].

Currently, there are no confirmed active arsenic species which reach the

Table 5-1 Respiratory Bioavailability of Arsenic in Humans and Animals

Compound	Species	Exposure method	Arsenic dosage	Experimental method	Exposure duration	Retention in the lung [AS] (%)	Arsenic Recovery (%) WB[1]	Urine	Feces	BF^2(%)	Ref
Inorganic											
[3]$NaAsO_2$	Human	Smoking	200-300 μc	External counting;	4 d	12-25				75-85[5]	a
		Nebulizing		Urinary and fecal	10 d			45	2.5		
				excretion	14 d	10-15				85-90[5]	
[3]$NaAsO_2$	Hamster	Intratracheal	100 μg	External counting;	3 d	0.06	25	55	15	80	b
[3]Na_3AsO_4			(2 mg/kg)	Urinary and fecal		0.02	15	56	16	71	
[4]As_3S_3				excretion		1	15	61	15	75	
[4]$Pb_3(AsO_4)_2$						46	60	33	6	47	
Organic											
[3]Dimethyl-arsinic acid	Rat	Intratracheal	33 μg	External counting; Urinary and fecal excretion	15 min	5					c
					24 h		24	60	8	92	

[1] WB = Whole Body; [2] BF = Bioavailability Factor; [3] Water soluble; [4] Slightly water soluble; [5] Data from the deposited compounds within the respiratory tract

Adapted from: a. Holland *et al.*, 1959; b. Marafante and Vahter, 1987; c. Stevens *et al.*, 1977

Table 5-2 Oral Bioavailability of Arsenic in Humans

Compound	Number of subjects	Arsenic carrier	Arsenic dosage	Experimental method	Exposure duration	Bioavailability factor (%)	Ref.
<u>Inorganic</u>							
[4]Liq. Arsenicals [As (III)]	11	Water	8.5 mg	Urinary and fecal excretion	10 d	97[1] 52[2]	a
[4]$NaAsO_2$	4	Water	125 µg x 5 250 µg x 5 500 µg x 5 1,000 µg x 5	Urinary excretion	14 d	54[2] (9)[3] 73[2] 5[3] 74[2] (14)[3] 64[2] (17)[3]	b
[4]As_2O_3 [6]As_2Se_3	1 1	Water Powder	0.76 mg x 5 12 mg	Urinary excretion	5 d	69-72 0	c
<u>Organic</u>							
[4]Fish arsenic	2	Meal	10 mg	Urinary and fecal excretion	-	99.7	d
[5]Arsanilic Acid	4	Water Meal [Chick (fed As)]	1.3-3.0 mg 160-210 g of pate	Urinary and fecal excretion	6 d	26 36	e

[1] the values from fecal excretion data that are considered as the upper bound of bioavailability factor; [2] the values from urinary excretion data (parent compound and its metabolites) that are considered as the lower bound of bioavailability factor; [3] the values that are the fraction of the parent compound in urine; [4] Water soluble; [5] Slightly water soluble; [6] Water insoluble

Adapted from: a. Bettley and O'Shea, 1975; b. Buchet *et al.*, 1981a; c. Mappes, 1977; d. Tam and Lacroix , 1980; e. Calesnick *et al.*, 1966

Table 5-3 Oral Bioavailability of Arsenic in Animals

Compound	Species	Exposure methods	Arsenic carrier	Arsenic dosage	Experimental method	Exposure duration	Bioavailability factor (%)	Ref.
Inorganic								
[4]$NaAsO_2$ [4]Na_2HAsO_4	Mouse	Gavage	Water	0.4-4 mg/kg	Urinary and fecal excretion; Whole body counting	2 d	98[1] 97[2]	a
[4]Na_2HAsO_4	Rabbit	Oral	Soil in capsule	0.78 mg As/kg 0.95 mg As/kg 3.9 mg As/kg	Urinary and fecal excretion vs. intravenous level	5 d	24	b
		Gavage	Water	1.9 mg As/kg			50	
[4]As_2O_3	Rat Rabbit	Feed	Diet	29 mg/kg	Urinary and fecal excretion	4 d 7 d	31[1] 37[1]	c
[4]As_2O_5	Hamster	Gavage	Water	0.01 μg (0.1 μg/kg)	Urinary and fecal excretion; Whole body counting	1 d	34[1]	d
[4]$NaAsO_2$ [4]Na_3AsO_4 [5]As_2S_3 [5]$Pb_3(AsO_4)_2$	Hamster	Oral	Water	2 mg/kg	Urinary and fecal excretion; Whole body counting	3 d	51[1](43)[3] 88[1](81)[3] 17[1](13)[3] 31[1](24)[3]	e

[1] the values from fecal excretion data that are considered as the upper bound of bioavailability factor; [2] the values from urinary excretion data (parent compound and its metabolites) that are considered as the lower bound of bioavailability factor; [3] the values that are the fraction of the urinary excretion data plus whole body retention; [4] Water soluble; [5] Slightly water soluble

Adapted from: a. Vahter and Norin, 1980; b. Freeman *et al.*, 1993; c. Ariyoshi and Ikeda, 1974; d. Charbonneau *et al.*, 1980; e. Marafante and Vahter, 1987

(Continued)

Compound	Species	Exposure methods	Arsenic carrier	Arsenic dosage	Experimental method	Exposure duration	Bioavailability factor (%)	Ref.
[4]As_2O_5	Dog	Oral	Gelatin capsule	0.3 µg (0.02-0.03 µg/kg)	Urinary and fecal excretion; Whole body counting	1 d	95[1]	f
[4]As_2O_5	Dog	Oral	Soil	0.6 µg/kg	Urinary and fecal excretion vs. intravenous level	5 d	8	g
[4]As_2O_3	Monkey	Gavage	Meal	1 µg/kg	Urinary and fecal excretion	14 d	98[1] 73[2]	h
Organic								
[4]Methylarsine sulfide	Rabbit	Feed	Diet	36 mg/kg	Urinary and fecal excretion	7 d	64	i
[4]Fish arsenic	Monkey	Feed	Meal	1 µg/kg	Urinary and fecal excretion	14 d	90	h
[4]Dimethylarsinic acid	Rat	Oral	-	0.1µg/kg	Urinary and fecal excretion; Whole-body retention	24 hr	66	i

Adapted from : f. Hollins *et al.,* 1979; g. Groen *et al.,* 1994; h. Charbonneau *et al.,* 1978; i. Stevens *et al.,* 1977

Table 5-4 Dermal Bioavailability of Arsenic in Animals *in vivo* and *in vitro*

Compound	Species	Exposure methods	Arsenic carrier	Arsenic concentration	Experimental method	Exposure duration	Bioavailability factor (%)	Ref.
H_3AsO_4	Monkey	Abdominal skin	Soil	0.00004 μg As/cm^2 0.6 μg As/cm^2	Urinary excretion versus intravenous levels (*in vivo*)	7 d	4.5 3	a
			Water	0.000024 μg As/cm^2 2.1 μg As/cm^2			6.4 2	
H_3AsO_4	Human	Cadaver skin	Soil	0.00004 μg As/cm^2	Diffusion cell (*in vitro*)	24 h	0.8	a
			Water	0.000024 μg As/cm^2			1.9	

Adapted from: a. Wester *et al.*, 1993

target tissues. Some studies indicated that the detoxification process was likely to be saturated at about 200 to 250 μg As/d ingested for adults [Buchet *et al.,* 1981b; USEPA, 1990c; North, 1992]. Whether methylated species can act as carcinogens is currently not clear. Hence, non-methylated arsenic is generally assumed to be an active species, arising when metabolic methylation becomes saturated, namely, at a methylation threshold of approximately 250 μg/d.

The USEPA Science Advisory Board [SAB, 1989] recommended that the risk assessment for arsenic in drinking water be based on the delivered dose of non-detoxified arsenic to active sites. Most recent bioavailability factors from either humans or animals are obtained from measurement of urinary or fecal excretion of arsenic. Results from fecal monitoring represent the amount of inorganic arsenic excretion prior to metabolism in the liver. However, the hepatic first-pass effect will dramatically influence the degree of absorption of inorganic arsenic into the systemic circulation. Thus, these bioavailability factors for inorganic arsenic are used as the upper bound. Because organic arsenic compounds do not experience presystemic metabolism, the measured bioavailability values for organic arsenic compounds based on fecal excretion appear to be close to their actual bioavailability factors. Values estimated from urinary excretion are a lower bound for the bioavailability factor. However, some studies only provide the total arsenic amount in urine without distinguishing parent compounds from their metabolites. Hopenhayn-Rich *et al.* [1993] reviewed the relationship between the exposure levels of inorganic arsenic and the non-methylated arsenic levels in urine from humans. Their results revealed that the levels of non-methylated arsenic in urine were fairly constant, approximately 20%, regardless of any increase of absorbed dose. These results raise questions about a methylation threshold hypothesis. In brief, uncertainties for health risk assessment for inorganic arsenic still involve substantial inadequacies of reliable knowledge of toxicokinetics and metabolisms of action.

E. CONCLUSIONS

Humans are routinely exposed to arsenic at low levels by inhalation, ingestion and dermal exposure. Inhalation of high levels of airborne arsenic primarily takes place in occupational settings or possibly at residential areas near contaminated industrial locations. Ingestion of inorganic arsenic through consumption of contaminated drinking water would pose a potential hazard for general populations. Exposure to arsenic via ingestion of soil and foods is less significant because of the generally low content of arsenic available to metabolic processes in these media. Dermal exposure is considered to be a minor route for the general population.

Either inorganic or organic arsenic is bioavailable via the exposure routes. Water-soluble inorganic arsenic is extensively and rapidly absorbed in the respiratory tract. Respiratory bioavailability for water soluble arsenic deposited within the human lung may be as high as 85% to 90%. However, the net absorption in the respiratory tract strongly depends on the deposition fraction for arsenic-bearing aerosols. The deposition fraction of water-soluble inorganic arsenic has been reported to be about 40% in humans. USEPA assumes a value of about 30% absorption for airborne arsenic [USEPA, 1984b]. In small rodent animals, the respiratory bioavailability factors for slightly water soluble inorganic arsenic range from 47% to 75%.

The oral bioavailability factors for water soluble inorganic and organic arsenic range from 50% to 100%. Currently, upper bound values (90% to 100%) are recommended to be used as the oral bioavailability factor for arsenic risk assessment and toxicity studies by WHO. For slightly water soluble organic arsenic, the oral bioavailability factors range from 26% to 36% in humans. Only 13% to 24% of slightly water soluble inorganic arsenic can be absorbed in the gastrointestinal tract in small rodent animals. For ingestion of arsenic contaminated soil, the dissolution rate in the gastrointestinal tract plays a key role in determining oral bioavailability. About 60% of arsenic may leach from an arsenic-bearing soil matrix into acidic aqueous solution (pH 2). Oral bioavailability of soil-bound arsenic varies with soil types, ranging from 8% to 24% in rabbits.

The degree of bioavailability is strongly governed by water solubility of arsenic. Arsenic pentoxide is highly water soluble. But chromated copper arsenate, a common wood preservative chemical, and soil-bound arsenic may have a relatively low water solubility in an acidic environment. The bioavailability for this type of arsenic is expected to be lower than that for arsenic pentoxide.

The dermal bioavailability of water-soluble arsenic is 2% to 6% in monkeys *in vivo*, and 0.8% in the human skin *in vitro*. 3% to 4.5% of soil-bound arsenic penetrates monkey skin *in vivo*, as compared with 1.9% through human skin *in vitro*.

F. RECOMMENDATIONS

For respiratory bioavailability studies, it is essential to determine the inhalable fraction of airborne arsenic aerosols and corresponding arsenic aerosol size distribution where inhalation exposure occurs at contaminated sites, so that the deposition fraction of arsenic inhaled within the respiratory tract can be estimated.

Because exposure to some inorganic arsenic via ingestion by humans is believed to lead to skin cancer, other internal tumors as well as systemic toxicity, the internal dose and delivered dose through the blood and other target tissues are essential for quantitative arsenic risk assessment. Accordingly, it is necessary to identify active arsenic species, to investigate the mechanism of arsenic metabolism, to determine the delivered dose through the blood and other target tissues, and to establish quantitative relationships between the dose of active species and arsenic toxicity or carcinogenicity.

II. CADMIUM

Cadmium occurs naturally in a metallic elemental form with an oxidation state of +2. There are a variety of cadmium salts found in the environment, such as cadmium acetate, cadmium chloride, cadmium nitrate, cadmium oxide, cadmium sulfate and cadmium sulfide [Dunnick and Fowler, 1988].

Cadmium has been found in air, water, soil and food. In industrial areas, cadmium oxide (cadmium fume) is the primary cadmium form in air [NRCC, 1979]. Airborne cadmium usually occurs in very small particulates (50% to 90% of suspended cadmium particles are less than 10 μm MMAD) and is removed by dry/wet deposition and precipitation [CEC, 1978; Elinder, 1985; WHO, 1992; ATSDR, 1993d].

In surface water, cadmium may exist as the hydrated ion and cadmium-organic substance complexes. A large amount of cadmium has been detected in sediment and suspended particles [Elinder, 1985]. Elevated soil cadmium may result from the deposition of airborne cadmium and from the use of sewage sludges and inorganic fertilizers [Thornton, 1992]. Commonly, the zinc ore complex (ZnS-CdS) is the main source of cadmium in soil and water from mining [NRCC, 1979].

Plants can easily take up cadmium from soil, especially under low soil pH conditions [Elinder, 1985, 1992]. Then, cadmium may bioaccumulate in all levels of the food chain, such as grasses, food crops, fish, poultry and cattle [ATSDR, 1993d].

Acute exposure to a high dose of cadmium has been associated with lung damage after inhalation and gastrointestinal disorders after ingestion [Friberg *et al.*, 1986; Waalkes *et al.*, 1992; ATSDR, 1993d]. Cadmium is a cumulative toxic agent. With chronic exposure, the primary target organ is the kidney, regardless of exposure route. The lung, bone, cardiovascular system, liver and reproductive system can also be affected, resulting in progressive lung disease, Itai-Itai disease, and hypertension [Friberg *et al.*, 1986; ATSDR, 1993d]. Cadmium and its compounds (cadmium chloride, cadmium oxide, cadmium sulfate and cadmium sulfide) may contribute to the development of lung cancer following inhalation exposure in humans and animals [Takenaka *et al.*, 1983; Thun *et al.*, 1985; Oberdörster, 1986, 1989a, 1989b; Glaser *et al.*, 1989]. Based on animal and limited human data, cadmium has been classified as a probable human carcinogen [IARC, 1976, 1982c; IRTS, 1993].

Cadmium is readily absorbed from the respiratory tract. In the gastrointestinal tract, bioavailability of cadmium is very low. Limited data indicate absorption of cadmium through the skin.

A. EXPOSURE ROUTES AND RECEPTORS

The major exposure route for workers is via inhalation exposure to cadmium particulates from tobacco manufacturing, smelting, soldering and electroplating [WHO, 1992; ATSDR, 1993d]. Occasionally, workers may ingest cadmium dust from their hands, cigarettes and contaminated food [Adamsson *et al.*, 1979; Friberg *et al.*, 1986].

For the smoking population, cigarette smoke is an important exposure source because cigarettes contain a high level of cadmium (about 1-2 μg Cd/cigarette), of which about 10% of cadmium particulates are inhaled. Cadmium in the respiratory tract is highly bioavailable [Friberg *et al.,* 1974; Elinder, 1985]. The primary source of cadmium for the non-smoking population is via ingestion of food, especially grain, cereal products, potatoes, meat (liver and kidney), shellfish and fruits [ATSDR, 1993d]. The daily intake of cadmium from food was estimated as about 30 μg/d. In most cases, exposure to cadmium from drinking water is negligible because of low levels of cadmium in drinking water [WHO, 1992]. Young children may be exposed to cadmium dust or soil through hand-to-mouth activity.

Because skin absorption of cadmium appears to be very low, the skin exposure route is regarded as insignificant for both workers and the general population.

B. TOXICOKINETICS (FOLLOWING ABSORPTION)

1. Distribution

Once absorbed, cadmium in blood is present in free form or bound to plasma proteins and distributed rapidly to other organs and tissues [Nordberg, 1972]. Cadmium is distributed in various organs and tissues (spleen, pancreas, heart, lungs, testes, thymus and muscle) in humans and animals [ATSDR, 1993d]. The highest amount of cadmium was found in the kidney and liver in humans [Elinder *et al.,* 1976].

2. Metabolism and Excretion

Cadmium may directly interact with metallothionein, a low molecular weight protein. This interaction will decrease the toxicity of cadmium. The cadmium-metallothionein complex is gradually accumulated by the kidney [Friberg *et al.,* 1974].

Cadmium is slowly eliminated from the body with a half-time of about 25 to 30 years [Friberg *et al.,* 1976, 1986; Elinder, 1985]. Most absorbed cadmium is excreted in urine in the parent form. Ingested cadmium that is unabsorbed will be eliminated in the feces.

C. BIOAVAILABILITY

1. Inhalation

Inhaled cadmium aerosols will be deposited in the lungs and slowly cleared. Carcinogenic potential of cadmium compounds depends on their *in vivo* solubility in lung fluid and bioavailability of cadmium ions in the respiratory tract. To become effective, cadmium compounds have to be dissolved within the respiratory tract [Oberdörster, 1989a, 1992]. Cadmium oxide (slightly water-soluble) and cadmium chloride (water-soluble) had a higher dissolution rate in lung fluid than cadmium sulfide [Oberdörster, 1979; Hadley *et al.,* 1980; Glaser *et al.,* 1986; König *et al.,* 1992].

Data on deposition, clearance and retention of cadmium are not available for humans. In rats, about 12% of inhaled cadmium oxide was deposited within the respiratory tract after 5 consecutive 30 min exposures [Boisset *et al.,* 1978]. About 19% of the deposition fraction of cadmium chloride was found in Syrian hamsters [Henderson *et al.,* 1979]. Retention fraction of cadmium ranged from 5% to 20% in rats, mice, rabbits, guinea pigs, dogs and monkeys, following exposure to cadmium oxide fumes for 10 to 30 min [Barrett *et al.,* 1947]. The deposited cadmium compounds (cadmium oxide and cadmium chloride) were cleared slowly in rats with a clearance half-time of 60 - 67 d [Boisset *et al.,* 1978; Oberdörster, 1979]. Clearance of cadmium chloride and cadmium sulphide was biphasic in rats [Klimisch, 1993]. Clearance of cadmium sulfide was very slow [Rusch *et al.,* 1986; Oberdörster, 1992].

Respiratory bioavailability of cadmium in humans was commonly estimated from total body burden of cadmium among smokers. The daily amount of cadmium inhaled from one pack of 20 cigarettes was estimated to be 2 to 4 μg (average 3 μg) [Friberg *et al.*, 1974]. Accumulated cadmium levels in the kidney, liver and lung, which were assumed to account for 50% of total body burden, were measured among smokers. The daily absorption amount is the ratio of total body burden to exposure days (packs per day x number of years of smoking, i.e. pack-years). The respiratory bioavailability factor of cadmium can be estimated by the daily absorption amount divided by the amount of cadmium inhaled:

$$BF = \frac{\text{total body burden of cadmium}}{\text{No. of pack-years x 365 d/yr x the amount of cadmium inhaled/d}}$$

Thus, based on the total body burden of cadmium among smokers, the bioavailability factor for cadmium aerosols ranged from 27% to 95% [Lewis *et al.*, 1972; Elinder *et al.,* 1976; Ellis *et al.*, 1979].

In animal studies, the bioavailability factor was mainly derived from the accumulated amount of cadmium in various organs and tissues, particularly in the kidney, liver and lung. This accumulated cadmium amount represents a minimum estimate of absorption fraction. Harrison *et al.* [1947] found about 40% of the inhaled cadmium in the kidney, liver and other organs in dogs following exposure to cadmium chloride. Henderson *et al.* [1979] reported about 25% to 35% of cadmium in the kidney, liver and skull at 3 wk post-exposure (cadmium chloride) in hamsters. In another study, rats were exposed to cadmium oxide for 30 min (total 5 d) [Boisset *et al.,* 1978]. About 60% of the deposited cadmium was absorbed. Because the total deposition fraction was 12%, the total bioavailability factor was 7.2% of exposure levels. The bioavailability factor of cadmium chloride in rats was about 41% [Moore *et al.,* 1973]. When rats were exposed to cadmium oxide via intratracheal instillation, the respiratory bioavailability was higher (at least 60%) than that via inhalation [Hadley *et al.*, 1980].

2. Ingestion

Absorption of cadmium in the gastrointestinal tract commonly involves two steps: (1) cadmium is trapped in the mucosal cells with a small part of cadmium bound to endogenous metallothionein within the mucosal cells and (2) cadmium penetrates the gut-blood barrier (the rate-limiting step) [Foulkes, 1980, 1988, 1989; Andersen, 1989, 1992]. Oral bioavailability of cadmium is usually estimated by measurement of fecal cadmium excretion because cadmium does not directly undergo metabolism and excretion of absorbed cadmium into feces via bile is very low. Alternatively, measurement of total body burden and whole-body counting is commonly used because absorbed cadmium is very slowly eliminated from the body.

Oral bioavailability of cadmium in humans has been extensively studied. Human volunteers ingested cadmium mixed with calf kidney suspension. The oral bioavailability factor of cadmium was 6% and 2.5% at a single dose of 100 μg [Rahola *et al.,* 1973a] and 50 μg [Shaikh and Smith, 1980], respectively. Following oral administration of cadmium incorporated into crab meat, about 2.7% of administered cadmium was absorbed from the gastrointestinal tract [Newton *et al.*, 1984]. An average of 4.6% ingested cadmium chloride in water was bioavailable [McLellan *et al.*, 1978]. In another study, oral bioavailability of cadmium in liquid (nutrient solution and lemon juice) was found to be 7% [Yamagata *et al.*, 1975]. Based on the total body burden of cadmium in humans, Ellis *et al.* (1979) estimated less than 4% of cadmium taken from daily food was absorbed. Persons with iron deficiency had higher cadmium bioavailability (8.9%) than did persons with normal iron stores (2.6%) [Flanagan *et al.*, 1978].

In small rodents (rats and mice), oral bioavailability of cadmium ranged from 0.2% to 8% (the highest value was 7% to 8%) [Decker *et al.,* 1957, 1958; Suzuki *et al.,* 1969; Cotzias *et al.,* 1961; Moore *et al.,* 1973; Engström and Nordberg, 1979; Müller *et al.,* 1986; Andersen and Nielsen, 1989; Schäfer *et al.*, 1990]. About 1% to 3% of ingested cadmium was absorbed in monkey [Friberg *et al.*, 1971, 1974].

A number of factors may affect cadmium absorption, such as age, gender, pregnancy, and dietary factors. A higher gastrointestinal absorption of cadmium had been found in newborn and young rats [Sasser and Jarboe, 1977; Kostial *et al.*, 1979, 1983]. Pregnant mice more easily absorbed cadmium from the gastrointestinal tract than nonpregnant mice [Bhattacharyya *et al.*, 1986]. Deficiency of zinc, iron, calcium and copper increased oral bioavailability of cadmium, while ingestion of cadmium at a high dose decreased oral bioavailability of these trace elements [Valberg *et al.,* 1976; Flanagan *et al.,* 1978; Nordberg *et al.,* 1985; Schäfer *et al.,* 1990]. Cadmium absorption was also found to be increased after oral administration of a very high dose of cadmium [Lehman and Klaassen, 1986; Andersen *et al.*, 1988].

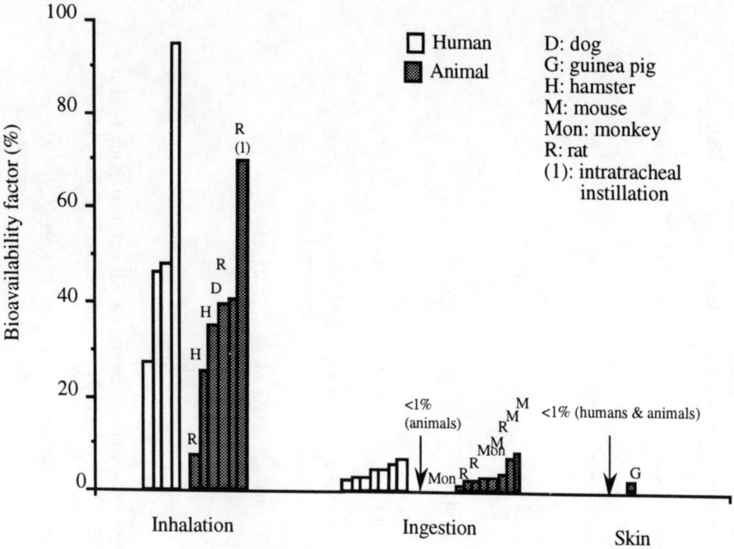

Figure 5-3 Bioavailability factors of cadmium in humans and animals by three exposure routes [Adapted from sources in Table 5-5, a-h (inhalation); Table 5-6, a-g (ingestion); Table 5-7, a-k (ingestion); and Table 5-8, a-c (skin)]

3. Dermal

Data on dermal bioavailability of cadmium are limited. One *in vitro* study indicated that dermal bioavailability of cadmium from water and soil was very low (0.1% to 0.6% for cadmium in water and 0.01% to 0.07% in soil) [Wester *et al.,* 1992]. Dermal absorption of cadmium in guinea pigs was generally less than 1% (maximum absorption of 1.8% at a very high concentration of 0.239 M) [Skog and Wahlberg, 1964]. Kimura and Otaki (1972) reported that dermal bioavailability factors were 0.4% to 0.6% in rabbits and 0.1% to 0.6% in mice.

4. Summary

Cadmium can be bioavailable through three exposure routes (Table 5-5, 5-6, 5-7, 5-8, and Figure 5-3). Cadmium is readily absorbed via inhalation in humans and animals. Oral absorption of cadmium is low in humans and animals. Cadmium penetrates human skin (*in vitro*) and animal skin (*in vivo*) to a limited degree.

D. UNCERTAINTIES

Humans may be exposed to cadmium by inhalation of cadmium fume (cadmium oxide), cadmium dust and fine cadmium particulates from cigarettes. A large proportion of cadmium aerosols is less than 10 μm and easily inhalable. The deposition fraction of cadmium aerosols within the

Table 5-5 Respiratory Bioavailability of Cadmium in Humans and Animals

Compound	Species	Exposure methods	Cadmium dosage	Experimental method	Exposure duration	Post-exp. duration	Bioavailability factor (%)	Ref.
Cd (aerosol)	Human	Smoking	4 µg/pack cigarette/day	Estimation based on body burden[1]	30-pack-year	-	27[2]	a
Cd (aerosol)	Human	Smoking	3 µg/pack cigarette/day	Estimation based on body burden[1]	25-pack-year	-	46	b
Cd (aerosol)	Human	Smoking	3 µg/pack cigarette/day	Estimation based on body burden[1]	27-pack-year	-	48-95	c
$CdCl_2$ (aerosol)	Dog	Inhalation	280-360 mg Cd/m^3	Mass-balance	30 min	10 wk	40[3]	d
$CdCl_2$ (aerosol)	Rat	Whole-body	1.8 mgCd/m^3	Whole-body counting	4 h	32 d	41	e
CdO_2 (Liquid)	Rat	Whole-body	9.4 mg Cd/m^3	Mass-balance	30 min x 5 d	6-8 wk	7.2 (60 of the deposited dose)	f
$CdCl_2$ (aerosol)	Hamster	Nose-only	2 - 17 mg Cd/m^3	Mass-balance	30 min	23 wk	25-35	g
CdO_2 (Liquid)	Rat	Intratracheal instillation	15 µg	Mass-balance	-	2 wk	70	h

[1] Total body burden is mainly based on the cadmium amount accumulated in the kidney, liver and/or lung, which accounts for 50% of total body burden; [2] Absorption rate was calculated by Friberg *et al.*, 1974 based on the study from Lewis *et al.*, 1972; [3] Absorption rate was calculated by Friberg *et al.*, 1974 based on the study from Harrison *et al., 1947*.

Adapted from: a. Lewis *et al.*, 1972; b. Elinder *et al.*, 1976; c. Ellis *et al.*, 1979; d. Harrison *et al.*, 1947; e. Moore *et al.*, 1973; f. Boisset *et al.*, 1978 g . Henderson *et al.*, 1979; h. Hadley *et al.*, 1980

Table 5-6 Oral Bioavailability of Cadmium in Humans

Compound	Number of Subjects	Cadmium carrier	Nutritional status	Cadmium dosage	Experimental method	Exposure duration	BF (%)[1]	Ref.
$^{115m}Cd\text{-}(NO_3)_2$	5	Calf kidney cortex	Fed	100 µg	Whole-body counting and fecal excretion	20 d	6	a
^{115m}Cd	3	Rice	Fed	150 µg	Whole-body counting and fecal excretion	6 d	7	b
$^{115m}CdCl_2$	10	Cereals	Iron deficiency	25 µg	Whole-body counting	1 wk	8.9	c
$^{115m}CdCl_2$	12	Cereals	Normal iron stores	25 µg	Whole-body counting	1 wk	2.3	d
Diet cadmium	20	Fed	Fed	75-100 µg	Estimation based on total body burden	52 y	<4	e
$^{115m}CdCl_2$	14	Water	Fed	22-29 µg	Fecal excretion	2-6 wk	4.6	e
$^{115m}CdCl_2$	12	Calf kidney cortex	Fed	50µg	Whole-body counting and fecal excretion	56-808 d	2.5	f
$^{115m}CdCl_2$	7	Crab sandwich	Fed	24-166 µg	Whole-body counting	87 d	2.7	g

[1] BF = Bioavailability Factor

Adapted from: a. Rahola *et al.*, 1973a; b. Yamagata *et al.*, 1975; c. Flanagan *et al.*, 1978; d. Ellis *et al.*, 1979; e. McLellan *et al.*, 1978; f. Shaikh and Smith, 1980; g. Newton *et al.*, 1984

Table 5-7 Oral Bioavailability of Cadmium in Animals

Compound	Species	Exposure methods	Cadmium carrier	Cadmium dosage	Experimental method	Exposure duration	Bioavailability factor (%)	Ref.
$^{115m}Cd\text{-}(NO_3)_2$	Rat	Gavage	Liquid	2 mg (6.6 µg/g)	Mass-balance	1 d	1-2	a
$CdCl_2$	Rat	Gavage	Drinking water	10 µg/g	Mass-balance	1 y	<1	b
$^{115m}CdCl_2$	Rat	Gavage	-	40 µC	Whole-body counting	60 d	2.3	c
$^{109}CdCl_2$	Rat	Gavage	Liquid	179 µg - 1.4 mg/kg	Whole-body counting	14 d	2.2-3.5	d
[1]Cd-Ac, Cd-MT, Cd-GSH	Rat	Gavage	-	2.5 µg/g/d (x 28 times)	Mass-balance	42 d	0.3	e
$^{109}CdCl_2$	Mouse	Gavage	-	40 µC	Mass-balance	4 h	0.5-8	f
$^{115}CdCl_2$	Mouse	Gavage	-	80 µg	Mass-balance	11 h	7	g
$^{109}CdCl_2$	Mouse	Gavage	-	1 µg/kg 15-750 µg/kg 3.75 mg/kg	Mass-balance	5 d	0.5 1.4 3.2	h
$^{109}CdCl_2$	Mouse	Gavage	Drinking water	500 µg/kg	Mass-balance	15 d	0.25	i

[1] Cd-Ac = Cd-acetate, Cd-MT = Cd-metallothionein, Cd-GSH = Cd-glutathione
Adapted from: a. Decker *et al.*, 1957; b. Decker *et al.*, 1958; c. Moore *et al.*, 1973; d. Shäfer *et al.*, 1990; e. Müller *et al.*, 1986; f. Cotzias *et al.*, 1961; g. Suzuki *et al.*, 1969; h. Engström and Nordberg, 1979; i. Andersen and Nielsen, 1989

(Continued)

Compound	Species	Exposure methods	Cadmium carrier	Cadmium dosage	Experimental method	Exposure duration	Bioavailability factor (%)	Ref.
$^{115}CdCl_2$	Monkey	Gavage	-	0.17-1.7 mg/kg	Whole-body counting	10 d	2.9	j
$^{115}CdCl_2$	Monkey	Gavage	-	1 µg/kg	Whole-body counting	10 d	1	k

Adapted from: j. Friberg et *al.*, 1971; k. Friberg *et al.*, 1974

Table 5-8 Dermal Bioavailability of Cadmium in Humans and Animals

Compound	Species	Cadmium carrier	Exposure region	Skin area	Cadmium concentration	Experimental method	Exposure duration	Bioavailability factor (%)	Ref.
$^{115}CdCl_2$	Human	Water	Cadaver (500 µm thickness)	-	0.58 µg/cm^2	Diffusion cell (*in vitro*)	16 h	0.1-0.6	a
$^{115m}CdCl_2$	Human	Soil	Cadaver (500 µm thickness)		0.52 µg/cm^2	Diffusion cell (*in vitro*)	16 h	0.01-0.07	a
$^{115}CdCl_2$	Guinea pig	Water	Back	-	0.239 M 0.005 M 0.080 M 0.398 M	Disappearance	5 h	1.8 <1 <1 <1	b
$CdCl_2$	Rabbit	Water Ointment	Back	10 cm^2	0.6 mg/cm^2 1.2 mg/cm^2	Mass-balance	2 wk	0.4-0.6 0.4-0.6	c
$CdCl_2$	Hairless mouse	Water Ointment	Back	10 cm^2	0.06 mg/cm^2	Mass-balance	1 wk	0.2-0.8	c

Adapted from: a. Wester *et al.* ,1992; b. Skog and Wahlberg, 1964; c. Kimura and Otaki, 1972

human respiratory tract has not been thoroughly investigated. However, data on respiratory bioavailability of cadmium aerosols are available for fine cadmium particulates from cigarettes in humans and for cadmium chloride in animals. Absorption of cadmium oxide and cadmium sulfide from the respiratory tract has not been studied in humans and animals.

Most oral absorption data are based on cadmium chloride mixed with different types of food. The actual species of cadmium in major diet sources have not been characterized. Moreover, mechanisms of interaction of cadmium with other metals during oral absorption processes are not fully understood.

E. CONCLUSIONS

Humans may be exposed to cadmium through three exposure routes. In occupational settings, the primary exposure route is via inhalation of cadmium aerosols. For smokers in the general population, exposure to fine cadmium aerosols in tobacco is a significant route. Ingestion of cadmium from food is the main exposure route for non-smokers in the general population. Because cadmium is not easily absorbed through the skin, the dermal exposure route is negligible for both workers and the general population.

Cadmium can be bioavailable after inhalation, ingestion and dermal exposure in humans and animals. The respiratory bioavailability factors of cadmium aerosols from cigarettes range from 27% to 95% among smokers. The deposition fraction of cadmium aerosols within the respiratory tract is 5% to 20% in small rodents. The respiratory bioavailability factors in small rodents are 7% for cadmium oxide (slightly water soluble) and 25% to 40% for cadmium chloride (water soluble). Absorption of water soluble cadmium from the gastrointestinal tract is low (2% to 7% in humans and 0.2% to 8% in animals). Dermal bioavailability is extremely low (less than 1%) in humans (*in vitro*) and in animals (*in vivo*).

F. RECOMMENDATIONS

Most absorption data through the three exposure routes are derived from administration of cadmium chloride in humans and animals. However, bioavailability for other cadmium species relevant to adverse human health effects needs to be investigated via all three exposure routes.

Deposition and mechanical clearance of cadmium aerosols and soil-bound cadmium within the respiratory tract should be determined. Studies on the interaction of cadmium in the gastrointestinal tract are needed.

III. CHROMIUM AND CHROMIUM COMPOUNDS

Chromium (Cr) is an element commonly found on earth which exists naturally in its elemental form and in oxidation states ranging from -2 to +6 [NAS, 1974; USEPA, 1985a, 1990b; Lide, 1993]. However, only elemental chromium, chromium (III) and chromium (VI) forms are stable and biologically important [Hartford, 1979; USEPA, 1984a; Hathaway *et al.*, 1991; ACGIH, 1991].

Because of its widespread distribution, chromium is commonly detected in soil, water, food and air. Chromium (VI) compounds, e.g. $Cr_2O_7^{-2}$ and CrO_4^{-2}, are the primary forms of atmospheric chromium which are mobile, reactive and bioavailable [USEPA, 1984a]. Both chromium (III) and chromium (VI) are stable in water under thermodynamic equilibrium and in sandy soils containing low levels of organic matter [Bartlett and James, 1988; Bartlett, 1991; USEPA, 1990b]. These two forms of chromium are likely to be interconvertible in natural water and in soils, a process called chromium cycling [Roberts and Anderson, 1975; Schroeder and Lee, 1975; Eary and Rai, 1987, 1988; Rai *et al.*, 1989; Bartlett, 1991]. Chromium compounds in airborne particulates can deposit on plant surfaces, and be taken up by plant roots growing on contaminated soils [NAS, 1974]. Chromium content in foodstuff can vary considerably. In general, the chromium content of most foods is very low [NAS, 1974; IARC, 1980]. Larger quantities of chromium have been reported in meats, vegetables, unrefined sugar and crustaceans in comparison with fish and fruit [NAS, 1974].

Chromium (III) is considered to be an essential trace nutrient for humans [Mertz, 1975; Anderson, 1981, 1989; USEPA, 1984a; WHO, 1988; Iyengar, 1989; ATSDR, 1989a; Burg and Liu, 1993]. However, some evidence has revealed that chromium (III) compounds may slowly react with nucleic acids and proteins *in vitro* or *in vivo*, thereby exhibiting mutagenicity and genotoxicity [Cupo and Wetterhahn, 1985; Elias *et al.,* 1986; Sugden *et al.,* 1990; Snow, 1991].

Chromium (VI), mainly produced by industrial processes, has been classified as a human carcinogen via inhalation exposure. Specifically, exposure to certain water-insoluble chromium (VI) compounds (particularly in industrial chromate production) and water-soluble chromium (VI) compounds was found to be associated with lung cancer in humans [IARC, 1980, 1987, 1990; USEPA, 1984a; WHO, 1988; Cohen *et al.,* 1993; Langard, 1993]. Data reviewed from IARC (1980) also indicate that a few water-soluble chromium (VI) compounds (including chromium trioxide) and some water-insoluble compounds (calcium chromate, lead chromate and zinc chromate) may be related to an increased risk of lung cancer in animals. Evidence on the carcinogenicity of chromium compounds via ingestion and dermal exposure is currently lacking.

Water-soluble chromium compounds can be rapidly absorbed in the respiratory system. They are slightly bioavailable in the gastrointestinal system. Quantitative data on skin absorption are lacking.

A. EXPOSURE ROUTES AND RECEPTORS

For occupational exposure, inhalation and dermal contact are both primary routes [WHO, 1988]. Exposure to both chromium (VI) and chromium (III) often occurs. Permissible exposure limits in air are 0.5 mg/m^3 for chromium (III) and 0.05 mg/m^3 for chromium (VI) at the workplace [Carson *et al.,* 1986; Sittig, 1991]. Inhaled water-soluble chromium compounds can readily penetrate through the air-blood barrier. In contrast, water insoluble compounds remain in the respiratory tract. Workers may contact chromium products through the skin.

The general population is mainly exposed to chromium compounds by ingestion of food, water or soil containing chromium. The levels of chromium compounds in foodstuffs and drinking water are generally very low and their compounds are only slightly absorbed in the gastrointestinal tract. Thus, the ingestion route from diet and drinking water is generally considered to be of little significance for most humans.The permissible exposure limit for chromium in drinking water is 50 μg/L [MSSC, 1989]. Chromium (VI) in groundwater is rarely detected, because of the common reduction of chromium (VI) to (III) when these compounds leach from soil into groundwater. Accordingly, the potential hazard from groundwater appears to be negligible.

On the other hand, the residents living near industrial and contaminated areas may be exposed to higher levels of chromium (VI) [USEPA, 1984a]. Under this circumstance, inhalation may become significant for human exposure. For example, a study from Hudson County, New Jersey, where soil was contaminated with chromium processing waste, indicated elevated chromium concentrations in outdoor air, indoor air and house dust [Lioy *et al.*,1992]. In particular, chromium in household dust was considered to be the primary source of exposure. The waste was used at numerous locations within the County as fill, resulting in chromium (VI) and chromium (III) being present at the surface at many of these locations. These locations represented a significant source of emission contributing to the elevated chromium levels in the air and house dust.

Soil ingestion is a potential exposure route for young children with pica behavior (eating non-food materials) living near chromium-contaminated sites. However, the hazard posed by soil ingestion is minimal unless the concentrations of soil chromium are extremely high. The safe levels for soil chromium suggested by the New Jersey Department of Environmental Protection were less than 1000 ppm for total chromium and 75 ppm for chromium (VI) [Paustenbach *et al.,* 1991a].

Occasionally, people may have dermal contact with contaminated soil or water during outdoor activities. To date, however, no special groups with an exceptional risk of exposure outside specific occupational environments have been identified [USEPA, 1984a].

B. TOXICOKINETICS (FOLLOWING ABSORPTION)

1. Distribution

Chromium compounds absorbed (except chromates) are rapidly removed from the systemic circulation, distributed in various tissues, or eliminated via urine [Mertz, 1969; WHO, 1988]. In humans, chromium and its compounds following inhalation are mainly distributed in the lung, lymph nodes, kidney, liver, bladder and bone [IARC, 1980]. Through ingestion, high chromium levels were found in hilar lymph nodes, lung, spleen, liver, kidney and heart [Teraoka, 1981].

2. Metabolism and Excretion

Chromium (VI) has been found to be reduced to chromium (III) with *in vivo* and *in vitro* studies [Wiegand, 1984]. Chromium (VI) compounds readily enter red blood cells by simple diffusion, where chromium (VI) is strongly bound to hemoglobin, while chromium (III) compounds have limited ability to cross blood cell membranes, but do have an affinity for serum protein [Mertz, 1969; Seiler and Sigel, 1988].

Excretion of chromium from the human body, following inhalation or ingestion, occurs primarily via urine. In most cases, chromium (VI) compounds are rarely detected in the urine because of the reduction of chromium (VI) to chromium (III) or the formation of chromium protein complexes [USEPA, 1984a]. Chromium (III) is excreted to a lesser degree compared with chromium (VI) because chromium (III) tends to form less excretable hydrate complexes. Data on excretion specifically following dermal exposure have not been located. Following the administration of intravenous chromium (VI) and (III) compounds, there was about 3.5% to 8.4% of chromium (VI) and 0.1% to 0.5% of chromium (III) found in the bile of rats [ATSDR, 1989a].

C. BIOAVAILABILITY

1. Inhalation

a. Humans

Several studies have revealed that chromium (VI) and (III) compounds could be absorbed following inhalation exposure in occupational or general populations, as evidenced by increased chromium in the blood and urine [Baetjer *et al.,* 1959; NAS, 1974; Gylseth *et al.*, 1977; Cavalleri and Minoia, 1985; Randall and Gibson, 1987; Stern, 1992]. The ICRP Task Group [Morrow *et al.,* 1966] classified the oxides and hydroxides of chromium as a slow clearance group, and other chromium compounds as a fast clearance group. Consequently, water-insoluble compounds tend to remain in the respiratory tract, while water-soluble compounds are likely to be absorbed in various regions of the respiratory tract. Inhaled water-soluble chromium (VI) compounds are rapidly absorbed according to studies on biological monitoring of occupational exposures [Franchini *et al.,* 1984]. However, quantitative data on the deposition, retention and clearance of chromium compounds for humans are generally lacking.

b. Animals

Once water-insoluble chromium (III) aerosols are inhaled, they are primarily cleared by mechanical clearance processes. One study from Visek *et al.* [1953] reported that chromic chloride ($CrCl_3$) dust, administered by intratracheal instillation in rats, was primarily excreted in the feces (55%). The results indicated that a large amount of inhaled chromium (III) was ultimately swallowed into the gastrointestinal tract. In another study [Sanders *et al.,* 1971], chromic oxide (Cr_2O_3) dust was introduced to hamsters by an inhalation chamber for 4 h at a dose of 0.5 to 1.0 mg/m^3. Over 90% of chromium (III) particles with the size of 0.33 μm median diameter were ingested by pulmonary macrophages.

Absorption of chromium (III) from the respiratory system is very low. Three studies showed that the bioavailability factors were less than 10% in small rodents following intratracheal administration of water-soluble chromium chloride at a single dose ranging from 100 to 200 μg by using a mass-balance technique [Baetjer *et al.,* 1959; Visek *et al.,* 1953; Wiegand *et al.*, 1984]. The majority of compounds inhaled were eliminated by mechanical clearance into the gastrointestinal tract. Baetjer *et al.* [1959] reported only 4% of administered chromium in the blood and other tissues and 69% in the lungs at 10 min post-exposure. At 24 h, there was 6% present in the urine and 45% remaining in the lungs. In this case, the cumulative urinary value accounted for the absorption fraction from both the respiratory tract and the gastrointestinal tract. In rats, the respiratory bioavailability factor was found to be less than 5% [Visek *et al.*, 1953]. Rabbit experiments indicated that 9% of chromium (VI) compounds could be absorbed from the lungs after 4 h dosing [Wiegand *et al.*, 1984].

Water-soluble chromium (VI) compounds are more readily absorbed than chromium (III) compounds. Experiments reported by Baetjer *et al.* [1959] showed only 15% of potassium dichromate (CrVI), after 10 min exposure, remained in the lung, with 20% present in the red blood cells, and 5% in the liver. Overall, the bioavailability factor of chromium (VI) was approximately 29%. In rabbits, the overall bioavailability factor of chromium (VI) was 46% after 4 h dosing [Wiegand *et al.,* 1984].

Weber [1983] used the whole body or lung counting and mass-balance techniques to determine the absorption and distribution of a chromium (VI) compound (sodium dichromate) in rats. Following the intratracheal administration of the compound (50 μg), the retention fraction in the lung was 43%, 36%, and 12% at 6 h, 24 h and 40 d, respectively. Most of the compound within the lung was localized in type II alveolar cells which form part of the air-blood barrier. These cells can synthesize and release a surfactant to regulate the surface tension for alveoli. However, the lack of data on urinary excretion of chromium (VI) precludes estimating a bioavailability factor for this study.

Bragt and his co-workers [Bragt and Van Dura, 1983; Bragt *et al.,* 1986] found that the clearance of chromates (sodium, zinc or lead chromate dusts) from the lungs and blood was directly associated with the degree of water solubility of the salts. 20% of sodium chromate (water-soluble) and 26% of zinc chromate (less water-soluble) were absorbed from the lung during 10 d exposure. Lead chromate (water-insoluble) was only slightly bioavailable (2%).

Zinc chromate (VI) dust was reported to be rapidly absorbed in rats [Langård *et al.*, 1978]. The dust contained 20.4% chromium (VI) with 76% as an inhalable fraction. After 100 min inhalation exposure with a dose of 7.35 mg/m^3 in an inhalation chamber, the blood chromium levels were increased 5-fold. However, the compound was slowly eliminated from the blood.

Chromium mixtures administered by inhalation generally have a low bioavailability [Salem and Katz, 1989]. Whetlerite dust (2.02% total chromium, 0.98% insoluble chromium (VI), 0.30% soluble chromium (VI) and 0.74% insoluble chromium (III)) was intratracheally introduced to rats at a single dose of 320 µg/g. After 24 to 48 h following dosing, only 17% of the total chromium dose was recovered in the lung, liver, kidney, spleen and blood.

2. Ingestion

a. Humans

Chromium (III) compounds in the diet, water or tablets are only slightly bioavailable in humans. Doisy *et al.* [1968] examined a number of normal, elderly and diabetic subjects who ingested chromic chloride at 150 µg/d with meals. After three days exposure, only 0.69% chromium was taken up by normal subjects, and 0.92% in elderly subjects. Donaldson and Barreras [1966] reported that the absorption fraction of chromic chloride in water for fasting subjects was 0.4% according to the fecal excretion values at 6 d after dosing 20 ng chromium. In another study, 0.4% of chromium was found in the urine following the administration of 200 µg/d chromium chloride tablets [Anderson *et al.,* 1983]. Offenbacher *et al.* [1986] studied two subjects who ingested chromium at 37 µg/d. They observed that the average absorption fraction was 1.8%. Another study found that the intake of low diet chromium (10 µg/7 d) had a high absorption (2%), while the intake of higher diet chromium had a lower absorption fraction (0.5%) [Anderson and Kozlovsky, 1985].

In general, chromium (VI) compounds appear to be absorbed rapidly and readily. A study from Donaldson and Barreras [1966] showed a 10.6% bioavailability factor for fasting subjects who were orally administered 20 ng of sodium chromate.

b. Animals

Data from animal experiments revealed that the degree of absorption of chromium (III) compounds was also low. Rats fed by chromic chloride in a buffer solution only absorbed less than 0.5% of chromium compounds after 4 d exposure [Visek *et al.,* 1953]. With the administration of chromic chloride by gavage, the bioavailability factor in rats was 2% to 3% [Mertz *et al.*, 1965]. Sullivan *et al.* [1984] observed that the extent of absorption of chromic chloride was higher in 2 d old rats (1.2%) than in adult rats (0.1%). Data on chromium (VI) absorption in rats and mice showed that bioavailability factors were slightly higher (2.3% to 2.4%) than that of chromium (III) compounds [Ogawa *et al.*, 1976; Donaldson and Barreras, 1966]. Moreover, mice in fasting state could absorb more chromium (III) and (VI) (11%) than mice with a normal diet [Ogawa *et al.*, 1976]. Rats in fasting state also absorb more chromium (VI) (6%) than rats with a normal diet (3%) [MacKenzie *et al.*, 1959].

Bioavailability of plant chromium complexes was studied by Starich and Blincoe [1983]. In their experiment, alfalfa and brewer's yeast were exposed to either chromium (III) or (VI) (CrO_4^{-2}) *in vitro*. Then, the plant chromium complexes were fed to rats for 10 d. The bioavailability factors of plant chromium complexes were 35% for alfalfa seedings, 37% for alfalfa extract and 29% brewer's yeast. Other studies suggested that 10% to 25% of the Brewer's yeast chromium was bioavailable [Underwood, 1977].

c. Others

In a soil solubility study, less than 0.5% to 1% of total soil chromium could be dissolved in an acid solution (pH 2 and pH 5 within 24 h) [Paustenbach *et al.*, 1991a; Sheehan *et al.*, 1991]. As a result, only a very small fraction of soil chromium was judged to be available for absorption in the gastrointestinal tract.

3. Dermal

Investigations have indicated some absorption of chromium (III) and (VI) through the skin in humans [Samitz and Shrager, 1966; Baranowska-Duttiewicz, 1981a; Kelly *et al.*, 1982; Lindberg and Vesterberg, 1983; Randall and Gibson, 1987; Gammelgaard *et al.*, 1992]. In principle, chromium VI compounds can easily pass through the skin without the formation of complexes [Bagdon and Hazen, 1991]. Chromium VI compounds can also be converted to chromium (III) compounds at certain applied concentrations [Gammelgaard *et al.*, 1992]. In contrast, chromium (III) compounds are likely to bind non-specific proteins on the skin with resulting difficulty in penetration of the skin. Samitz and Shrager [1966] reported that chromium (III) nitrate only slightly penetrated the intact human skin, while chromium (III) sulfate failed to pass through the skin.

The dermal permeability of chromium (III) and (VI) compounds has been studied in humans *in vivo* and *in vitro*. In one *in vivo* study, Na_2CrO_4 (Cr VI) solution at different concentrations (0.01 M, 0.1 M and 0.2 M) was applied on the forearm skin in human volunteers [Baranowska-Dutkiewicz, 1981a]. The dermal permeability was 3.5 x 10^{-3} cm/h at 0.01 M level after 15 min dosing and 1 x 10^{-3} cm/h at 0.2 M after 60 min dosing. The dermal permeability of chromium (III) and (VI) compounds through human skin *in vitro* is lower than that *in vivo*. Gammelgaard *et al.* [1992] reported only 0.01 x 10^{-3} cm/h of *in vitro* dermal permeability for a chromium (VI) compound ($K_2Cr_2O_7$) in humans.

In animal experiments, absorption of chromium compounds varied with the concentrations of the compounds at the applied site [Wahlberg and Skog, 1965]. Chromic chloride and sodium chromate ranging from 0.00048 to 1.689 M were applied to the skin of guinea-pigs. In the lowest (0.00048 and 0.005 M) and highest (0.753 and 1.689 M) concentrations, the absorption fractions were less than 1% after 5 h exposure. For the remaining concentrations (0.017 to 0.398 M), the absorption fractions of chromium (VI) were higher than chromium (III).

Dermal absorption of chromium is pH dependent. With higher pH of applied solution, the extent of absorption of chromium (III) and (VI) compound will increase. Czernielewski *et al.* [1965] observed that guinea-pigs with pretreatment of alkali on the skin had a higher degree of absorption of chromium (VI) (12.5%) than those without alkali treatment (1.30%).

When chromium compounds are bound to soil, dermal bioavailability will be low. Based on a solubility study, an estimated 1% of total soil chromium may leach from the soil matrix and be dissolved in sweat on the skin [Paustenbach *et al.*, 1991b].

4. Summary

Overall, either chromium (III) or chromium (VI) is bioavailable in humans and animals through inhalation, ingestion and dermal exposure (Table 5-9, 5-10, 5-11 and 5-12). Available quantitative results are summarized in Figure 5-4. The extent of absorption is higher via inhalation than via ingestion or dermal exposure. Chromium (VI) (other than water-insoluble compounds) is more readily absorbed than chromium (III) by any of these routes. Species differences are significant but generally small.

D. UNCERTAINTIES

At contaminated sites, chromium (VI) compounds (CrO_3) may be present in surface soil and lagoons. Soil-bound chromium may be transported off-site to residential areas by wind. As a result, residents in the vicinity of chromium contaminated sites may be exposed to airborne chromium by inhalation.

The respiratory tract following inhalation of chromium (VI) has been recognized as the only target tissue where carcinogenicity arises. However, the mechanisms of chromium carcinogenesis have not been clarified and are under review. Some studies indicated that the occurrence of lung cancer was related to the genotoxicity in alveolar tissue that was contributed by the reduced

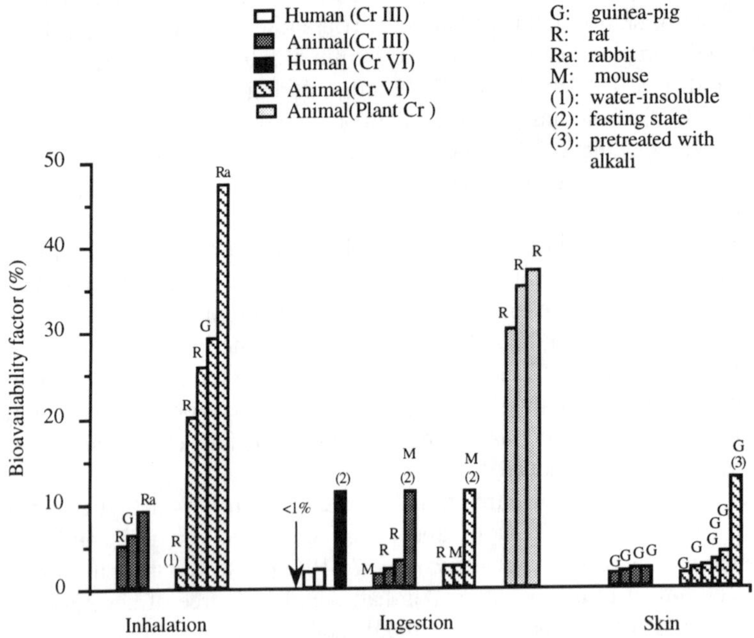

Figure 5-4 Bioavailability factors of chromium compounds in humans and animals by three exposure routes [Adapted from sources in Table 5-9: a, b, c, d, (inhalation); Table 5-10: a, b, c, d, e (ingestion); Table 5-11: b, c, d, f (ingestion); and Table 5-12: a, b (skin)]

forms (Cr III or Cr V) of chromium (VI) [USEPA, 1990c]. Therefore, an understanding of not only the extent of absorption, but also deposition, retention and mechanical clearance of chromium-containing aerosols are important for cancer risk assessment. Factors, such as particle size, inhalable fraction and solubility of soil chromium in the respiratory tract (under neutral pH conditions), strongly influence these processes. To date, the respiratory bioavailability has only been determined for some water-soluble and a few water-insoluble chromium (VI) compounds in animals. Extrapolation of chromium (VI) bioavailability from animals to humans without an appreciation of metabolism carries substantial uncertainty.

The quantitative data for CrO_3 absorption are lacking. However, CrO_3 is highly water-soluble and reactive. It may leach from soil particles trapped in the respiratory tract and be rapidly absorbed into the systemic circulation in humans following inhalation.

Chromium (VI) can leach from surface soil into groundwater or be released into surface water by runoff and fallout. Suspended chromium dust may deposit on locally grown plants. Chromium (VI) is readily reduced to chromium (III) in water and plants. Consequently, resident exposure to

Table 5-9 Respiratory Bioavailability of Chromium in Animals

Compound	Species	Exposure method	Cr dosage	Experimental method	Exposure duration	Retention in the lung [Cr] (%)	Cr Recovery (%) B[1]	U[2]	F[3]	Other	TC[4]	BF[5] (%)	Ref
Cr(III)													
$CrCl_3$	Guinea-pig	Intratracheal	200 µg	Mass-balance	10 min	69	4				31	4[6]	a
					24 h	45		6			55	6	
					60 d	12					88		
$CrCl_3$	Rat	Intratracheal	40-250 µg	Mass-balance	7 d			7	55			<5	b
$CrCl_3$	Rabbit	Intratracheal	100 µg	Mass-balance	4 h	85	0.83	7.9		0.51	15	9	c
Cr(VI)													
$K_2Cr_2O_7$	Guinea-pig	Intratracheal	200 µg	Mass-balance	10 min	15	20			5	88	25[6]	a
					24 h	11	9	13		7	89	29	
					60 d	1.6					98		
$Na_2Cr_2O_7$ + Na_2CrO_4	Rabbit	Intratracheal	100-500 µg	Mass-balance	4 h	47	6.1	15.5		24	53	47	c
Na_2CrO_4	Rat	Intratracheal	69 µg	Mass-balance	10 d		0.25	20	20			20	d
$ZnCrO_4$			66 µg				0.75	25	50			26	
$PbCrO_4$			38 µg				0	2	75			2	

[1] B = Blood; [2] U = Urine; [3] F = Feces; [4] TC = Total Clearance; [5] BF = Bioavailability Factor; [6] the value from blood and other tissues

Adapted from: a. Baetjer *et al.*, 1959; b. Visek *et al.*, 1953; c. Wiegand *et al.*, 1984; d. Bragt and van Dura, 1983

Table 5-10 Oral Bioavailability of Chromium in Humans

Compound	Number of subjects	Chromium carrier	Chromium dosage	Experimental method	Exposure duration	Bioavailability factor (%)	Ref.
Cr(III)							
$CrCl_3$	15 (normal) 7 (elderly)	Meal	150 µg/d	Urinary excretion	72 h	0.69 (0.3-1.3) 0.92 (0.4-1.5)	a
$CrCl_3$	42	Tablet	200 µg/d	Urinary excretion	24 h	0.4	b
$CrCl_3$	6 (fasting)	Water	Single: 20 ng	Urinary and fecal excretion	6 d	0.4 (Minimal 0.5)	c
Dietary chromium	2	Diet	37 µg/d	Urinary and fecal excretion	6 d	1.8 (Minimal 0.8)	d
Dietary chromium	32	Diet	10-40 µg/d	Urinary excretion	7 d	2 (10µg) 0.5 (40µg)	e
Cr(VI)							
Na_2CrO_4	6 (fasting)	Water	Single: 20 ng	Urinary and fecal excretion	6 d	11 (Minimal 2.1)	c

Adapted from: a. Doisy *et al.*, 1968; b. Anderson *et al.*, 1983; c. Donaldson and Barreras, 1966; d. Offenbacher *et al.*, 1986; e. Anderson and Kozlovsky, 1985

Table 5-11 Oral Bioavailability of Chromium in Animals

Compound	Species	Exposure methods	Chromium carrier	Chromium dosage	Experimental method	Exposure duration	Bioavailability factor (%)	Ref.
<u>Cr(III)</u>								
$CrCl_3$	Rat	Feed	Buffer solution	1.25-2.5 mg	Fecal excretion	4 d	<0.5	a
$CrCl_3$	Rat	Gavage	Acidic solution	Single: 0.15-10 μg/100g	Whole-body counting	4-10 d	2-3	b
$CrCl_3$	Rat	Gavage	Saline	Single: 1 ng	Urinary and fecal excretion	7 d	2 (Minimal 1.4)	c
$CrCl_3$	Mouse	Oral	-	-	Whole-body counting; Urinary and fecal excretion	48 h	1.4	d
$CrCl_3$	Mouse (fasting)	Oral	-	-	Whole-body counting; Urinary and fecal excretion	48 h	11	d
$CrCl_3$	Rat (adults)	Gavage	Acidic solution	1 μc	Whole-body counting	-	0.1	e
$CrCl_3$	Rat (2-day old)	Gavage	Acidic solution	1 μc	Whole-body counting	-	1.2	e

Adapted from: a. Visek *et al.*, 1953; b. Mertz *et al.*, 1965; c. Donaldson and Barreras, 1966; d. Ogawa *et al.*, 1976; e. Sullivan *et al.*, 1984

(Continued)

Compound	Species	Exposure methods	Chromium carrier	Chromium dosage	Experimental method	Exposure duration	Bioavailability factor (%)	Ref.
<u>Cr(VI)</u>								
Na_2CrO_4	Rat	Gavage	Saline	Single: 1 ng	Urinary and fecal excretion	7 d	2.3 (Minimal 0.8)	c
Na_2CrO_4	Mouse	Oral	-	-	Whole-body counting; Urinary and fecal excretion	48 h	2.4	d
Na_2CrO_4	Mouse (fasting)	Oral	-	-	Whole-body counting; Urinary and fecal excretion	48 h	11	d
<u>Mixtures</u>								
Chromium complex	Rat	Feed	Alfalfa. seedings labeled extract *Brewer's* yeast	-	Urinary and fecal excretion	10 d	35 37 30	f

Adapted from: f. Starich and Blincoe, 1983

Table 5-12 Dermal Bioavailability of Chromium in Animals

Compound	Species	Exposure region	Chromium carrier	Chromium concentration	Experimental method	Exposure duration	Bioavailability factor (%)	Ref.
Cr(III)								
$CrCl_3$	Guinea-pig	Skin	Water	0.017 M	External counting	5 h	2.2	a
				0.080 M			1.6	
				0.126 M			1.8	
				0.239 M			2.0	
Cr(VI)								
Na_2CrO_4	Guinea-pig	Skin	Water	0.017 M	External counting	5 h	2.6	a
				0.080 M			2.1	
				0.239 M			2.3	
				0.261 M			4.0	
				0.398 M			2.8	
Na_2CrO_4	Guinea-pig	Dorsal area	Saline	15 µg	Mass-balance	24 h	1.3 13 (alkali-pretreated)	b

Adapted from: a. Wahlberg and Skog, 1965; b. Czernielewski *et al.*, 1965

chromium (VI) from drinking water and locally grown plants should be generally negligible. At operating industrial sites, the levels of soil-bound chromium (VI) may be high. However, levels of these compounds in residential areas should be lower. Residents, particularly children with pica behavior, may be exposed to soil chromium by ingestion. In fact, the oxidizing potential of chromium (VI) ions relies upon pH. In an acidic environment (e.g. gastric juice), chromium (VI) is very efficiently reduced to chromium (III) that is less readily absorbed in the gastrointestinal tract (Petrilli and De Flora, 1988). Thus, human gastric juice is considered to be a significant barrier against the absorption of chromium (VI) from the gastrointestinal tract. Furthermore, no evidence currently supports the carcinogenicity of chromium (VI) by ingestion. Therefore, the health hazard by soil ingestion should be minimal.

Oral bioavailability in humans and animals is well studied. Chromium (III) is likely to be absorbed in the small intestine by passive diffusion [Dowling *et al.,* 1989]. The mechanism of absorption for chromium (VI) in the gastrointestinal tract is poorly understood.

Because the majority of chromium compounds absorbed are eliminated via urine, some results of oral bioavailability for chromium are derived from chromium urinary excretion data, by making an assumption that the urinary excretion is equal to total absorption by ingestion [Franchini *et al.*, 1984; WHO, 1988]. However, quantitative study on distribution of chromium is currently incomplete. In particular, absorbed chromium (VI) effectively passes through cell membranes in the body and may be partially and slowly excreted via urine. Thus, chromium urinary data represent a minimum chromium absorption potential, but probably not recent exposure [USEPA, 1990c]. Bioavailability from fecal excretion generally reflects net absorption because (1) the unabsorbed chromium accounts for most of total chromium excretion in feces, (2) the biliary excretion of chromium accounts for a very small part of chromium in feces and (3) a hepatic first-pass effect is not observed for chromium. Oral bioavailability values are mainly derived from the water-soluble compounds contained in water and diet. However, quantitative data for water-insoluble or soil chromium (VI) are not available. Solubility studies reveal a very low solubility rate for soil-bound chromium (VI) in an acidic environment. In animal studies, plant chromium complexes formed *in vitro* are highly absorbed, but these levels of plant complexes do not reflect realistic plant intake levels from the environment.

Dermal exposure for residents near contaminated sites likely represents a minor route. Residents will only occasionally contact soil-bound or waterborne chromium (VI) with their skin. Furthermore, soil chromium (VI) is not substantially dissolved in sweat on skin (1%). Likewise, the dermal bioavailability factors in animals have been shown to be low. Quantitative values for humans are limited. Finally, there is currently no evidence indicating carcinogenicity of chromium (VI) through dermal exposure.

The general population is often exposed to chromium mixtures in the environment. Information on the interaction during absorption between chromium and other xenobiotics is limited.

E. CONCLUSIONS

In an occupational setting, inhalation and dermal exposure are major exposure routes for chromium. For the general population, ingestion of foodborne, waterborne and soil-bound chromium compounds is a primary exposure route. In most cases, however, this route is less significant because of the low concentration of chromium in these media and low bioavailability in the gastrointestinal tract. Residents living near heavily chromium-contaminated surface soils have some potential to be exposed to these compounds by inhalation, dermal contact or soil ingestion. Contaminated surface soils may represent a limited source of chromium contamination for off-site release and human exposure.

Both chromium (VI) and (III) are bioavailable to humans and animals. Chromium (VI) is more readily absorbed than chromium (III). In accordance with the carcinogenic risk by inhalation and toxic effects mainly from chromium (VI), the evaluation of chromium (VI) bioavailability is important.

Bioavailability factors for water-soluble chromium (VI) by ingestion are approximately 11% in fasting humans. Quantitative bioavailability factors by inhalation and dermal exposure for humans are not yet available.

Respiratory bioavailability factors from animal studies with small rodents are 20% to 47% for water-soluble chromium (VI) and 2% for water-insoluble chromium (VI). Only 4% to 9% of chromium (III) can be absorbed from the lungs. The oral bioavailability of chromium (VI) is about 2% in small rodents and 11% in fasting mice. The oral bioavailability of chromium (III) ranges from <1% to 2%. The values by dermal exposure to chromium (VI) vary with the concentrations of the compounds applied on the skin, ranging from 1.3% to 4% in small rodents. With alkali treatment of the skin, the dermal bioavailability increases to 13%. The dermal bioavailability of chromium (III) is about 2%. The solubility of soil chromium (VI) is 0.5% to 1% under acidic condition (pH 2 and pH 5). Based on the pH-dependent profile involved in chromium absorption, it is possible to reduce the extent of absorption of chromium through alteration of pH in chromium contaminated media.

Chromium carcinogenicity in the respiratory tract is currently considered to be a local effect. Consequently, there is no need for estimation of an internal dose for the purpose of chromium cancer risk assessment.

F. RECOMMENDATIONS

Evaluation of carcinogenic risk by inhalation of chromium (VI) compounds (including CrO_3) in humans and animals requires more knowledge on parameters relevant to estimating the respiratory bioavailability of chromium (VI) at specific sites. These include the respiratory bioavailability factor for each specific chromium form, fractions of deposition, retention and clearance and physicochemical properties of airborne chromium at the sites, such as particle size distribution in air, inhalable fraction and dissolution rate of chromium particulates in the respiratory tract (pH 7.4 environment). Clarifying the mechanisms of chromium carcinogenesis and determining the biologically active chromium will aid in assessing the potential risk for humans arising from exposure to chromium.

Oral bioavailability of chromium (VI) in humans and animals has been well-documented. But values from measurement of chromium urinary excretion are not reliable. Quantitatively studying the distribution of chromium will reduce the errors arising from this method. Data on the levels of chromium (VI) in surface soil, groundwater and locally grown plants at specific sites are required for accurately assessing exposure to the general population because residents may be exposed to high level chromium (VI).

Bioavailability studies indicate that absorption of chromium via the main exposure routes greatly depends on valence, water solubility, pH and chromium speciation. For quantitative risk assessment, it is necessary to determine valence states, chemical forms and water solubility of chromium compounds at chromium contaminated sites. The values of bioavailability and toxicity that are selected for calculating average daily doses, if possible, should be consistent with or close to those from chromium contaminated sites. For instance, the bioavailability factors for water-soluble chromium (VI) may be used for CrO_3.

With exposure to chromium mixtures, interaction between chromium and other xenobiotics may alter the absorption profiles of chromium. Thus, studies on the bioavailability for chromium mixtures are needed.

IV. COPPER

Copper is widely distributed in the environment with a variety of concentrations ranging from levels that may be nutrient deficient to those that are toxic for humans. There are oxidation states from 0 to +3, but the free copper (II) ions have the main biological significance [Seiler and Sigel, 1988; ATSDR, 1990b].

The information about the form of copper in air is limited. Copper may exist as oxides, sulfate or carbonate [ATSDR, 1990b]. In water, copper occurs as Cu^{2+}, $Cu(HCO_3)^+$ and $Cu(OH)_2$. Copper (II) also forms a number of stable complexes with ligands (inorganic or organic), or is bound to particulate matter in water [Stiff, 1971]. Large amounts of copper from various sources are released to soils and sediments [USEPA, 1980a]. The forms of copper in these media depend on soil or sediment conditions, such as pH, Eh, types of soils and the amount of organic matter in soil and sediment [USEPA, 1979a, 1980a]. Copper is essential for normal growth and development of plants. In natural conditions, copper rarely accumulates to high levels in plants [ATSDR, 1990b]. Concentrations of plant copper may become elevated with an increase in soil copper levels.

Humans take copper as a trace element from their daily diet. For example, adults require approximately 1.5 to 3 mg/d for normal health [NAS, 1989]. As an essential nutrient, copper plays an important role in human metabolism [Harris, 1991]. Because there is limited evidence of severe toxic effects in humans following exposure, copper is usually not considered to be a particularly hazardous industrial substance [Carson *et al.*, 1986; ATSDR, 1990b; ACGIH, 1991].

Absorption of copper in the gastrointestinal tract is a homeostatic process. Hence, inhalation and dermal exposure routes may be of great concern. Currently, quantitative estimates of copper bioavailability by inhalation and dermal contact are limited.

A. EXPOSURE ROUTES AND RECEPTORS

Humans can be exposed to copper by inhalation, ingestion and dermal contact because copper is widely distributed in various environmental media. In principle, only the free copper (II) ion or weakly complexed copper, as an available form, is a potential health hazard for humans [ATSDR, 1990b]. In effect, the majority of copper in the environment is tightly bound to organic or mineral matter with these bound forms being less toxic for humans.

In the occupational environment, workers may be exposed to high levels of copper fumes and dusts via inhalation [ACGIH, 1991]. Dermal exposure may occur for workers who deal with soluble cupric salts.

Although people ingest a large amount of copper daily from various foods, absorption of copper in the gastrointestinal tract tends to be restricted to levels that satisfy metabolic requirements. The average daily intake of dietary copper is less than 2 mg [ATSDR, 1990b]. Accordingly, ingestion of copper from foods does not appear to be a significant exposure route for humans. An excess amount of free copper (II) ions may be present in drinking water when copper pipes are used for distribution of corrosive water. In such cases, ingestion of drinking water may become an important exposure route for the general population.

Copper sulfate may be used to control blooms of algae. In such cases, humans may be exposed dermally to high levels of free copper during bathing or swimming. Additionally, residents living near copper smelters may be exposed to high levels of copper-containing soil or dust.

Exposure to copper via inhalation or dermal contact may be substantial in very special circumstances for either the general or occupational population. Quantitative information about bioavailability by these two exposure routes is limited.

B. TOXICOKINETICS (FOLLOWING ABSORPTION)

1. Distribution

Once absorbed, the distribution of dietary copper takes place in two stages [Goode *et al.*, 1989]. First, copper is bound to the albumin and amino acids in the portal blood. Upon arrival at the liver, the copper released is transported to the hepatocytes where it is incorporated into various copper proteins, such as ceruloplasmin [Marceau *et al.*, 1970]. Then, ceruloplasmin as well as a small fraction of albumin-copper or amino acid-copper is released into the systemic circulation and taken up by extrahepatic organs. Alternatively, the copper complexes are excreted via the biliary tract if a large amount of copper is present in the liver. Consequently, copper levels in the liver usually remain relatively constant, making the liver a major homeostatic organ for copper [Harris, 1991].

The highest concentrations of copper are found in the liver, brain, heart and kidney [Harris, 1991]. The lung, intestine, spleen, bone and muscle contain relatively low copper levels. The total body burden of copper is typically about 80 to 120 mg. Approximately 10% of the total copper burden is enriched in the liver [Seiler and Sigel, 1988].

2. Metabolism and Excretion

Copper will exist in a free form in the human body. Absorbed copper is soon complexed with a variety of proteins, amino acids and peptides. Copper becomes part of the molecular structure of these proteins and acts as a co-factor of some enzymes. In the cytosol of hepatocytes, 80% of copper is incorporated into hepatocuprein, copper-chelatin, metallothionein and other specific copper proteins [NAS, 1977b]. Of all the important copper proteins, ceruloplasmin, a blue plasma glycoprotein, is considered to be the main transfer agent for copper because it accounts for 70% to 90% of copper in plasma [Harris, 1991].

Based on excretion studies via oral and intravenous exposure, copper is observed to be primarily eliminated via bile in fecal excretion. Biliary copper is seldom reabsorbed in the intestine because of being readily bound to proteins. The rest of the fecal copper results from the salivary, gastric, pancreatic, and jejunal secretions, the intestinal mucosae, as well as non-absorbed dietary copper [NAS, 1977b]. Urinary excretion of copper plays a minor role, being only 2% to 4% of the total copper excretion [Triebig and Schaller, 1984].

C. BIOAVAILABILITY

1. Inhalation

Absorption of copper may result from inhalation of copper fumes and dusts in an occupational environment [Triebig and Schaller, 1984]. However, there are no quantitative bioavailability data via inhalation documented for humans or animals. One study indicated that the retention of copper within the lung declined sharply at 24 h after the intratracheal administration of whetlerite dust in rats [Salem and Katz, 1989].

2. Ingestion

Oral bioavailability of dietary copper has been well studied in humans and animals because it is a trace nutrient. In principle, the absorption and excretion of dietary copper are regulated in accordance with the daily requirements for copper in humans and animals. Therefore, either toxicity or deficiency of copper rarely occur from oral exposure to copper unless individuals have copper-related diseases, such as Wilson's disease or Menkes' disease.

The main absorption sites for copper are located in the stomach and duodenum in humans [Bearn and Kunkel, 1955; Weber *et al.*, 1969]. Based on animal experiments, there are two mechanisms involved in the absorption of copper. One includes a specific carrier for copper ions in the epithelial cells of the gastrointestinal tract [Harris, 1991]. Other metal ions, such as cadmium, iron and zinc, may compete for the enzymatic carrier [King *et al.*, 1978; Kies and Harms, 1989; Greger, 1992]. The second mechanism is associated with the absorption of an amino acid-copper complex which is formed by the binding of metallothionein in the intestinal mucosa [Mason, 1979; Harris, 1991].

Various factors can influence the oral bioavailability of copper [Johnson, 1989; Johnson *et al.,* 1992]. Besides other metal ions, age, gender, pregnancy, diseases, types of food and ascorbic acid can contribute to variable absorption of copper.

Well-documented quantitative human bioavailability data for ingestion have been reported in the literature [Cartwright and Wintrobe, 1964; Sternlieb, 1967; Weber *et al.,* 1969; Strickland *et al.*, 1972; Dekaban *et al.,* 1974; King *et al.*, 1978; Nageswara Rao, 1980; Turnlund *et al.*, 1982, 1983, 1985, 1988a, 1988b, 1989; Jacob *et al.,* 1987; Johnson *et al.,* 1988, 1992; Turnlund, 1989]. In these studies, volunteers took dietary copper and labeled

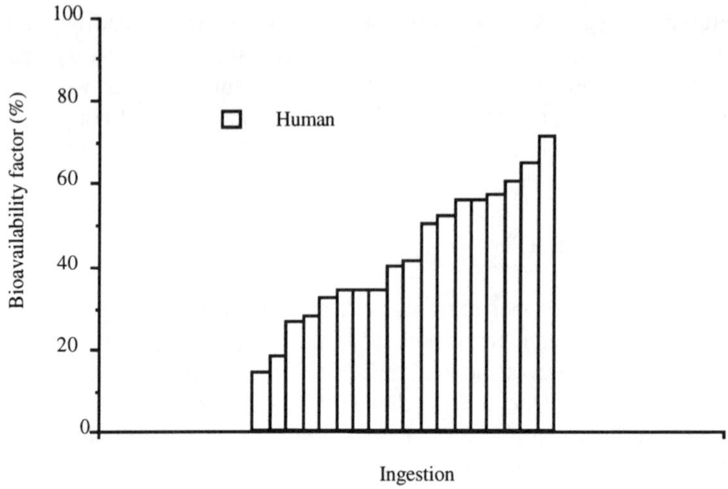

Figure 5-5 Oral bioavailability factors of copper in humans [Adapted from source: Table 5-13: a-j]

copper solutions by both oral and intravenous routes at various levels and duration. Quantitative results were usually obtained by using a whole-body counting technique or fecal excretion monitoring. In effect, copper absorption might be underestimated by using the fecal excretion method. Fecal copper represents the portion of nonabsorbed copper and endogenous losses of copper. Absorbed copper excretion in feces is estimated to be approximately 9% to 16% [Mason, 1979; Turnlund, 1989].

From a variety of studies in humans, the average values of absorption of dietary copper range from 14% to 71% (Table 5-13 and Figure 5-5). The variation of bioavailability is strongly dependent on the daily intake of dietary copper (Figure 5-6). For exposures to near daily required levels (around 1.5 to 3 mg/d), about 20% to 40% of dietary copper was absorbed. With an increase of administered copper to 7.5 mg/d, the oral bioavailability declined to 14%, whereas the oral bioavailability increased to 50% to 70% at a deficient daily copper intake level of around 1 mg/d.

3. Dermal

Absorption of copper via dermal exposure has been reported [Bentur *et al.*, 1988; Hostynek *et al.*, 1993]. Little information on quantitative dermal absorption is available for humans and animals. A few *in vivo* and *in vitro* studies revealed that copper combined with various carriers could penetrate the skin [Walker and Reeves, 1977; Beveridge *et al.*, 1984].

D. UNCERTAINTIES

Water-soluble copper (II) could be released to surface soils and tightly bound to organic matter at specific contaminated sites. Residents near

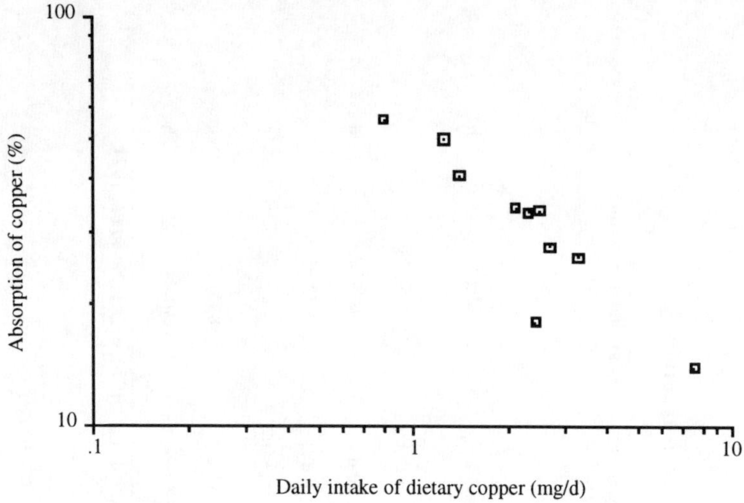

Figure 5-6 Bioavailability factor vs. daily intake of dietary copper (Adapted from Nageswara Rao 1980, Johnson *et al.* 1988, Turnlund, 1989).

waste sites could be exposed to soil-bound copper in airborne dust. However, the copper forms, concentration and particle size distribution in the air are generally not known at contaminated sites. Based on data from other studies, most airborne copper may be present as copper-sulfur complexes .

There have been a few reports of adverse health effects by inhalation of fumes and dusts in occupational settings. Toxicity by inhalation for the general population is poorly understood. Furthermore, quantitative data on the bioavailability via inhalation in both humans and animals are very limited.

Although copper may contaminate surface and groundwaters, and surface soils, the occurrence of toxic effects via ingestion is dependent on the amount of free copper (II) ions or weakly bound complexes in these media. The ratio of free copper (II) ions to the complexes is not known at most contaminated sites. In bioavailability studies, differences in absorption for different copper (II) compounds from the gastrointestinal tract are also not well characterized.

Toxicity and absorption of copper by dermal exposure has received little attention. Thus, many uncertainties are involved in the assessment of dermal exposure but this exposure route does not appear to be important for soil-bound copper.

E. CONCLUSIONS

Although the ingestion of copper from foods and drinking water is a major exposure pathway for the general population, toxicity rarely occurs in healthy humans because of homeostasis. A potential health hazard posed by

Table 5-13 Experiments of Oral Bioavailability of Copper in Humans

Compound	Number of subjects	Exposure method	Copper dosage	Experimental method	Exposure duration	Bioavailability factor (%)	Ref.
Dietary copper	235	Oral	2-5 mg	Measurement of body fluid	-	32	a
^{64}Co	49	Oral/i.v.	2 mg	Measurement of serum level	2 d	40	b
Copper acetate	7	Oral/i.v.	Single: 300 µg	Whole-body counting	4 d	60 (15-97)	c
Copper acetate	4	Oral/i.v.	0.4 - 4.5 mg	Whole-body counting	20 d	56 (40-70)	d
Cupric chloride	1	Oral/i.v.	-	Whole-body counting	8 d	52	e
Formula copper	8	Oral	3.2 mg/d	Fecal excretion	10-12 d	57	f
Dietary copper	6	Oral	2.4 mg/d	Fecal excretion	11 d	18 (7-37)	g
Dietary copper	7	Oral	1.25 mg/d	Fecal excretion	5.5 mon	50-65	h

Adapted from: a. Cartwright and Wintrobe, 1964; b. Sternlieb, 1967; c. Weber *et al.*, 1969; d. Strickland *et al.*, 1972; e. Dekaban *et al.*, 1974 f. King *et al.*, 1978; g. Nageswara Rao, 1980; h. Johnson *et al.*, 1988

(Continued)

Compound	Number of subjects	Exposure method	Copper dosage	Experimental method	Exposure duration	Bioavailability factor (%)	Ref.
Formula copper or dietary copper or both forms	4 - 11	Oral	0.8 mg/d	Fecal excretion	42 - 98 d	56	i
			1.4 mg/d			41	
			2.1 mg/d			34.4	
			2.3 mg/d			33.5	
			2.5 mg/d			33.8	
			2.7 mg/d			27.5	
			3.3 mg/d			25.8	
			7.5 mg/d			14	
Dietary copper	127	Oral	1.3 mg/d in men	Whole-body counting	21 d	64 (20-59 y)	j
			1.1 mg/d in women			71 (20-59 y)	

Adapted from: i. Turnlund, 1989 (Data from seven studies); j. Johnson *et al.*, 1992

oral exposure is minor with the exception of exposure to excessive free copper (II) in drinking water. Inhalation and dermal exposure may be significant for occupational settings. Residents living near industrial areas may be exposed to high levels of airborne copper.

Humans require a certain amount of copper for daily metabolic needs. This is usually satisfied by ingestion of dietary copper (about 1.5 to 3 mg/d). Absorption and excretion of dietary copper are regulated. The mean values of bioavailability factor range widely from 14% to 71% in humans, depending on the daily intake of dietary copper. Quantitative bioavailability data for inhalation and dermal exposure are not currently available.

F. RECOMMENDATIONS

Compared to chromium and arsenic, copper poses a relatively low health risk for humans. The health risk for ingestion of copper may be negligible. In contrast, exposure by inhalation may be a greater concern. For the evaluation of respiratory and dermal bioavailability, the following data are needed: the forms of copper in the environmental media, particle size distribution, inhalable fraction, the rate of deposition-retention-clearance within the respiratory tract and the permeability of copper through the skin.

V. MERCURY

Mercury occurs naturally in the environment, but can be influenced by anthropogenic activity. There are three general forms of mercury in the environment: elemental, inorganic and organic mercury [WHO, 1976, 1990, 1991; ATSDR, 1993e]. Elemental mercury is a silver-white, volatile liquid at room temperature. Elemental mercury vapor can be readily dissolved in biological fluids [Hursh, 1985]. Inorganic mercury compounds include mercuric Hg^{++} and mercurous Hg_2^{++} states. Organic mercury compounds are covalently bound to carbon, such as methylmercury ($HgCH_3$) and phenylmercury (HgC_6H_5). All these forms of mercury are biologically significant.

There is a global biogeochemical cycle for mercury. Mercury vapor, the primary form of mercury in air (accounting for 60% of airborne mercury), may be transported to remote areas by physical processes [USEPA, 1984c]. Airborne mercury (vapor and mercury aerosols) may be removed by photolysis, wet or dry deposition, precipitation and sorption to soil, sediment and surface water [USEPA, 1984c]. Mercury compounds in surface water and soil undergo biotransformation and chemical reactions (reduction and oxidation). Elemental mercury may be oxidized to inorganic mercury, while inorganic mercury may reduced to volatile elemental mercury [WHO, 1991]. Inorganic mercury in these media may also be methylated by microorganisms under both aerobic and anaerobic conditions. Methylated mercury may be demethylated to volatile elemental mercury. Then, volatile mercury will evaporate to the air. Methylmercury may be taken up by aquatic organisms, particularly in fish, and accumulated in the food chain [WHO, 1989d, 1990]. Eventually, mercury enters the human body through the food chain.

Inhalation of mercury vapor, with chronic exposure, may primarily affect the central nervous system in humans [Goyer, 1991]. Ingestion of mercuric compounds may lead to gastrointestinal and kidney damage. Exposure to methylmercury for a long time may cause permanent damage to the brain, kidney and developing fetus [ATSDR, 1993e]. Neurotoxic effects, such as parathesia, ataxia, tremor, coma and death, are the main adverse human health effects [Clarkson, 1993].

Mercury vapor is rapidly absorbed into the human body via inhalation exposure, but elemental mercury is not absorbed in the gastrointestinal tract or across the skin. Inorganic compounds are bioavailable in the gastrointestinal tract and through skin. Organic mercury compounds can enter the human body by all three exposure routes.

A. EXPOSURE ROUTES AND RECEPTORS

The workers in various occupational groups (e.g. chemical industries, instrument manufacture, mining, smelting and refining, health services and business services) may be exposed to mercury vapor via inhalation [Berlin, 1986; Magos, 1988; WHO, 1990, 1991; ATSDR, 1993e]. In some occupational settings, the workers may also inhale aerosols of inorganic mercury salts and organic mercury compounds [Berlin, 1986]. Dermal exposure may also be a possible route for many workers [ATSDR, 1993e].

The general population may be exposed to elemental, inorganic and organic mercury from various environmental media through different exposure routes. The elderly, young children and people with renal disorders are at high risk. The main exposure to mercury for the general population is via ingestion of mercury-contaminated food, drinking water and dental or medical treatments. Organic mercury (e.g. methylmercury) is usually enriched in aquatic food chains, affecting fish. Carnivorous fish consumption is the most significant source of exposure to organic mercury for the general population [ATSDR, 1993e]. Ingestion and inhalation of elemental mercury from dental amalgam filling may contribute to the daily intake of total mercury for persons with extensive dental restorations or those engaged in applying the dental treatments [Brune and Evje, 1985]. Occasionally, persons may ingest elemental mercury or inhale mercury vapor from broken thermometers.

Residents living near industrial areas using mercury may be exposed to mercury (vapor or aerosols) via inhalation. Exposure through the skin may occur for people using mercury-containing creams or soaps [ATSDR, 1993e].

B. TOXICOKINETICS (FOLLOWING ABSORPTION)

1. Distribution

Mercury vapor and inorganic mercury compounds are distributed to all organs in the body, with the highest level found in the kidney [Takahata *et al.*, 1970]. Mercury vapor in the body fluid can readily penetrate the placental barrier and the blood-brain barrier and mercury will be retained in the brain for a long time [Clarkson, 1989]. Inorganic mercury compounds cross the blood-brain barrier and the placental barrier only slightly. Organic mercury compounds have a high affinity for the brain following ingestion exposure in humans [Aschner and Aschner, 1990].

2. Metabolism and Excretion

Elemental mercury is oxidized to divalent mercury by catalase in the red blood cells, liver, brain and lung [Magos *et al.,* 1978; Clarkson, 1989]. The oxidized divalent mercury is reduced to metallic mercury in the liver and kidney [WHO, 1976, 1991]. Absorbed organic mercury compounds can be converted to divalent inorganic mercury in tissues that undergo an oxidation-reduction cycle [Dunn and Clarkson, 1980].

All forms of mercury are primarily excreted from the body via urine and feces in humans [WHO, 1976; ATSDR, 1993e]. A small amount of mercury vapor from tissues can be exhaled from the lung. Biliary excretion through feces is the primary route following exposure to inorganic and organic mercury compounds [ATSDR, 1993e]. After exposure to a high level of inorganic mercury compounds, the urinary excretion route may dominate

[WHO, 1991]. Methyl mercury has also been shown to be excreted via hair in rats [Farris *et al.,* 1993].

C. BIOAVAILABILITY

1. Inhalation

Humans can rapidly absorb mercury vapor following inhalation exposure [Cherian *et al.,* 1978; Barregard *et al.*, 1992]. Because mercury vapor is highly lipophilic, the inhaled mercury vapor in the pulmonary region may readily cross the air-blood barrier into the systemic circulation [Nielsen-Kudsk, 1969; Berlin, 1986; Clarkson, 1989]. Thus, the deposition coefficient for mercury vapor in the respiratory tract determines the respiratory bioavailability factor, as no additional resistance to absorption occurs. Human volunteers were exposed to mercury vapor by nasal inhalation at mercury vapor concentrations of 50 to 350 $\mu g/m^3$. The bioavailability factors for mercury vapor ranged from 74% to 77% [Teisinger and Fiserova-Bergerova, 1965; Nielsen-Kudsk, 1965a; Hursh *et al.*, 1967]. By oral inhalation, the bioavailability factor decreased to 20% [Teisinger and Fiserova-Bergerova, 1965]. Also, the absorption of mercury vapor was reduced by oral administration of alcohol in humans [Nielsen-Kudsk, 1965b]. This reduction was attributed to the inhibition of mercury vapor oxidation in red blood cells caused by alcohol [Magos *et al.,* 1973].

Rats were exposed to mercury vapor at concentrations of 1 to 1.4 mg/m^3. About 50% to 91% of mercury vapor was absorbed from the respiratory tract [Gage, 1961; Hayes and Rothstein, 1962]. The respiratory bioavailability factor in mice was found to be 77% for mercury vapor [Magos, 1967].

Quantitative data on absorption of inorganic mercury salt aerosols are limited. In animal experiments, absorption of mercury chloride aerosol from the respiratory tract was lower than mercury vapor [Friberg and Vostal, 1972]. About 45% of mercury oxide aerosols deposited within the respiratory tract were cleared within one day in dogs [Morrow *et al.*, 1964].

Inhalation of some organic mercury compounds could produce intoxication in humans [Friberg and Vostal, 1972], but quantitative data were not available. In small rodents, methylmercury and dimethylmercury were rapidly and completely absorbed from the lung into the systemic circulation [Östlund, 1969; Fang, 1980].

2. Ingestion

Absorption of mercury in the gastrointestinal tract may vary in humans and animals, depending on its form. Elemental liquid mercury has been found to be poorly absorbed following ingestion exposure [Friberg and Nordberg, 1973].

Oral bioavailability of inorganic mercury is low in humans and animals. Inorganic mercury salts are less bioavailable than organic mercury compounds. The mechanism of inorganic mercury absorption is similar to that for cadmium [Foulkes and Bergman, 1993]. Human volunteers ingested protein-bound divalent mercury at a single dose of 6 μg mercury [Rahola *et al.*, 1973]. The oral bioavailability factor was about 7%. Miettinen [1973] reported that about 85% of protein-bound inorganic mercury which was ingested by human volunteers was eliminated in feces during the first four to five days after dosing. Consequently, the fraction of ingested mercury absorbed would have been less than 15%.

In mice, the absorption fraction of dietary mercury chloride was only 2% [Clarkson, 1971]. When mice ingested mercury chloride dissolved in water, the oral bioavailability factor increased to the range of 20% to 25% [Nielsen and Andersen, 1989, 1990; Nielsen, 1992].

Absorption of organic mercury is almost complete in humans and animals. Aberg *et al.* [1969] and Miettinen [1973] reported that about 95% of methylmercury was absorbed in humans after oral administration. The oral bioavailability factor for organic mercury mixed with food in monkey was about 95% [Berlin *et al.,* 1973]. After ingestion of methylmercury in water or incorporated into food, about 97% to 100% of mercury was absorbed from the gastrointestinal tract in mice [Clarkson, 1971; Nielsen and Andersen, 1991a, 1991b; Nielsen, 1992].

Absorption of ingested mercury is influenced by many factors (e.g. age and dietary factors). Young rats had higher absorption for inorganic mercury than adults [Kostial *et al.,* 1978, 1983; Walsh, 1982]; no difference between young and adult rats was found for absorption of organic mercury [Walsh, 1982]. An increase in pH of the gastrointestinal fluid tended to increase the absorption of inorganic mercury from the gastrointestinal tract in rats [Endo *et al.*, 1990]. Exposure to elemental mercury with a metal mixture(cadmium and manganese in slag) in rats had no influence on the absorption of mercury [Kostial *et al.*, 1989].

3. Dermal

All forms of mercury are bioavailable to a limited degree via dermal exposure. Dermal absorption of mercury vapor was estimated to be 2.2% in humans following exposure to mercury vapor (0.61 to 2.41 ng/cm^2/h) in air for 0.25 to 0.75 h [Hursh *et al.*, 1989]. However, mercury chloride could penetrate human skin *in vitro* [Wahlberg, 1965]. By using disappearance measurements, about 2% to 4.5% of mercury chloride was absorbed through the Guinea pig skin [Friberg *et al.*, 1961; Skog and Wahlberg, 1964]. Dermal bioavailability of methylmercury ranged from 2.25 to 5.9% in Guinea pigs (Friberg *et al.*, 1961, Skog and Wahlberg, 1964).

4. Summary

Absorption of mercury in humans and animals varies, depending on the form (Table 5-14, 5-15, 5-16, Figure 5-7). Mercury vapor and some organic mercury compounds are readily absorbed after inhalation exposure. Organic mercury compounds are almost completely absorbed in the gastrointestinal tract, while absorption of inorganic mercury compounds is relatively low.

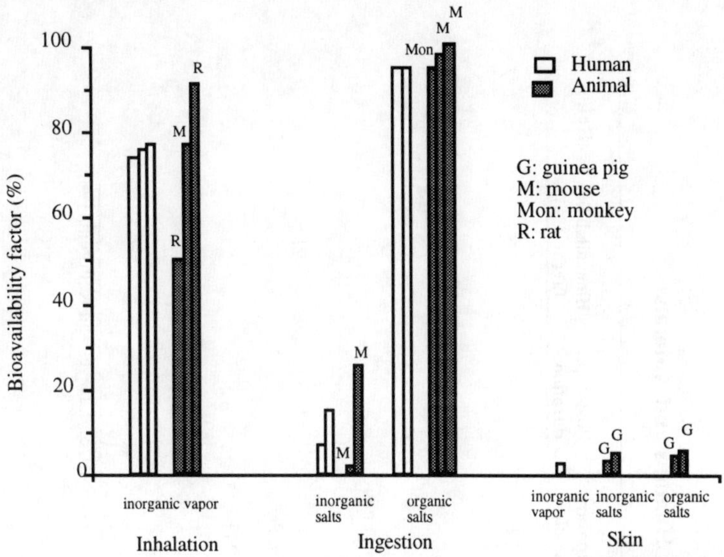

Figure 5-7 Bioavailability factors of mercury in humans and animals by three exposure routes [Adapted from sources in Table 5-14, a-f (inhalation); Table 5-15, a-f (ingestion); and Table 5-16, a-d (skin)]

Mercury vapor, inorganic mercury and organic mercury can be absorbed to a limited degree through the skin.

D. UNCERTAINTIES

Information on respiratory bioavailability is available for mercury vapor in humans and animals. Absorption of other forms of mercury has not been extensively studied. Data on physiochemical properties and environmental fate for inorganic mercury aerosols and organic mercury are limited.

Oral bioavailability for inorganic and organic mercury from food and water has been extensively studied in humans and animals. However, there is no direct information about oral absorption from soil-bound mercury.

Quantitative data on dermal absorption for inorganic and organic mercury are limited in humans.

E. CONCLUSIONS

Humans may be exposed to all forms of mercury via different exposure routes. In occupational settings, the principal exposure route is by inhalation of mercury vapor. For the general population, exposure to organic mercury from ingestion of food is the most important route of environmental exposure. Absorption of mercury in humans and animals varies, depending

Table 5-14 Respiratory Bioavailability of Mercury in Humans and Animals

Compound	Species (No. of subjects)	Exposure methods	Mercury dosage	Experimental method	Exposure duration	Post-exp. duration	Bioavailability factor (%)	Ref.
Vapor	Human (5)	Nasal inhalation Oral inhalation	70 $\mu g/m^3$	Measurement of total intake	7 h	-	76 20	a
Vapor	Human (4)	Nasal inhalation	50-350 $\mu g/m^3$	Measurement of total intake	-	-	77	b
Vapor	Human (5)	Nasal inhalation	62-100 $\mu g/m^3$	Whole-body counting	14 - 24 min	-	74	c
Vapor	Rat	Whole-body exposure	1.4 mg/m^3	Whole-body counting	5 h	15 d	91	d
Vapor	Rat	Whole-body exposure	1.0 mg/m^3	Whole-body counting	24 h	-	50	e
Vapor	Mouse	Whole-body exposure	33-900 $\mu g/m^3$	Mass-balance	4 h	-	77	f

Adapted from: a. Teisinger and Fiserova-Bergerova, 1965; b. Nielsen-Kudsk, 1965a; c. Hursh *et al.* , 1976; d. Hayes and Rothstein, 1962; e. Gage, 1961; f. Magos , 1967

Table 5-15 Oral Bioavailability of Mercury in Humans and Animals

Compound	Species (No. of subjects)	Exposure methods	Mercury carrier	Mercury dosage	Experimental method	Exposure duration	Bioavailability factor (%)	Ref.
Inorganic mercury								
$^{203}Hg\text{-}(NO_3)_2$	Human (8)	Oral	Liver paste	Single: 6 µg	Whole-body counting	74 d	7	a
$^{203}Hg\text{-}(NO_3)_2$	Human (10)	Oral	Liver paste	3.8-14 µC	Whole-body counting	4-5 d	15	b
$HgCl_2$	Mouse	Oral	Food	0.05 ppm	Whole-body counting	21 d	2	c
$HgCl_2$	Mouse	Gavage	Water	0.2-2 µg/kg	Whole-body counting	2 wk	20-25	d
Organic mercury								
CH_3HgNO_3	Human (3)	Oral	Water	10 µg	Whole-body counting	240 d	95	e
CH_3Hg-fish	Human (15)	Oral	Fish	2 µC	Whole-body counting	4 d	94	b
CH_3HgCl	Mouse	Oral	Food	0.05 ppm	Whole-body counting	21 d	98	c
CH_3HgCl	Mouse	Oral	Water	50µg - 6.25 mg/kg	Whole-body counting	14 d	97-100	f
CH_3HgNO_3	Monkey	Oral	Food	2 µC/kg	Whole-body counting	85 d	95	g

Adapted from: a. Rahola *et al.*, 1973b; b. Miettinen, 1973; c. Clarkson, 1971; d. Nielsen and Andersen, 1989, 1990, Nielsen, 1992; e. Aberg *et al.* , 1969; f. Nielsen and Andersen, 1991a, 1991b, Nielsen, 1992; g. Berlin *et al.*, 1973

Table 5-16 Dermal Bioavailability of Mercury in Humans and Animals

Compound	Species	Mercury carrier	Exposure region	Skin area	Mercury concentration	Experimental method	Exposure duration	Bioavailability factor (%)	Ref.
Vapor	Human	-	Forearm	-	0.88-2.14 μg/cm^2	Disappearance (external counting)	0.5 - 0.75 h	2.2	a
$HgCl_2$	Human	Liquid	Abdominal	-	0.005-0.239 M	Diffusion cell (*in vitro*)	5 h	0.62 x 10^{-5} cm/h[1]	b
$HgCl_2$	Guinea pig	Liquid	Abdominal	3.1 cm^2	2.5-5.1 mg/cm^2	Disappearance (external counting)	5 h	3.3-4.5	c
$HgCl_2$	Guinea pig	Liquid	Back	-	0.04-0.16 M	Disappearance (external counting)	5 h	2.0-3.2	d
[2]Me-Hg-DCD	Guinea pig	Liquid	Abdominal	3.1 cm^2	2.5-5.1 mg/cm^2	Disappearance (external counting)	5 h	2.2-5.9	c
CH_3HgCl	Guinea pig	Liquid	Back	-	0.04-0.08M	Disappearance (external counting)	5 h	3.4-4.5	d

[1] Permeability; [2] Me-Hg-DCD = Methyl Mercury Dicyandiamide
Adapted from: a. Hursh *et al.*, 1989; b. Wahlberg, 1965; c. Friberg *et al.*, 1961; d. Skog and Wahlberg, 1964

on mercury forms and exposure routes. Mercury vapor is easily absorbed into the human body via inhalation (bioavailability factor of 75%). The respiratory bioavailability factor of mercury vapor is 50% to 90% in small rodents. Elemental mercury is poorly absorbed from the human gastrointestinal tract. Oral absorption for inorganic mercury is low in humans (7% to 15%) and in animals (2% to 25%). Absorption of organic mercury in the gastrointestinal tract is almost complete in humans (95%) and animals (95% to 100%). Oral bioavailability of mercury is affected by age and dietary factors. All forms of mercury only slightly penetrate human skin (dermal bioavailability factor of 2.2% for mercury vapor) and animal skin (2% to 4.5% for inorganic mercury and 2.2.% to 5.9% for organic mercury).

F. RECOMMENDATIONS

Data on respiratory bioavailability are available for mercury vapor. The physicochemical properties, environmental fate, deposition and clearance for inorganic and organic mercury aerosols should also be studied.

VI. LEAD

Lead is a heavy bluish-gray, water-insoluble metal, having inorganic and organic forms [ATSDR, 1991b, 1993c; Sittig, 1991; Lide, 1993]. Some of its salts (nitrate, chlorate and acetate) are water soluble and others (sulfate, chromate, phosphate, sulfide, and oxide) are insoluble in water. The nitrate, sulfide, and phosphate salts are soluble in acid. Toxicity of inorganic lead compounds is caused by lead *per se* [Seiler and Sigel, 1988]. Lead can form stable organic compounds (tetra-alkayl) that are soluble in most organic solvents, but are insoluble in water and decompose into free lead and organic radicals at high temperature.

Lead in the environment occurs naturally and anthropogenically. Lead levels in environmental media are highly variable. Generally, lead tends to accumulate near discharge points [WHO, 1977 1989c]. Lead is found in air, soil, sediment, surface water, groundwater and biota. In surface water, lead is likely to form insoluble sulfates and carbonates [Carson *et al.,* 1986]. However, drinking water levels may be elevated by local dissolution in acidic and soft waters [USEPA, 1980c]. Generally, lead is strongly adsorbed onto soil particles and sediment. Plants can be enriched in lead by atmospheric deposition and root absorption from soil [Seiler and Sigel, 1988]. The global average daily intake of lead, estimated by the United Nations Environment Program (UNEP) (1991), was 80 μg/d from food and 40 μg/d from drinking water.

Adverse human health effects produced by lead include: anemia, kidney diseases, brain damage, impaired function in peripheral nerve system, high blood pressure (in middle-aged men), reproductive abnormalities, developmental defects, abnormal vitamin D metabolism and even death under certain circumstances [ATSDR, 1991b, 1993c]. Low-level lead exposure may decrease intelligence quotient scores, impair the function of learning, attention and classroom performance and reduce the growth rate of young children [Ellenhorn and Barceloux, 1988; Demers, 1991]. Lead can be absorbed via inhalation and ingestion. Organic lead compounds can easily penetrate the human skin.

A. EXPOSURE ROUTES AND RECEPTORS

Humans may be exposed to lead via inhalation, ingestion and dermal contact in the occupational and general environments. Lead toxicity in adults mainly occurs in the occupational environment, while children can experience lead poisoning from non-occupational exposure. High risk groups include young children (<6 years old), the elderly, pregnant women, smokers, alcoholics and people who have genetic diseases that are related to heme synthesis and other diseases (e.g. kidney dysfunction) and various nutritional deficiencies (e.g. zinc, iron, and calcium) [ATSDR, 1993c].

Occupational populations may inhale airborne lead (dust and fumes) during mining, smelting, refining and other operations at the workplace

[Sittig, 1991]. Lead exposure also occurs via ingestion during eating or smoking in a lead-contaminated workplace [Baxter *et al.*, 1985]. Dermal exposure is much less significant than inhalation and ingestion. Only alkyl lead compounds and lead naphthenate in the occupational setting were reported to be slightly absorbed through the skin [Lauwerys, 1983; ATSDR, 1991b].

General populations may be exposed to airborne lead via inhalation of dusts or aerosols outdoors and indoors. However, direct inhalation of lead dusts generally accounts for only a small part of the total human exposure. Health and Welfare Canada estimated that about 23% of daily exposure for an urban adult and 1% to 2% for a child arose from daily inhalation [MOE, 1987]. Although inhalation is a minor exposure route for children, they may be more vulnerable than adults for exposure to airborne particles. Children may exhibit higher deposition efficiencies for inhaled airborne particles than adults because of the physiological characteristics of their respiratory tract [Xu *et al.,* 1986]. A potential hazard for the general population apparently arises from exposure to lead through ingestion of contaminated drinking water, food and various types of dusts or soils. Direct ingestion of soil/dust-bound lead, lead-containing paints and other lead-containing materials is of great concern for young children. Because lead has a long residence time in soils, lead-contaminated soil or dust can become a substantial contributor to blood lead for the general population. There is evidence indicating a positive correlation between soil lead from urban areas and blood lead [Steele *et al.,* 1990]. This type of relationship between soil lead and blood lead was not observed at mining sites.

General population exposure to inorganic lead by skin may occur through the use of hair-darkening cosmetics (containing lead acetate), but this exposure route is generally considered to be insignificant because of the low dermal bioavailability of lead acetate in humans. Cohen and Rose [1991] reported that human exposure to lead acetate from a hair coloring agent only accounted for about 0.5% of the total absorption of lead from the environment.

B. TOXICOKINETICS (FOLLOWING ABSORPTION)

1. Distribution

The distribution of lead in the human body has been well characterized. In a simplified scheme, absorbed lead is distributed in blood, soft tissue and bone [Rabinowitz *et al.*, 1976; Rabinowitz, 1991]. Once absorbed, 96% to 99% of blood lead is associated with red blood cells [ATSDR, 1991b, 1993c]. About 80% of this erythrocyte (red blood cell) pool is bound to hemoglobin, with a particular affinity for fetal hemoglobin, and the rest is bound to other proteins, mainly albumin. A half-life for lead circulating in the blood is approximately one month [Rabinowitz *et al.,* 1976].

Bone lead levels increase with age at rates dependent on the skeletal site and the degree of lead exposure. Approximately 95% of the total body burden is present in the bone of human adults, compared with 73% in the bone of children [Barry, 1975]. A half-life for bone lead can range from 10 to 30 years. Teeth can also accumulate lead. Tooth lead levels vary considerably with tooth types. Lead will accumulate in certain soft tissues, including brain, kidney, lungs, liver, heart and spleen [Barry, 1975].

2. Metabolism and Excretion

Inorganic lead is not metabolized during phase I processes (such as oxidation, reduction and hydrolysis), but does undergo phase II metabolic processes (biosynthetic reactions, such as conjugation with glucuronic acid and sulfate) [USEPA, 1986]. Alkyl lead compounds may be oxidized in the liver [Gerson, 1990].

Lead is mainly eliminated via urine and feces. A "normal" value for lead in urine is ≤0.05 mg/L in the general population [ATSDR, 1991b, 1993c]. Exhalation of alkyl lead compounds following inhalation is a major route of elimination.

C. BIOAVAILABILITY

1. Inhalation

Human exposure to lead often results from inhalation of lead aerosols. Prior to any evaluation of bioavailability, determination of the deposition of lead in the respiratory tract is crucial. The deposition fraction of lead aerosols is very dependent on the particle size distribution. At a particle size less than 0.05 μm and inhalation rate of 15 breath/min, the deposition fractions of inhaled lead ranged from 35% to 70%, but decreased to 10% to 30% for a particle size range between 0.05 to 0.5 μm [Booker *et al.*, 1969; Hursh and Mercer, 1970; Chamberlain *et al.*, 1975; Wells *et al.*, 1977; James, 1978; Morrow *et al.*, 1980; Gross, 1981; Chamberlain, 1985].

Respiratory bioavailability studies have been conducted in human volunteers [Booker *et al.*, 1969; Hursh and Mercer, 1970; Chamberlain *et al.*, 1975]. In the studies from Booker *et al.* [1969] and Hursh and Mercer [1970], radio-labeled lead (^{212}Pb) obtained from thoron decay was administered to volunteers by oral-only inhalation exposure. The absorption and retention of lead in the body was detected by means of external whole-body counting with a correction for lead levels in the blood and excreta. The absorption fractions of the deposited lead were 48% (24 h) and 77% (72 h). Wells *et al.* [1977] reported that about 50% of the deposited lead from exhaust aerosols was absorbed from the lungs into the systemic circulation.

Tetraethyl lead from exhaust aerosols was inhaled by volunteers for 15 minutes. About 50% of the deposited lead and 5% of the ingested lead (swallowed to the gastrointestinal tract) was absorbed into the blood [Chamberlain *et al.*, 1975]. In another study, after exposure to tetraalkyl lead vapors, about 37% to 51% of the compounds were deposited within the human respiratory tract [ATSDR, 1993c]. About 68% to 80% of the deposited compounds were absorbed.

2. Ingestion

Isotope absorption studies in adults and lead balance studies in children have clarified the understanding of human lead bioavailability in the gastrointestinal tract. Absorption of lead occurs in the small intestine and is affected by a variety of factors, such as age, solubility of lead salts and lead-containing soil or dust, nutritional status (feeding and fasting), dietary types (lipid content and milk) and dietary deficiency of essential elements (iron, calcium, zinc, copper, selenium, chromium and manganese).

Generally, children tend to more easily absorb lead from the gastrointestinal tract than do adults. The metabolic balance technique was utilized to determine the oral bioavailability of lead for children [Alexander *et al.*, 1974; Ziegler *et al.*, 1978]. Infants and young children were given lead with lead-containing food by feeding milk or formula and other foods. The daily intake and urinary and fecal excretion data were recorded. The net absorption and retention of lead were calculated from these measurements. The oral bioavailability factors ranged from 42% to 53% for normal feeding status.

In adult studies, radio-isotope lead was orally administered to volunteers at various doses (single or multiple doses). The absorption and retention of lead in the blood were determined by whole-body counting, sometimes, with a correction of lead levels measured in the blood and excreta. In feeding studies, about 4% to 13% of lead ingested with meals was absorbed [Harrison *et al.*, 1969; Rabinowitz *et al.,* 1976, 1980; Blake *et al.*, 1983; Chamberlain, 1985]. When the volunteers ingested the diet lead which was incorporated into foods, the bioavailability factor ranged between 3% to 10% [Rabinowitz *et al.,* 1980; Heard and Chamberlain, 1982, 1983]. With slight fasting (light meal), about 16% to 21% of lead in water and 8.5% to 19% of lead in beverages (beer, tea and coffee) were bioavailable [Hursh and Suomel, 1968; Blake, 1976; Moore *et al.,* 1979; Watson *et al.*, 1980; Heard and Chamberlain, 1983; Newton *et al.*, 1992]. In a fasting status, the oral bioavailability was markedly increased, ranging from 31% to 70% [Rabinowitz *et al.,* 1980; Flanagan *et al.*, 1982; Heard and Chamberlain, 1982, 1983; Blake and Mann, 1983; Chamberlain, 1985].

Deficiency of calcium, iron, zinc, copper, phosphorus, vitamin D, dietary lipids and certain milk contents (particularly lactose) were reported to have very strong effects on lead absorption, resulting in an enhancement of lead absorption in humans [Heard and Chamberlain, 1982; USEPA, 1986; Chowdhury and Chandra, 1987; ATSDR, 1993c]. A study showed that the bioavailability factor of lead in human volunteers declined from 69% without supplementation of any minerals to 10% with supplementation of calcium (200 mg) and phosphorus (140 mg) [Heard and Chamberlain, 1982]. Watson *et al.* [1980] found that iron deficiency resulted in elevated lead absorption in humans. However, another study showed that there was no strong relationship between iron deficiency and increased lead absorption [Flanagan *et al.*, 1982].

Numerous studies on oral bioavailability have been performed in animals [Forbes and Reina, 1972; Garber and Wei, 1974; Barltrop and Khoo, 1975;

Barltrop and Meek, 1975; Barltrop *et al.,* 1975; Quarterman and Morrison, 1978; Quarterman *et al.*, 1978; Pounds *et al.*, 1978; Aungst *et al.*, 1981; Baldini *et al.,* 1989; Kapoor *et al.*, 1989; Clapp *et al.,* 1991; Dieter and Matthews, 1993; Hayashi *et al.*, 1993; Yannai and Sachs, 1993; Freeman *et al.*, 1994]. These studies, either qualitative or quantitative, indicated that lead absorption was also affected by age, nutritional status, dietary factors and carrier vehicles.

Lead in the environment often exists as a mixture of lead salts, soil/dust-bound lead or lead incorporated into a food matrix. The oral bioavailability depends on the solubility of these lead forms in the gastrointestinal tract. Some lead compounds are easily dissolved in gastric acid. A study revealed that the solubility of lead salts increased as pH declined from pH 7 to pH 3, but as pH fell below 3, the solubility of lead decreased [Cheng *et al.,* 1991]. The solubility under physiological conditions of soil-bound lead from mining sites is lower than that from urban areas because of differences in chemical form [Steele *et al.,* 1990; Davis *et al.*, 1992; Ruby *et al.*, 1992]. For exposure, Davis *et al.* [1992] reported only 6% of mine-waste-impacted lead was dissolved in rabbit small intestine. The low bioavailability of lead from mining sites was attributed to the low solubility of lead minerals.

3. Dermal

Absorption of inorganic lead compounds through the skin has been considered to be much less significant than other exposure routes because inorganic lead compounds have the ability to form complexes with proteins on the skin to prevent absorption [Hostynek *et al.*, 1993]. A recent study by Stauber *et al.* [1994] indicated that inorganic lead compounds could readily and rapidly absorb through the human skin *in vivo.* The rapid absorption phase was via sweat and hair follicles. Likewise, organic lead compounds are lipid-soluble and readily bioavailable.

Lead nitrate was applied to the human forearm skin (< 5 cm^2) for 24 h [Stauber *et al.,* 1994]. By measurement of surface disappearance, about 29% of lead nitrate was found to penetrate human skin. Lead naphthenate in lubricating oil was applied to human forearm skin for one hour at a dose of 32 µg/ml [Rasetti *et al.*, 1961]. The dermal permeability was 2×10^{-3} cm/h to 3×10^{-3} cm/h. In another study, the rate of absorption of lead acetate (containing in hair-darkening formulation) was reported to be 0.038 µg/cm^2 in 12 h for human volunteers [Moore *et al.*, 1980].

With *in vitro* study, dermal bioavailability varied with the forms and speciation of lead [Bress, 1991]. The bioavailability factors were 6.3% for tetrabutyl lead, 1.3% for lead nuolate (oleate and linoleate), 0.3% for lead naphthenate and 0.05% for lead acetate.

In an animal study, tetraethyl lead was found to be rapidly absorbed through rat skin [Laug and Kunze, 1948]. The dermal bioavailability factor was approximately 6.5%, with a dermal permeability of 6×10^{-6} cm/h . With the application of lead acetate to rat skin, between 1.5% to 3% and 4.1% to

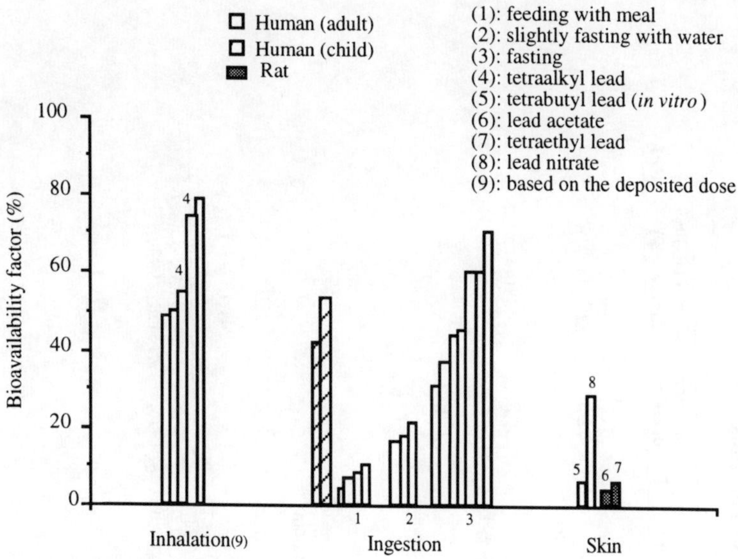

Figure 5-8 Bioavailability factors of lead in humans and animals by three exposure routes [Adapted from sources in Table 5-17, a-e (inhalation); Table 5-18, a-j, m-p (ingestion); and Table 5-19, a, c, d (skin)]

4.4% of the compound was absorbed in one week and eight week exposure periods, respectively [Pounds, 1979].

4. Summary

Lead can be rapidly bioavailable after inhalation and ingestion (Table 5-17, Table 5-18, and Figure 5-8). The deposition fraction of lead within the respiratory tract is inversely proportional to particle size distribution of lead aerosols. The oral bioavailability is influenced by nutritional status, age, lead species and forms, particle size and solubility of lead soil, and dietary types. Tetra-alkyllead compounds more easily penetrate the human skin than inorganic lead compounds (Table 5-19).

D. UNCERTAINTIES

Generally, the respiratory and oral bioavailability of lead are well characterized in humans. Respiratory bioavailability factors have been derived from lead deposited within the respiratory tract. Thus, determination of the deposition fraction of lead is critical for lead risk assessment. Estimation of deposition fraction requires information on the particle size distribution for lead in air at lead-contaminated sites. However, such data are generally not available.

Most oral bioavailability studies are related to administration of isotope lead to humans. In reality, humans are exposed to lead with soil or dust, food

Table 5-17 Respiratory Bioavailability of Lead in Humans

Compound	Exposure methods	Lead carrier	Lead dosage	Experimental method	Exposure duration	Bioavailability factor (%)	Ref.
Inorganic							
^{212}Pb	Oral-only	Aerosol (0.05-0.5 μm)	1.3-1.6 μCi	Whole-body counting Blood and excreta measurement	72 h	77[1] (34-60)[2]	a
^{212}Pb	Oral-only	Aerosol (0.02-0.2 μm)	0.66-2.3 μCi	Whole-body counting Blood and excreta measurement	24 h	48[1] (44)[2]	b
^{203}Pb	Oral-only	Exhaust aerosol (0.2-1 μm)	0.5-2.5 μCi	Whole-body counting Blood and excreta measurement	48 h	50[1] (35)[2]	c
Organic							
Tetraethyl lead	Oral-only	Exhaust aerosol (0.2-1 μm)	0.5-5 mg/ m^3	Whole-body counting Blood and excreta measurement	72 h	55[1]	d
Tetraethyl lead	Oral-only	Vapor	1 mg/ m^3	-	-	74[1] (37-51)[2]	

[1] data from the deposited dose of lead within the respiratory tract.; [2] deposition fraction

Adapted from: a. Booker *et al.*, 1969; b. Hursh and Mercer, 1970; c. Wells *et al.*, 197; d. Chamberlain *et al.*, 1975; e. ATSDR, 1993c

Table 5-18 Oral Bioavailability of Lead in Humans

Compound	Number of subjects	Lead carrier	Nutritional status	Lead dosage	Experimental method	Exposure duration	Bioavailability factor (%)	Ref.
$^{203}PbCl_2$	Adult (2)	Meal	Fed	300 μg	Whole-body counting	4 d	4	a
$^{203}PbCl_2$	Adult (6)	Meal	Fed	-	Whole-body counting Blood and excreta measurement	7 d	7.5	b
$^{204}Pb(NO_3)_2$	Adult (4)	Meal	Fed	80 μg/d	Fecal excretion	20 d	8	c
$^{204}Pb(NO_3)_2$	Adult (5)	Meal	Fed	79-204 μg/d (1-124 d)	Blood and excreta measurement	1-124 d	10	d
$^{203}PbCl_2$	Adult (3)	Meal	Fed	5 μCi x 2 d	Whole-body counting Blood and excreta measurement	5 d	13	e
Diet lead	Child (12)	Meal	Fed	30-130 μg/d (3 d)	Fecal and urinary excretion Lead intake	3 d	42	f
Diet lead	Child (86)	Meal	Fed	1 μg/d x 3 d	Fecal and urinary excretion Lead intake	3 d	53	g
$^{203}PbCl_2$	Adult (6)	Water	Light meal	10 μCi	Whole-body counting	12 d	16	h
$^{203}PbCl_2$	Adult (10)	Water	Light meal	10 μCi	Whole-body counting	4 d	18	i
$^{203}PbCl_2$	Adult (11)	Water	Light meal	-	Whole-body counting	-	21	j
$^{203}PbCl_2$	Adult (3)	Beer	Light meal	5 μCi	Whole-body counting	2 d	8.5	k

Adapted from: a. Blake *et al.*, 1983; b. Chamberlain *et al.*, 1985; c. Rabinowitz *et al.*, 1980; d. Rabinowitz *et al.*, 1976; e. Harrison *et al.*, 1969; f. Ziegler *et al.*, 1978; g. Alexander *et al.*, 1974; h. Watson, 1980; i. Blake, 1976; j. Moore *et al.*, 1979; k. Hursh and Suomel, 1968

(Continued)

Compound	Number of subjects	Lead carrier	Nutritional status	Lead dosage	Experimental method	Exposure duration	Bioavailability factor (%)	Ref.
$^{203}PbCl_2$	Adult (9)	Beer	Light meal	212 μg x 28 d	Whole-body counting Blood measurement	6 d	14	l
$^{203}PbCl_2$	Adult (4)	Beer Tea/coffee	Light meal	1 μCi	Whole-body counting	2 d	19 14	m
Diet lead	Adult (6)	Meal	Fed	100 μg	Whole-body counting	7 d	3	n
Diet lead	Adult (9)	Meal	Fed	1 μCi	Whole-body counting Blood measurement	2 d	5.5	m
Diet lead	Adult (4)	Meal	Fed	80 μg/d	Fecal excretion	20 d	10	c
$^{204}PbSO_4$ $^{204}Pb(NO_3)_2$	Adult (4)	Water	Fasted	79 μg/d 74 μg/d	Fecal excretion	20 d	31 37	c
$^{203}PbCl_2$	Adult (6)	Water	Fasted	1 μCi	Whole-body counting Blood measurement	2 d	44	m
$^{203}PbCl_2$	Adult (6)	Water	Fasted	75 μg	Whole-body counting	-	45	b
$^{203}PbCl_2$	Adult (8)	Water	Fasted	100 μg plus Ca (200 mg) and P (140 mg)	Whole-body counting Blood measurement	7 d	60 10	n
$^{203}PbCl_2$	Adult (85)	Water	Fasted	-	Whole-body counting	15 d	60	o
$^{203}PbCl_2$	Adult (4)	Water	Fasted	300 μg	Whole-body counting	4 d	70	p

Adapted from : l. Newton *et al.*, 1992; m. Heard and Chamberlain, 1983; n. Heard and Chamberlain, 1982; o. Flanagan *et al.*, 1982; p. Blake and Mann, 1983

Table 5-19 Dermal Bioavailability of Lead in Humans and Animals

Compound	Species	Exposure region	Skin area	Lead concentration	Experimental method	Exposure duration	Bioavailability factor (%)	Ref.
Lead nitrate	Human	Forearm	< 5 cm^2	4.4 mg/cm^2	Surface disappearance (*in vivo*)	24 h	29	a
Lead naphthenate in lubricating oil	Human	Forearm	-	32 ng/ml	-	1 h	2-3 x 10^{-3} cm/h[1]	b
Tetrabutyl lead Lead nuolate Lead naphthenate Lead acetate	Human	Abdominal	1.3 cm^2	10 mg	Diffusion cell (*in vitro*)	24 h	6.3 1.3 0.3 0.05	c
Tetraethyl lead	Rat	Dorsal	2.9 cm^2	106 mg	Mass-balance	24 h	6.5	d
Lead acetate	Rat	-	-	-	-	1 wk 4 wk 8 wk	1.3-3 3.3-3.9 4.1 - 4.4	e

[1]Permeability

Adapted from: a. Staber et *al.,* 1994; b. Rasetti *et al.*, 1961 (cited by Hostynek *et al.* 1993); c. Bress, 1991; d. Laug and Kunze, 1948; e. Pounds, 1979 (cited by Hostynek *et al.*,1993) and drinking water. The solubility of soil-bound lead and dissolution rate of lead salts in the gastrointestinal tract determine the variability of oral bioavailability of lead. Such relevant information including the form of lead at contaminated sites is usually not available.

The mechanisms of lead toxicity and interaction with other agents have not been clarified. Currently, a relationship between environmental lead levels and blood lead levels is commonly used for lead risk assessment. In some cases, environmental lead levels in certain media have not shown a linear relationship to blood lead levels [USEPA, 1991b]. One of the contributors for this non-linear correlation may be caused by variable bioavailability of lead with different carrier vehicles or different lead forms.

E. CONCLUSIONS

The primary exposure routes for workers are through inhalation of lead dust and fumes, ingestion of lead-contaminated food and dermal contact with organic lead compounds. General populations are mainly exposed to lead via ingestion of lead-contaminated food, drinking water and soil, and inhalation of lead soil or dust either outdoors or indoors. Direct ingestion of lead soil and paint is a major potential risk for young children with pica behavior. The potential hazard by dermal exposure to inorganic lead is negligible for the general population.

Inorganic lead compounds may be absorbed via inhalation and ingestion exposure. The respiratory bioavailability factors range from 48% to 77%. The deposition fractions are about 35% to 70% for lead aerosols less than 0.05 μm and 10% to 30% for ones between 0.05 to 0.5 μm. The oral bioavailability factors in a normal feeding status are 42% to 53% in children and 4% to 13% in adults, while the values increase to 30% to 70% in adults in a fasting status and 16% to 21% in a slight fasting status with drinking water. Only 3% to 10% of dietary lead is bioavailable. About 30% of lead nitrate could penetrate the human skin. The dermal bioavailability factors of some lead compounds were low in humans and rats *in vivo* and *in vitro*. Lead absorption, especially in the gastrointestinal tract, is strongly affected by various factors. Thus, control of these factors will aid in reducing lead absorption and, subsequently, in reducing lead toxicity for humans. These control methods may include a supplementation of certain minerals (e.g. Ca, Fe, Cu and P) in daily diet and hygienic education for young children.

F. RECOMMENDATIONS

In order to determine the deposition fraction of lead, the specific information on the particle size distribution and inhalable fraction of soil-bound lead aerosols at contaminated sites is needed. The solubility of lead-containing soil from contaminated sites under physiological conditions is required. Determining the speciation of lead in water and dissolution of inorganic lead in the gastrointestinal fluid is also important.

VII. 2,3,7,8-TETRACHLORODIBENZO-p-DIOXIN (TCDD)

2,3,7,8-Tetrachlorodibenzo-p-dioxin (TCDD), popularly referred to as dioxin, is the most toxic among the polychlorinated aromatic compounds, the polychlorinated dibenzo-p-dioxins. TCDD is a colorless solid with a low vapor pressure, an extremely low water solubility and a very large octanol-water partition coefficient [Schroy *et al.,* 1985; Shlu, 1988; WHO, 1989e; Fiedler and Schramm, 1990; Eitzer and Hites, 1991; USEPA, 1992c]. TCDD is generally regarded as a synthetic chemical, a byproduct formed during the manufacture of certain chlorinated compounds and herbicides. It is not produced naturally in substantial quantity except as a combustion byproduct [ATSDR, 1989e].

TCDD released to the environment partitions mainly between soil (85%) and sediment (14%) with less than 1% distributing into air, water, suspended sediment and biota [Travis and Hattemer-Frey, 1991]. The potential removal processes for TCDD in air are wet and dry deposition, photolysis and reactions with OH, HO_2 and NO_3 radicals and with ozone [Atkinson, 1991]. The few data available indicate that TCDD is unlikely to occur in drinking-water supplies [Jobb *et al.,* 1990]. TCDD in surface water may be removed by volatilization and photodegradation rather than by biodegradation [ATSDR, 1989e]. TCDD is readily bound to and immobilized in most soils. However, soil-bound TCDD may undergo volatilization, photodegradation and biodegradation, but the rates of transport and transformation of TCDD in soil are very slow [USEPA, 1992c]. The half-life of TCDD in different soils ranged from 18 months to greater than 100 years [Gough, 1991]. Accumulation in vegetation involves deposition, root intake, and air-leaf transfer [Travis and Hattemer-Frey, 1991]. The bioaccumulation of TCDD has been found in several aquatic species to a substantial degree [WHO, 1989e]. Also, recent studies have shown bioaccumulation of TCDD in tissue and eggs of foraging animals [Stephens *et al.,* 1990; USEPA, 1992c].

In humans, TCDD may cause severe skin lesions (chloracne), liver damage and gastrointestinal disorders [Fingerhut *et al.,* 1991; ATSDR, 1989e]. In the accident area around Seveso, Italy where a few kilogram of TCDD were accidentally released, there was only equivocal evidence of an increased incidence of cardiovascular deaths and a few specific cancers among the populations in exposed areas as compared to populations in a reference area [Bertazzi *et al.*, 1989, 1993].

TCDD is toxic to many animals at extremely low doses [Iverson and Grant, 1991; Van den Heuvel and Lucier, 1993]. TCDD in animals can produce adverse effects on immune function, reproductive development, behavior and metabolism of glucose and lipid [Skene *et al.,* 1989; Iverson and Grant, 1991; Birnbaum, 1993]. TCDD is a potent tumor promoter for liver and skin cancer in small rodents [Poland *et al.*, 1982; Graham *et al.*, 1988; Lucier *et al.,* 1991]. Although TCDD is classified as a possible human carcinogen based on animal data, evidence on the carcinogenic effect in humans from environmental exposure to TCDD remains uncertain [Dickson and Buzik, 1993].

TCDD is readily absorbed in humans via ingestion and dermal exposure. Data on respiratory bioavailability for humans are absent. TCDD is bioavailable through all three exposure routes in animals.

A. EXPOSURE ROUTES AND RECEPTORS

The general population and workers who are involved in the use of certain herbicides and chlorinated phenol compounds may be exposed to TCDD through all three exposure routes. Commonly, workers and people living near heavily TCDD-contaminated areas exhibit higher levels of TCDD in liver and adipose tissue than found in the general population [Beck *et al.,* 1988; Ryan and Norstrom, 1991]. Dermal exposure to TCDD has been a major exposure route for occupational groups.

Although the general population may inhale TCDD vapor or aerosols from the environment, the inhalation exposure route is considered to be only marginally important because air concentrations of TCDD are usually extremely low [Paustenbach *et al.,* 1986; Ryan and Norstrom, 1991]. Only 1% of the total daily intake is estimated to result from inhalation of TCDD vapor and aerosols in air [Travis and Hattemer-Frey, 1991]. Young children have a potential risk from dermal exposure to TCDD adsorbed on soil in highly TCDD contaminated areas. The uptake dose from dermal contact to TCDD in soil accounts for less than 1% of total daily intake [Travis and Hattemer-Frey, 1991].

The primary exposure route for the general population is via ingestion of food (fish, meat and dairy products), accounting for greater than 90% of the total daily intake [USEPA, 1985c, 1988; Furst *et al.*, 1991; Travis and Hattemer-Frey, 1991; Henry *et al.*, 1992]. Ingestion of soil-bound TCDD may be significant for people, especially for young children, living near TCDD-contaminated areas. Exposure to TCDD via drinking water is negligible for the general population because of the extremely low water solubility of TCDD.

B. TOXICOKINETICS (FOLLOWING ABSORPTION)

1. Distribution

Following ingestion of TCDD in corn oil by a human volunteer, the largest portion of TCDD absorbed was distributed into the fat [Poiger and Schlatter, 1986]. Data from two human autopsies showed that high levels of TCDD were present in the fat and liver [Ryan *et al.*, 1985]. In rats, mice and guinea pigs, the distribution of TCDD was found in the liver and fat after ingestion exposure [USEPA, 1984c].

2. Metabolism and Excretion

There are no available data on metabolism of TCDD in humans. In rats, TCDD could undergo hydroxylation and conjugation with glucuronide and sulfate [Poiger *et al.*, 1982; Olson *et al.*, 1983]. The rate of metabolism was very slow.

Fecal excretion is a major route for elimination of TCDD for both humans and experimental animals [USEPA, 1985c; Poiger and Schlatter, 1986; ATSDR, 1989e]. Urinary excretion accounted for a small part of the total excretion dose of TCDD. The half-life for elimination of TCDD from the human body was 5.8 years [Poiger and Schlatter, 1986].

C. BIOAVAILABILITY

1. Inhalation

Data on respiratory bioavailability of TCDD in humans are not available. Nessel *et al.* [1992] studied respiratory bioavailability of TCDD in rats which were intratracheally instilled with TCDD aerosols (10 m) at a concentration of 74.4 ppm (in 10 mg soil). Absorption of TCDD from the respiratory tract was determined by measurement of TCDD concentrations in the liver. A respiratory bioavailability factor of 100% was reported.

2. Ingestion

One study on absorption of TCDD from the gastrointestinal tract was conducted with one human volunteer in a fasting condition [Poiger and Schlatter, 1986]. Following dosage of 1.14 ng/kg of TCDD in corn oil, about 13% of the administered TCDD was excreted from feces during the first week. Most of this amount represented unabsorbed TCDD. Thus, oral bioavailability of TCDD could be no more than 87% in this case.

In a few animal studies, oral bioavailability of TCDD was estimated by measurement of cumulative fecal and urinary excreta and total body burden. Piper *et al.* [1973] reported oral bioavailability of about 70% for rats after oral administration of TCDD in acetone-corn oil at a single dose of 50 g/kg. In another study, 84% and 86% of TCDD in acetone-corn oil was absorbed in rats following a single exposure (1 g/kg) and repeated exposure (0.1 to 1 g/kg), respectively [Rose *et al.*, 1976]. Olson *et al.* [1980] gave TCDD in olive oil to hamsters (650 g/kg) by oral administration. Approximately 74% of TCDD was absorbed from the gastrointestinal tract.

When TCDD mixed with food was given to rats, the oral bioavailability factor decreased to about 50% to 60% [Fries and Marrow, 1975]. Nolan *et al.* [1979] reported that only about 50% of TCDD (without indication of the carrier vehicle) was absorbed in guinea pigs.

Absorption of TCDD in the gastrointestinal tract is affected by the various carrier vehicles. Comparing liver levels of absorbed TCDD in different animal species, oral bioavailability of TCDD with soot, fly-ash and soil was found to be lower than that of TCDD with soot solvent extract, fly-ash solvent extract and corn oil [Bonaccorsi *et al.*, 1984; Kaminsky *et al.*, 1985; Lucier *et al.*, 1986; McMonnell *et al.*, 1984; Poiger and Schlatter, 1980; Silkworth *et al.*, 1982; Shu *et al.*, 1988; Umbreit *et al.*, 1985, 1986, 1987; van den Berg *et al.*, 1983, 1985, 1986, 1987].

Oral bioavailability of TCDD with different vehicles was determined by using the measurement of liver-TCDD levels and activity of arylhydrocarbon hydroxylase (AHH) induction, an extremely sensitive measure of exposure. Because TCDD has a long elimination half-life and about 70% of the total body burden was in the liver, estimation of TCDD in the liver could provide information on the comparative absorption of TCDD with various vehicles [Fries and Marrow, 1975; Poiger and Schlatter, 1980]. Based on this method, Poiger and Schlatter [1980] found that, after 10 to 15 h post-exposure, oral bioavailability of soil-bound TCDD was lower than that of TCDD in 50% ethanol solution.

Oral bioavailability of TCDD varied with different types of soil. Relative bioavailability (the ratio of TCDD in soil to TCDD in corn oil) of soil-bound TCDD from manufacturing sites was 0.005 in guinea pigs, while a corresponding estimate of soil-bound TCDD from a metal salvage yard was 0.21 [Umbreit *et al.*, 1986]. The absolute bioavailability of TCDD in corn oil for guinea pigs was not stated, but would be presumed to be relatively high. In rabbits, the relative oral bioavailability factor for TCDD with lab contaminated soil was 0.29 to 0.44, while bioavailability was 0.68 for TCDD with soil from Seveso (heavily TCDD contaminated area) [Bonaccorsi *et al.*, 1984]. Lucier *et al.* (1986) reported a relative bioavailability factor of about 0.25 for soil-bound TCDD at a dose of 1 μg/kg and 0.50 for soil-bound TCDD at a dose of 5 μg/kg in rats.

Shu *et al.* [1988a] studied oral bioavailability of soil-bound TCDD by comparing liver TCDD level in rats that were administered soil-bound TCDD to that of rats that were administered TCDD with corn oil (assuming 100% bioavailability). Because TCDD with corn oil was not completely absorbed from the gastrointestinal tract in rats, relative oral bioavailability of soil-bound TCDD was adjusted by the unabsorbed fraction of TCDD in corn oil. The unabsorbed fraction from Piper *et al.*'s study was 30% for TCDD in corn oil for rats. After adjusting for the unabsorbed fraction of 30%, an absolute oral bioavailability of about 43% for soil-bound TCDD was estimated by Shu *et al.* [1988a].

Because TCDD and other compounds have the ability to induce AHH in some animal species, AHH induction was used to calculate oral bioavailability for soil-bound TCDD [McConnell *et al.*, 1984; Lucier *et al.*, 1986]. By using this method, soil-bound TCDD was found to be highly bioavailable, but no quantitative data were reported.

3. Dermal

Data on *in vivo* dermal bioavailability in humans are not available. One dermal absorption study was conducted *in vitro* for human skin [Weber *et al.*, 1991]. TCDD in acetone more easily penetrated human skin (back and thigh region) than TCDD in mineral oil. The dermal bioavailability factor for TCDD ranged from 40% (6.5 ng/cm^2) to 56% (65 ng/cm^2). The value of TCDD in mineral oil was between 10% (65 ng/cm^2) and 14% (6.5 ng/cm^2).

Dermal absorption of TCDD was found to be dependent on different dosage levels of TCDD applied on the back skin of rats [Brewster *et al.*, 1989; Banks *et al.*, 1991]. At a low dose (0.32 μg/kg), the dermal bioavailability factor of TCDD delivered in acetone was about 40%. When the applied dose of TCDD increased, the absorption of TCDD through the skin decreased to 27% (3.2 μg/kg) and 17% (32-320 μg/kg) [Brewster *et al.*, 1989].

Dermal absorption of TCDD was influenced by age and carrier vehicles. Young rats more easily absorbed TCDD than older rats following dermal exposure [Banks *et al.*, 1990; Anderson *et al.*, 1993]. Soil-bound TCDD had low dermal bioavailability in rats [Poiger and Schlatter, 1980; Shu *et al.*, 1988b]. In the study by Poiger and Schlatter [1980], dermal absorption of soil-bound TCDD was determined by a comparison of TCDD levels in liver following application of TCDD in soil and in methanol to the rat skin. Only 0.05% to 2.2% of TCDD appeared in the liver 24 h after dermal exposure to soil-bound TCDD at various doses, whereas about 15% of the administered TCDD in methanol reached the liver. Relative dermal bioavailability (soil/methanol) ranged from 0.003 to 0.15. Shu *et al.* [1988b] estimated a relative dermal bioavailability factor of 0.13 for soil-bound TCDD in rats by comparing the level of TCDD in the liver following dermal administration of soil-bound TCDD to that following oral administration of TCDD in corn oil.

4. Summary

TCDD can be absorbed after inhalation, ingestion and dermal exposure in humans and animals. Quantitative data are available for humans following ingestion and dermal exposure routes and for animals through all three exposure routes (Table 5-20, 5-21a, 5-21b, 5-22, Figure 5-9). TCDD is essentially completely absorbed from the respiratory tract. TCDD is readily absorbed from the gastrointestinal tract. Dermal absorption is much lower for soil-bound TCDD than for TCDD in a solvent carrier.

D. UNCERTAINTIES

Data on respiratory bioavailability of TCDD aerosols were only available for rats by intratracheal administration. Because TCDD is a lipophilic compound, it appears to completely penetrate the air-blood barrier upon reaching the pulmonary region. The respiratory bioavailability of TCDD by inhalation exposure will be expected to be lower than that by intratracheal instillation. Thus, information on deposition and the mechanical clearance of TCDD-bearing aerosols within the respiratory tract will be useful for estimating the respiratory bioavailability of TCDD. However, such data are dependent on the characteristics of the aerosols carrying TCDD and no specific studies have been reported.

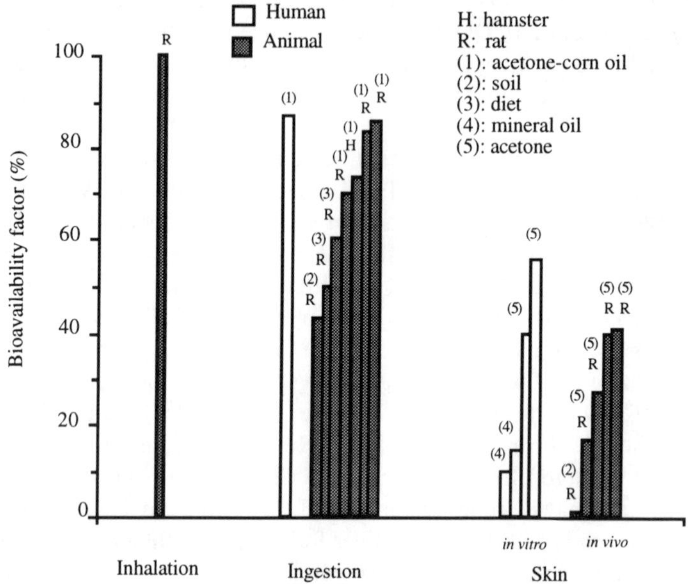

Figure 5-9 Bioavailability factors of 2,3,7,8-tetrachlorodibenzo-p-dioxin in humans and animals by three exposure routes [Adapted from sources in Table 5-20, a (inhalation); Table 5-21a, a, b, c, d, e, g (ingestion); and Table 5-22, a-d (skin)]

Data on oral bioavailability of TCDD are mainly derived from administration of TCDD in corn oil in one human volunteer and experimental animal studies. These oral bioavailability estimates are higher than those made by administration of TCDD with other vehicles. Although relative oral bioavailability factors (soil/corn oil) for TCDD with different types of soils are available in a few animal species (such as guinea pigs and rabbits), the oral bioavailability factors are difficult to estimate because of the lack of absolute bioavailability data for TCDD in corn oil for those species. Oral bioavailability of TCDD is dependent on types of soils and levels of TCDD in soil. Studies on the mechanism of absorption in the gastrointestinal tract have not been reported for soil-bound TCDD.

Dermal bioavailability of TCDD with acetone, mineral oil and soil has been studied *in vitro* using human skin and in rats. In animal studies, the dermal values for TCDD in acetone were obtained using disappearance measurements. The values for TCDD in soil were based on liver-TCDD measurements so that these estimates will be lower than those from disappearance measurements.

Table 5-20 Respiratory Bioavailability of 2,3,7,8-Tetrachlorodibenzo-p-dioxin in Animals

Compound	Species	Exposure methods	TCDD dosage	Experimental method	Exposure duration	Post-exp duration	Bioavailability factor (%)	Ref.
TCDD aerosols	Rat	Intratracheal	74.4 ppm in 10 mg soil (10 µm)	Measurement of TCDD level in liver	-	28 d	100	a

Adapted from: a. Nessel *et al.*, 1992

Table 5-21a Relative Oral Bioavailability of 2,3,7,8-Tetrachlorodibenzo-p-dioxin (TCDD) in Humans and Animals

TCDD carrier	Species	Exposure methods	TCDD dosage	Experimental method	Exposure duration	Relative Bioavailability factor	Ref.
Soil from manufacturing sites	Guinea pig	Gavage	12 µg/kg	Measurement of liver level	60 d	0.005 (soil/corn oil)	a
Soil from metal yard land			320 ng/kg			0.21 (soil/corn oil)	
Soil from lab	Rabbit	Gavage	40 ng/d x 7 d	Measurement of liver level	7 d	0.29 (soil/corn oil)	b
Soil from lab			80 ng/d x 7 d			0.44 (soil/corn oil)	
Soil from Seveso			80 ng/d x 7 d			0.68 (soil/corn oil)	
Soil	Rat	Gavage	1 µg/kg	Measurement of liver level	6 d	0.25 (soil/corn oil)	c
			5 µg/kg			0.50 (soil/corn oil)	

Adapted from: a. Umbreit *et al.* , 1986; b. Bonaccorsi *et al.*, 1984; c. Lucier *et al.*, 1986

Table 5-21b Oral Bioavailability of 2,3,7,8-Tetrachlorodibenzo-p-dioxin (TCDD) in Humans and Animals

TCDD carrier	Species (No. of subjects)	Exposure methods	TCDD dosage	Experimental method	Exposure duration	Bioavailability factor (%)	Ref.
Corn oil	Human (1) (fasting)	Oral	Single: 1.14 ng/kg	Measurement of fecal excretion	7 d	87	a
Acetone-corn oil	Rat	Gavage	Single: 50 μg/kg	Measurement of fecal excretion	2 d	70	b
Diet	Rat	Gavage	0.5 μg/kg or 15 μg/kg (x 42 d)	Measurement of total body burden	42 d	50-60	c
Acetone-corn oil	Rat	Gavage	Single: 1 μg/kg 0.1-1 μg/kg (x 5d/wk x 7 wk)	Measurement of urinary excretion	22 d 7 wk	84 86	d
Olive oil	Hamster	Oral	Single: 650 μg/kg	Measurement of urinary excretion	1 d	74	e
-	Guinea pig	Gavage	3.2-1,450 ng/kg	Mass-balance	22 d	50	f
Soil	Rat	Gavage	3.2-1,450 ng/kg	Measurement of liver level with a collection	1 d	43	g

Adapted from: a. Poiger and Schlatter, 1986; b. Piper *et al.*, 1973; c. Fries and Marrow, 1975; d. Rose *et al.*, 1976; e. Olson *et al.*, 1980; f. Nolan *et al.*, 1979; g. Shu *et al.*, 1988a

Table 5-22 Dermal Bioavailability of 2,3,7,8-Tetrachlorodibenzo-p-dioxin (TCDD) in Humans and Animals

TCDD carrier	Species	Exposure region	Skin area	TCDD concentration	Experimental method	Exposure duration	Bioavailability factor (%)	Ref.
Acetone	Human	Back and thigh	3.8 cm^2	6.5 ng/cm^2	Diffusion cell	16 h	40	a
Mineral oil				6.5 ng/cm^2	(*in vitro*)		14	
Acetone				65 ng/cm^2			56	
Mineral oil				65 ng/cm^2			10	
Acetone	Rat	Back	-	0.32 μg/kg	Disappearance measurement	3 d	20	b
				3.2 μg/kg			27	
				32-320 μg/kg			17	
Acetone	Rat	Back	1.8 cm^2	0.32 μg/kg	Disappearance measurement	5 d	41	c
Soil	Rat	Back	-	12.5-125 ng/kg	Measurement of liver levels with a correction	1 d	1.3	d

Adapted from: a. Weber *et al.*, 1991; b. Brewster *et al.*, 1989; c. Banks and Birnbaum, 1991; d. Shu *et al.*, 1988b

E. CONCLUSIONS

Humans may be exposed to TCDD through three exposure routes. In occupational settings, dermal exposure to TCDD is significant. The general population is primarily exposed to TCDD by ingestion of TCDD-contaminated food. Young children have a potential risk of exposure from soil-bound TCDD by ingestion.

Data are lacking on respiratory bioavailability for TCDD-aerosols in humans. In rats, TCDD-aerosols within the respiratory tract appear to be completely absorbed into the systemic circulation. The oral bioavailability factors of TCDD in corn oil are higher in humans (87%) and animals (74% to 86%) than that of TCDD in diet (50% to 60% in rats) and in soil (43% in rats). Oral absorption of soil-bound TCDD varies with types of soil and TCDD levels in soil. Also, dermal absorption of TCDD depends on different vehicles with an absolute dermal bioavailability factor of 40% to 56% for TCDD in acetone for humans (*in vitro*), 10% to 14% for TCDD in mineral oil for humans (*in vitro*) and only 1.3% for soil-bound TCDD in animals.

F. RECOMMENDATIONS

Data on respiratory bioavailability for TCDD are limited. Such data (deposition and mechanical clearance for TCDD aerosols within the respiratory tract) need to be investigated. Studies on oral bioavailability for TCDD with different types of soils are required because most estimates are based on relative bioavailability factors. Also, the mechanisms of absorption for soil-bound TCDD in the gastrointestinal tract and dermal bioavailability for soil-bound TCDD need to be determined.

VIII. POLYNUCLEAR AROMATIC HYDROCARBONS (PAHs)

Polynuclear aromatic hydrocarbons (PAHs) are a class of organic chemicals consisting of multiple fused-benzene rings. The different physicochemical properties of PAHs, such as the number and orientation of rings, molecular weight, water solubility and volatility, cause differences in their environmental characteristics, primary exposure routes and toxic effects. Some of the PAHs which will be discussed in this section include:

Two-ring aromatics

- Naphthalene (volatile, slightly water soluble)
- 2-methylnaphthalene (volatile)

Three-ring aromatics

- Acenaphthene (volatile)
- Anthracene (relatively water insoluble)
- Carbazole (relatively water insoluble)
- Fluorene (volatile)
- Phenanthrene (relatively water insoluble)

Four-ring aromatics

- Chrysene (hydrophobic and less volatile)
- Fluoranthene (hydrophobic and less volatile)
- Pyrene (hydrophobic and less volatile)

Five-ring aromatics

- Benzo[a]anthracene, B[a]A
(highly hydrophobic and non-volatile)
- Benzo[a]pyrene, B[a]P
(highly hydrophobic and non-volatile)
- Benzo[b]fluoranthene, B[b]F
(highly hydrophobic and non-volatile)
- Benzo[k]fluoranthene, B[k]F
(highly hydrophobic and non-volatile)
- Benzo[g,h,i]perylene, B[g,h,i]P
(highly hydrophobic and non-volatile)
- Dibenz[a,h]anthracene, DB[a,h]A
(highly hydrophobic and non-volatile)
- Indeno[1,2,3-c,d,]pyrene
(highly hydrophobic and non-volatile)

PAHs have been detected in soil, sediment, water (ground water, surface water, drinking water and waste water) and air [IARC, 1983; Menzin *et al.*, 1992]. Creosote vapors (naphthalene, 2-methylnaphthalene, acenaphthene and fluorene) may be released to air by volatilization from treated wood and creosote products [Heikkila *et al.*, 1987; ATSDR, 1990c, 1990d]. Particulates containing PAHs have been observed in the ambient atmosphere and indoors [IARC, 1983; Menzie *et al.*, 1992]. Airborne PAHs can undergo photochemical degradation [ATSDR, 1990e]. Volatilized PAHs in surface water are subject to volatilization and photolysis. Water insoluble PAHs and highly hydrophobic compounds tend to partition into organic matter in sediment. PAHs in soil may persist by adsorption to soil organic matter, giving rise to sorption, volatilization and biological or chemical degradation [Bulman *et al.*, 1985]. Some PAHs are detected in certain food groups, e.g. smoked meats and fish, vegetables (green leafy vegetables, tomatoes and potatoes), grains, fruits, fats and oils [ARC, 1983; ATSDR, 1990e; Bartle, 1991; Menzie *et al.,* 1992].

Toxic effects remain incompletely studied for most PAHs despite the substantial attention this group of compounds has received as environmental contaminants. Naphthalene and 2-methylnaphthalene can cause adverse health effects in the lung, blood and liver in humans [ATSDR, 1990d]. B[a]P has been reported to produce skin lesions in humans [ATSDR, 1990f]. Inhalation and dermal exposure to PAH-containing mixtures (e.g. coal tar, coke oven emissions, soot and crude oil) in occupational settings may be associated with an increased incidence of lung and skin cancer [ATSDR, 1990e].

Although PAHs are frequently described as carcinogenic, currently only the following are considered as possible human or animal carcinogens [IARC, 1983]: carbazole, chrysene, benzo[a]-anthracene, benzo[a]pyrene, benzo[b]fluoranthene, benzo[k]fluoranthene, benzo[g,h,i]perylene, dibenz[a,h]-anthracene, indeno[1,2,3-c,d,]pyrene. B[a]P can induce various tumors in animals through all exposure routes, so that IARC has classified B[a]P as a probable human carcinogen [IARC, 1987]. Other carcinogenic PAHs are primarily associated with skin cancer [IARC, 1983].

Toxicity profiles may be altered with exposure to PAH-containing mixtures. Some PAH components may serve as antagonistic or synergistic agents in animals. For instance, B[a]A was found to decrease the extent and rate of metabolism of B[a]P and B[a]P-DNA adduct formation in hamster embryo cells, while anthracene may enhance the mutagenicity of B[a]P [ATSDR, 1990e].

Limited information on bioavailability is available for a few carcinogenic PAHs. Overall, B[a]P has been most extensively investigated with regard to its toxicokinetics. Hence, B[a]P is often used as a surrogate for carcinogenic PAHs.

A. EXPOSURE ROUTES AND RECEPTORS

Inhalation and dermal exposure are primary pathways for workers using volatilized PAHs [ATSDR, 1990e; Withey *et al.,* 1993]. The general population may be exposed to these compounds from smoking, other combustion processes or heavily PAH-contaminated sites. High risk groups for exposure to PAHs include residents at high density areas, individuals who have nutritional deficiencies (e.g. iron, vitamins A and C), heavy smoking habits, genetic diseases (e.g. low DNA-repair capacities) and women of child-bearing age (maternal exposure) [ATSDR, 1990e].

Important exposure arises from active or passive inhalation of PAHs in tobacco smoke. Because of the persistence of hydrophobic PAHs in soil, residents living around hazardous waste sites or high density urban areas have some potential exposure to airborne PAHs (e.g. engine exhaust and PAH-containing aerosols) via inhalation.

Exposure to PAHs by ingestion of food is a potential exposure route for the general population, particularly for nonsmokers [Menzie *et al.,* 1992]. Carcinogenic PAHs have been commonly found in surface soils. Accordingly, exposure to soil-bound PAHs through ingestion may be a concern for the general population, particularly for young children with pica behavior [ATSDR, 1990c, 1990e]. Exposure to PAHs by ingestion of drinking water is substantially less significant because of the low water solubility of the PAHs [Menzie *et al.*, 1992]. Dermal contact with PAH-contaminated soil may be an exposure pathway for residents living around PAH contaminated sites.

B. TOXICOKINETICS (FOLLOWING ABSORPTION)

1. Distribution

B[a]P was found to be distributed in the lung, blood and intestines of rats after intratracheal administration [Weyand and Bevan, 1986]. Following oral exposure, B[a]P was distributed in the liver, lung and kidney in rats [ATSDR, 1990e]. A few studies have revealed that a very small fraction of PAHs was distributed to various organs and tissues.

2. Metabolism and Excretion

Lung and skin have both been found to be capable of metabolizing B[a]P [Shinohara and Cerutti, 1977; Boroujerdi *et al.*, 1981; Wester and Maibach, 1985; Noonan and Wester, 1987, 1989]. However, the liver has a much greater capacity for B[a]P metabolism in comparison with the lung and skin.

Metabolism of B[a]P occurs in multiple steps in the liver [IARC, 1983; Williams and Weisburger, 1990]. B[a]P is metabolized initially by cytochrome P-450 monooxygenase into arene oxides (e.g. 7,8-epoxide). Arene oxides may be catalyzed by epoxide hydrolase to form dihydrodiols before further conversion to diol epoxides (e.g. 7,8-dihydrodiol-9,10-epoxide). Arene oxides, dihydrodiols and diol epoxides may undergo conjugation with GSH, glucuronides and sulfate esters. Alternatively, arene oxides and dihydrodiols can form phenols and phenol diols by rearrangement or hydrolysis. Of all the metabolites, diol epoxides are considered to be ultimate carcinogens which may be covalently bound to DNA and proteins.

B[a]P is primarily excreted via feces (biliary excretion), but it is usually metabolized prior to elimination through the bile duct [IARC, 1983; ATSDR, 1990f]. Because of the hepatic presystemic metabolism, the biliary excretion rate could be enhanced following oral administration of B[a]P and the distribution of B[a]P in tissues may be lower relative to other exposure routes [IARC, 1983; ATSDR, 1990f]. Moreover, enterohepatic circulation has been observed for B[a]P and its metabolites [ATSDR, 1990f].

C. BIOAVAILABILITY

1. Inhalation

a. Humans

Some PAHs have been reported to be absorbed via inhalation exposure in humans. These compounds include: naphthalene, chrysene, B[a]P, B[a]A, B[b]F and DB[a,h]A [ATSDR, 1990d, 1990f, 1990g, 1990h, 1990i, 1990j]. Little is known of their quantitative bioavailability via inhalation in humans.

b. Animals

Chrysene

Grimmer *et al.* [1988] investigated urinary and fecal excretion of chrysene after intratracheal administration in rats. Based on cumulative urinary and fecal excretion, the respiratory bioavailability factor was estimated to be 53% and 41%, at single doses of 400 ng and 800 ng, respectively. About 16% of chrysene deposited within the respiratory tract was cleared to the gastrointestinal tract and was, therefore, not available for respiratory absorption.

Benzo[a]pyrene

Deposition fractions of pure B[a]P aerosols and B[a]P on ultrafine Ga_2O_3 particles (0.1 μm) in rats were examined by Sun *et al.* [1982]. Following inhalation of both forms of B[a]P aerosols for 30 min, 20% of pure B[a]P aerosols and 10% of B[a]P on Ga_2O_3 particles were deposited within the respiratory tract. B[a]P on Ga_2O_3 particles was mainly cleared by mucociliary transport, while pure B[a]P aerosols appeared to be absorbed directly to the systemic circulation. The rate of mechanical clearance for B[a]P on Ga_2O_3 particles was slower (one day) than pure B[a]P aerosols (1.5 h).

Pure B[a]P crystals and B[a]P-containing aerosols (0.1 to 1.0 μm and 15 to 30 μm) were intratracheally administered to mice [Cresia *et al.*, 1976]. The clearance rates of B[a]P from the lung varied depending on their aerosol size. About 95% of pure B[a]P crystals were cleared within 24 h. Only 50% of B[a]P-containing aerosols between 0.1 to 1.0 μm were cleared from the lung within 36 h.

The bioavailability factors by intratracheal administration of B[a]P to rats during 6 h periods of exposure ranged from 57% to 78% [Weyand and Bevan, 1986, 1987]. The values were 55% to 68% in guinea pig and 74% in hamsters [Weyand and Bevan, 1987]. When B[a]P on diesel carbonaceous particles was intratracheally administered to rats, the bioavailability factor declined to 20% after 6 hours dosing [Bevan and Ruggio, 1991].

The bioavailability factors from isolated perfused rat lungs were highly variable (24% and 87%) [Foth *et al.*, 1984; Törnquist *et al.*, 1988]. The highest value (87%) was obtained from experimental data with a correction for dissolution from air particulates [Törnquist *et al.*, 1988].

2. Ingestion

a. Humans

Naphthalene and B[a]P were reported to be bioavailable via ingestion exposure in humans [ATSDR, 1990d]. Hecht *et al.* [1979] investigated oral absorption of B[a]P in humans. After ingestion of B[a]P-containing broiled meat (9 µg B[a]P) by eight volunteers, only < 0.1 µg of B[a]P was detected in fecal excretion. This result indicated that B[a]P was readily absorbed from the gastrointestinal tract. Quantitative data were not reported for other PAHs in humans.

b. Animals

Anthracene

Anthracene was intraduodenally administered to rats at a single dose of 1 mg in corn oil. After 24 hours post-exposure, the oral bioavailability factor was found to be 71% [Rahman *et al.,* 1986].

Chrysene

Grimmer *et al.* [1988] investigated the urinary and fecal excretion of chrysene after oral administration in rats. Based on cumulative urinary and fecal excretion, the oral bioavailability factor was estimated to be between 60% and 87%. The highest bioavailability estimates based on fecal excretion represented upper bound estimates of bioavailability to the portal circulation.

Phenanthrene

Phenanthrene was intraduodenally administered to rats at a single dose of 1 mg in corn oil. After 24 h post-exposure, the oral bioavailability factor was 97% [Rahman *et al.,* 1986].

Benzo[a]pyrene

B[a]P was intraduodenally administered to rats at a single dose of 1 mg in corn oil. After 24 hours post-exposure, the oral bioavailability factor was 23% [Rahman *et al.*, 1986]. When rats were given a low dose of B[a]P in peanut oil (0.2 to 1 µg), the bioavailability factor was only about 10% [Foth *et al.*, 1988].

3. Dermal

a. Humans

Most PAHs have been observed to be absorbed through the skin in human exposure. Compounds tested included naphthalene, acenaphthene, anthracene, fluorene, fluoranthene, phenanthrene, pyrene, carbazole, chrysene, benzo[a]anthracene, benzo[a]pyrene, benzo[b]fluoranthene, benzo[k]-fluoranthene, benzo[g,h,i]perylene, indeno[1,2,3-c,d]pyrene and dibenz[a,h]-

anthracene [Storer *et al.*, 1984; ATSDR, 1990d; VanRooij *et al.*, 1993a and 1993b].

Based on a skin monitoring program combined with personal air and biological monitoring, the dermal bioavailability of PAHs was estimated for twelve coke oven workers [VanRooij *et al.*, 1993b]. The workers were exposed to pyrene, which was used as a marker compound at 21 to 166 μg/d. An average absorption fraction through the skin was 75%. The dermal bioavailability factor of B[a]P was estimated to be 51% based on the ratio of pyrene to B[a]P in the air and in skin contamination samples. The extent of absorption varied with anatomic sites and individuals [VanRooij *et al.*, 1993a]. The dermal bioavailability factor for humans was estimated to range from 20% to 56% for PAHs, following application of coal-tar ointment on various regions of the human skin at a low dose rate (e.g. about 1 ng pyrene/cm^2) [VanRooij *et al.*, 1993a]. The results from *in vitro* studies indicated that the dermal bioavailability factor of B[a]P was about 24% to 31% in humans [Storm *et al.*, 1990; Wester *et al.*, 1990].

b. Animals

<u>Naphthalene</u>

Naphthalene in pure, sandy soil and clay soil forms was applied to rat abdominal skin at a dose of 3.3 μg/cm^2 [Turkall *et al.*, 1994]. At 2 d, the dermal bioavailability factors of naphthalene in these three forms were 88% (pure), 89% (sandy soil) and 95% (clay soil).

<u>Anthracene</u>

Anthracene dissolved in organic solvents was applied to the dorsal skin (2 cm^2) in rats. At 6 d, 52% of the chemical was recovered in the urine, feces (via bile) and various tissues [Yang *et al.*, 1986a]. Results from an *in vitro* experiment showed about 56% of the chemical penetrated full-thickness skin [Yang *et al.*, 1986a].

<u>Phenanthrene</u>

Dermal penetration of phenanthrene was studied by *in vivo* and *in vitro* methods in hairless guinea pig [Ng *et al.*, 1991]. About 80% of dermal absorption was observed from the *in vivo* method at 24 h post-exposure. The *in vitro* penetration was 71% to 78% of applied chemical .

<u>Pyrene</u>

Pyrene was studied in the hairless guinea pigs by *in vivo* and *in vitro* methods [Ng *et al.*, 1992]. The bioavailability factor (*in vivo*) was 94% after 7 d dosing. Results from the *in vitro* method revealed about 40% to 69% of dermal penetration at the low and high doses during 24 h exposure.

Dibenz[a,h]anthracene

DB[a,h]A dissolved in benzene was administered to mouse skin [Heidelberger and Weiss, 1955]. At 2 d post-exposure, about 8% of the chemical was absorbed. Another study reported about 33% dermal absorption following dosing the chemical with acetone for 24 h in mice [Sander *et al.*, 1986].

Benzo[a]pyrene

The dermal bioavailability of B[a]P has been extensively studied in animals by *in vivo* and *in vitro* methods [Kao *et al.,* 1985; Sanders *et al.*, 1984, 1986; Yang *et al.,* 1986b, 1989a, 1989b; Storm *et al.*, 1990; Wester *et al.*, 1990; Ingram and Phillips, 1993]. B[a]P was studied *in vivo* by application on the dorsal skin of animals with B[a]P dissolved in organic solvents (commonly acetone) at low and high doses. Bioavailability factors were determined after one day to one week post-exposure by measurement of the cumulative biliary, urinary and fecal excretion data or by using the complete mass-balance technique.

In vitro studies were frequently conducted using a diffusion cell. A skin culture technique (i.e. viable skin) for determining the dermal absorption in various species was reported by Kao *et al.* [1985]. Values from this technique were lower (3% in humans and 0.1% to 10% in small rodents) than those from the diffusion cell technique. The presystemic metabolism of B[a]P on the skin was thought to account for the apparent low dermal bioavailability.

There was relatively good agreement between *in vivo* and *in vitro* experiments for the dermal bioavailability of B[a]P. Overall, the dermal bioavailability in animals was slightly higher than in humans. Values from a variety of studies varied according to the differences in species, applied dose and exposure duration. Results for B[a]P available in the literature are summarized as :

	Bioavailability ***in vivo***	**Bioavailability** ***in vitro***	**Reference**
Mouse	41%-83%	55%-60%	Sanders *et al.,*1984, 1986; Ingram and Phillips, 1993; Storm *et al.*, 1990
Rat	35%-46%	38%-55%	Yang *et al.*, 1986b, 1989a, 1989b; Storm *et al.*, 1990
Hairless guinea pig	73%	40%-67%	Ng *et al.,* 1992
Monkey	51%	-	Wester *et al.*, 1990
Human	-	24%-31%	Storm *et al.,* 1990; Wester *et al.*, 1990

Mixtures

Because people are exposed to PAH-containing mixtures rather than isolated individual PAH chemicals in the environment, the study of bioavailability for PAH-mixtures is important for quantitative risk assessment.

Cruzan *et al.* [1986] investigated dermal bioavailability by *in vivo* and *in vitro* methods for some PAHs from clarified slurry oil in rats. The results indicated good agreement between *in vivo* and *in vitro* experiments for these hydrophobic compounds. The bioavailability factors were 48% (*in vivo*) and 44% (*in vitro*) for carbazole, 21% (*in vivo*) and 13% (*in vitro*) for four-ring PAHs (phenanthrene and fluorene), 1% (*in vitro*) for chrysene, 5% (*in vivo*) and 3% (*in vitro*) for five-ring PAHs (B[a]P) and 8% (*in vitro*) for 2-methylnaphthalene.

In another study, pure B[a]P and a complex mixture of PAHs were applied to mouse skin [Dankovic *et al.*, 1989]. B[a]P in the mixtures was more readily absorbed (16%) than the pure B[a]P (5%). The bioavailability factors were 16% to 19% for B[a]P, benzo[a]anthracene and benzo[b]fluorene, 13% for chrysene and 8% for pyrene.

4. Summary

B[a]P and other selected PAHs can be absorbed by humans and animals through all three exposure routes (Table 5-23, 5-24 and 5-25). Dermal bioavailability of B[a]P is well documented from *in vivo* and *in vitro* studies in either humans or animals (Figure 5-10, 5-11). Dermal bioavailability for each compound in a complex mixture will be lower than a pure chemical exposure. Pure B[a]P is more easily absorbed from the respiratory tract to the systemic circulation than B[a]P aerosols. Oral bioavailability of B[a]P appears to be lower in comparison with other exposure routes.

D. UNCERTAINTIES

Applications of bioavailability for evaluating PAH cancer risk is a complex issue, depending on the mechanisms of carcinogenesis as well as exposure routes.

Lung tumors in animals are reported to be related to exposure to B[a]P via not only inhalation or intratracheal administration, but also other exposure routes [IARC, 1983; ATSDR, 1990f; Collins *et al.*, 1991]. Hence, the interaction between the B[a]P metabolites and DNA molecules may occur because of the metabolism of deposited B[a]P within the respiratory tract or B[a]P and its metabolites transported from the systemic circulation via other exposure routes. B[a]P may also cause various tumors (including gastric tumors, pulmonary adenomas and leukemia) via ingestion [IARC, 1983; ATSDR, 1990f; Collins *et al.*, 1991]. Application of B[a]P on animal skin has been observed to increase the incidence of skin cancer [IARC, 1983; ATSDR, 1990f; Collins *et al.*, 1991]. Consequently, both the quantitative analysis of DNA adduct formation and protein binding at the applied sites as

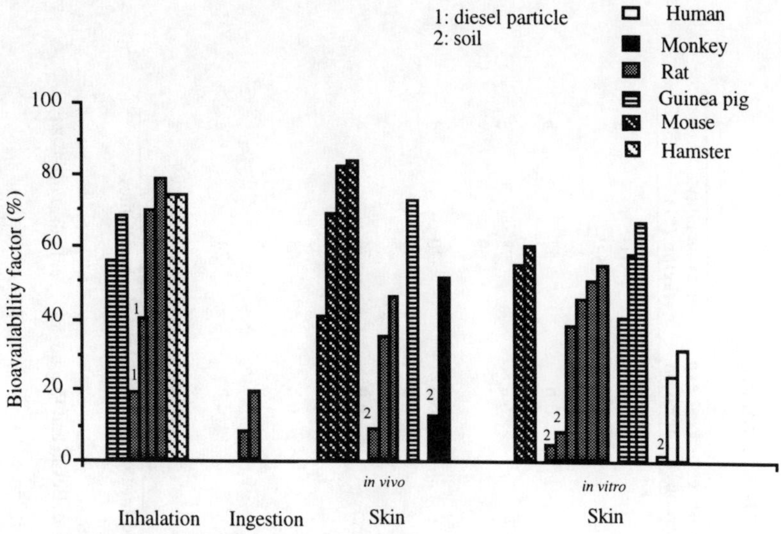

Figure 5-10 Bioavailability factors of benzo[a]pyrene in humans and animals by three exposure routes [Adapted from sources in Table 5-23: b, c ,d (inhalation); Table 5-24: a, c (ingestion); and Table 5-25: d, f, g, h, j, k, l (skin)]

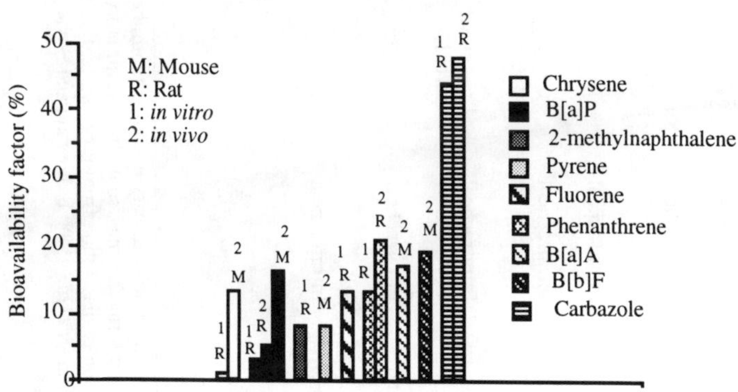

Figure 5-11 Dermal bioavailability factors for individual compounds after application of the PAH-containing mixtures to animal skin [Adapted from sources in Table 5-25: m, n (skin)]

Table 5-23 Respiratory Bioavailability of Some PAHs in Animals

Compound	Species	Exposure methods	PAHs dosage	Experimental method	Exposure duration	Post-exp duration	Bioavailability factor (%)	Ref.
Chrysene	Rat	Intratracheal	Single: 800 ng Single: 400 ng	Urinary and fecal excretion	-	3 d 5 d	41[1] 53[1]	a
B[a]P	Rat	Intratracheal	1 μg/kg	Mass balance	-	3 h	57	b
B[a]P	Rat	Intratracheal	0.16 μg 350 μg	Mass balance	-	6 h	78 69	c
B[a]P	Rat	Intratracheal	1 mg diesel particles	Mass balance	-	6 h 3 d	20 40	d
B[a]P	Hamster	Intratracheal	0.16 μg 350 μg	Mass balance	-	6 h	74 74	c
B[a]P	Guinea pig	Intratracheal	0.16 μg 350 μg	Mass balance	-	6 h	68 55	c
B[a]P	Rat	-	-	Isolated perfused lung	-	2 h	24	e
B[a]P	Rat	-	-	Isolated perfused lung	-	2.5 h	87[2]	f

[1] Data from the sum of fraction of urinary excretion (both the unaltered form and metabolites) and fraction of fecal excretion (metabolites, excluding the unaltered form that may come from respiratory mechanical clearance)
[2] Data with a correction of dissolution from urban air particulates and covalently bound DNA adducts.
Adapted from: a. Grimmer *et al.*, 1988; b. Weyand and Bevan, 1986; c. Weyand and Bevan, 1987; d. Bevan and Ruggio, 1991; e. Foth *et al.*, 1984; f. Törnquist *et al.*, 1988

Table 5-24 Oral Bioavailability of Some PAHs in Animals

Compound	Species	Exposure methods	PAHs dosage	Experimental method	Exposure duration	Post-exp duration	Bioavailability factor (%)	Ref.
Anthracene	Rat	Intraduodenal	1 mg in corn oil	Biliary and urinary excretion	-	24 h	71	a
Chrysene	Rat	Gavage	50 µg in corn oil	Urinary and fecal excretion	-	4 d	87[1] 60[2]	b
Phenanthrene	Rat	Intraduodenal	1 mg in corn oil	Biliary and urinary excretion	-	24 h	97	a
B[a]P	Rat	Intraduodenal	1 mg in corn oil	Biliary and urinary excretion	-	24 h	23	a
B[a]P	Rat	Gavage	0.2-1 µg in peanut oil	Blood measurement	-	24 h	10	c

[1] Data from the cumulative fecal excretion (the unaltered form) as the upper bound value.
[2] Data from the fecal excretion (metabolites) and the urinary excretion (both the altered form and metabolites).
Adapted from: a. Rahman *et al.*, 1986; b. Grimmer *et al.*, 1988; c. Foth *et al.*, 1988

Table 5-25 Dermal Bioavailability of Some PAHs in Humans and Animals (*in vivo* and *in vitro*)

Compound	Species	PAH carrier	Exposure region	Skin area	PAH application	Experimental method	Exposure duration	Bioavailability factor (%)	Ref
Naphthalene	Rat	None Sandy soil Clay soil	Costo-abdominal	1.3 cm^2	3.3 μg/cm^2	Mass-balance	48 h	88 (*in vivo*) 89 (*in vivo*) 95 (*in vivo*)	a
Anthracene	Rat	Acetone/ hexane	Dorsal	2 cm^2	9.3 μg/cm^2	Mass-balance Diffusion cell	6 d	52 (*in vivo*) 56 (*in vitro*)	b
Phenanthrene	Hairless Guinea pig	Acetone	Dorsal	4 cm^2	6.25 μg/cm^2	Urinary and fecal excretion	24 h	80 (*in vivo*)	c
			250 μm thick	1.7 cm^2	6.6 μg/cm^2 15.2 μg/cm^2	Diffusion cell Diffusion cell		78 (*in vitro*) 71 (*in vitro*)	
Pyrene	Hairless Guinea pig	Acetone	Dorsal	4 cm^2	6.2 μg/cm^2	Urinary and fecal excretion	7 d	94 (*in vivo*)	d
			200 μm thick	2 cm^2	6.8 μg/cm^2 37 μg/cm^2	Diffusion cell Diffusion cell	24 h	69 (*in vitro*) 41 (*in vitro*)	
DB[a,h]A	Mouse	Benzene	Sacral	-	55 μg	Mass-balance	2 d	8 (*in vivo*)	e
DB[a,h]A	Mouse	Acetone	Dorsal	1.8 cm^2	5.4 μg/cm^2 56 μg/cm^2 515 μg/cm^2	Mass-balance	24 h	82 (*in vivo*) 71 (*in vivo*) 33 (*in vivo*)	f

Adapted from: a. Turkall *et al.*, 1994; b. Yang *et al.*, 1986a; c. Ng *et al.*, 1991; d. Ng *et al.*, 1992; e. Heidelberger and Weiss, 1951; f. Sanders *et al.*, 1984, 1986

(Continued)

Compound	Species	PAH carrier	Exposure region	Skin area	PAH application	Experimental method	Exposure duration	Bioavailability factor (%)	Ref
B[a]P	Mouse	Acetone	Dorsal	1.8 cm^2	1.25-12.5 μg/cm^2 125 μg/cm^2	Mass-balance	24 h	83 (*in vivo*) 83 (*in vivo*)	f
B[a]P	Mouse	Acetone Mineral oil	Dorsal	5 cm^2	0.4 μg/cm^2	Urinary and fecal excretion	2 d	69 (*in vivo*) 82 (*in vivo*)	g
B[a]P	Mouse	Acetone	Dorsal	5 cm^2	2 μg/cm^2	Skin culture	24 h	10 (*in vitro*)	h
B[a]P	Sencar mouse	Acetone	200 μm thick	0.64 cm^2	3 μg/cm^2	Diffusion cell	24 h	55 (*in vitro*)	i
B[a]P	Balb/c mouse	Acetone	200 μm thick	0.64 cm^2	3 μg/cm^2	Diffusion cell	24 h	60 (*in vitro*)	i
B[a]P	Osb-mendel rat	Acetone	200 μm thick	0.64 cm^2	3 μg/cm^2	Diffusion cell	24 h	45 (*in vitro*)	i
B[a]P	Fuzzy rat	Acetone	200 μm thick	0.64 cm^2	3 μg/cm^2	Diffusion cell	24 h	55 (*in vitro*)	i
B[a]P	Rat	Acetone/ hexane	Dorsal	2 cm^2 2.4 cm^2	9.2 μg/cm^2 9-10 μg/cm^2	Mass-balance Diffusion cell	5 d	46 (*in vivo*) 50 (*in vitro*)	j
B[a]P	Rat	Acetone	Dorsal	5 cm^2	2 μg/cm^2	Skin culture	24 h	2 (*in vitro*)	h

Adapted from: g. Ingram and Phillips, 1993; h. Kao *et al.*, 1985; i. Storm *et al.*, 1990; j. Yang *et al.*, 1986b

(Continued)

Compound	Species	PAH carrier	Exposure region	Skin area	PAH application	Experimental method	Exposure duration	Bioavailability factor (%)	Ref
B[a]P	Rat	Crude oil Soil/crude	Dorsal	7 cm^2	0.09 mg/cm^2 9 mg/cm^2	Mass-balance	4 d	35 (*in vivo*) 9 (*in vivo*)	k
B[a]P	Rat	Crude oil Soil/crude Soil/crude Soil/B[a]P	Dorsal 350 μm thick	7 cm^2	0.09 mg/cm^2 9 mg/cm^2 56 mg/cm^2 9 mg/cm^2	Diffusion cell	4 d	38 (*in vitro*) 8 (*in vitro*) 1 (*in vitro*) 1.7-3.7(*in vitro*)	k
B[a]P	Guinea pig	Acetone	Dorsal	5 cm^2	2 $\mu g/cm^2$	Skin culture	24 h	0.1 (*in vitro*)	h
B[a]P	Hairless Guinea pig	Acetone	Dorsal	4 cm^2	7.8 $\mu g/cm^2$	Urinary and fecal excretion	7 d	73 (*in vivo*)	d
			200 μm thick	2 cm^2	8.0 $\mu g/cm^2$ 74$\mu g/cm^2$	Diffusion cell	24 h	67 (*in vitro*) 40 (*in vitro*)	
B[a]P	Hairless Guinea pig	Acetone	200 μm thick	0.64 cm^2	3 $\mu g/cm^2$	Diffusion cell	24 h	58 (*in vitro*)	i
B[a]P	Rabbit	Acetone	Dorsal	5 cm^2	2 $\mu g/cm^2$	Skin culture	24 h	1.5 (*in vitro*)	h
B[a]P	Marmoset	Acetone	Dorsal	5 cm^2	2 $\mu g/cm^2$	Skin culture	24 h	3.5 (*in vitro*)	h
B[a]P	Monkey	Acetone Soil	Abdominal	12 cm^2	10 ppm 0.04 g soil /cm with 10 ppm B[a]P	Urinary excretion	24 h	51 (*in vivo*) 13 (*in vivo*)	l

Adapted from: k. Yang *et al.*, 1989a and 1989b; l. Wester *et al.*, 1990

(Continued)

Compound	Species	PAH carrier	Exposure region	Skin area	PAH application	Experimental method	Exposure duration	Bioavailability factor (%)	Ref
B[a]P	Human	Acetone	Abdominal	0.64 cm^2	3 µg/cm^2	Diffusion cell	24 h	31 (*in vitro*)	i
B[a]P	Human	Acetone	Leg	5 cm^2	2 µg/cm^2	Skin culture	24 h	3 (*in vitro*)	h
B[a]P	Human	Acetone Soil	500 µm thick	1 cm^2	10 ppm/cm^2 0.04 g soil /cm with 10 ppm B[a]P	Diffusion cell	24 h	24 (*in vitro*) 1.4 (*in vitro*)	l
<u>Mixtures</u>									
Carbazole	Rat	CSO[1]	Dorsal 350 µm thick	1.8 cm^2 2.4 cm^2	20 mg/cm^2	Mass-balance Diffusion cell	4 d	48 (*in vivo*) 44 (*in vitro*)	m
2-Methyl-naphthalene	Rat	CSO[1]	350 µm thick	2.4 cm^2	20 mg/cm^2	Diffusion cell	4 d	8 (*in vitro*)	m
Phenanthrene	Rat	CSO[1]	Dorsal 350 µm thick	1.8 cm^2 2.4 cm^2	20 mg/cm^2	Mass-balance Diffusion cell	4 d	21 (*in vivo*) 13 (*in vitro*)	m
Fluorene	Rat	CSO[1]	350 µm thick	2.4 cm^2	20 mg/cm^2	Diffusion cell	4 d	13 (*in vitro*)	m
Chrysene	Rat	CSO[1]	350 µm thick	2.4 cm^2	20 mg/cm^2	Diffusion cell	4 d	1 (*in vitro*)	m
B[a]P	Rat	CSO[1]	Dorsal 350 µm thick	1.8 cm^2 2.4 cm^2	20 mg/cm^2	Mass-balance Diffusion cell	4 d	5 (*in vivo*) 3 (*in vitro*)	m
B[a]P	Mouse	Acetone	Dorsal	-	25 µg	Surface disappearance	24 h	5 (*in vivo*)	n

[1] CSO = Clarified Slurry Oil

Adapted from: m. Cruzan *et al.*, 1986; n. Dankovic *et al.*, 1989

(Continued)

Compound	Species	PAH carrier	Exposure region	Skin area	PAH application	Experimental method	Exposure duration	Bioavailability factor (%)	Ref
B[a]P	Mouse	COM[1]	Dorsal	-	8 mg COM	Surface disappearance	24 h	16 (*in vivo*)	n
B[a]A	Mouse	COM[1]	Dorsal	-	8 mg COM	Surface disappearance	24 h	17 (*in vivo*)	n
B[b]F	Mouse	COM[1]	Dorsal	-	8 mg COM	Surface disappearance	24 h	19 (*in vivo*)	n
Chrysene	Mouse	COM[1]	Dorsal	-	8 mg COM	Surface disappearance	24 h	13 (*in vivo*)	n
Pyrene	Mouse	COM[1]	Dorsal	-	8 mg COM	Surface disappearance	24 h	8 (*in vivo*)	n

[1] COM = Coal-derived Complex Organic Mixture

well as internal dose (either the parent compound and its metabolites) are important for assessing cancer risk for PAHs. Attempts have been made to determine quantitative data on the DNA and protein binding of B[a]P at the applied sites [Törnquist *et al.*, 1988; Ingram and Phillips, 1993]. However, such information remains limited for most PAHs.

Because the occurrence of lung tumors was found to be correlated with B[a]P aerosols rather than pure B[a]P via inhalation, aerosols are considered to exert synergistic action in the development of lung tumors. An understanding of the characteristics of aerosols and their deposition, retention and mechanical clearance within the respiratory tract will aid in assessing the cancer risk of PAHs. Parameters (inhalable fraction of B[a]P aerosols in the air and particle size distribution) which can be used to predict the deposition fraction are not available for contaminated sites. Data on the deposition, retention and clearance for B[a]P aerosols and other PAHs are limited.

Current quantitative values of respiratory bioavailability are obtained from intratracheal administration of a few pure individual PAH compounds to animals. Such data will generally overestimate the human bioavailability for PAH-containing organic mixtures commonly found in the environment.

Oral bioavailability studies have been conducted for a few PAHs in animals. Bioavailability values have been commonly determined by measurement of the extent of biliary, urinary and fecal excretion. B[a]P is metabolized prior to elimination via the bile duct. As a result, the oral bioavailability factor usually consists of the fractions of the unaltered (parent) form and its metabolites that may come from both the presystemic and systemic metabolism. To some extent, such results are likely to overestimate the oral bioavailability in humans.

Quantitative studies on the dermal bioavailability in several species have been performed *in vivo* and *in vitro*. Dermal absorption of B[a]P is dependent on skin metabolic capacities as well as B[a]P physicochemical properties [Kao *et al.*, 1985]. Human data are only available *in vitro*. However, *in vitro* bioavailability does not account for skin metabolism, so that such values tend to be an overestimation of bioavailability in humans.

Because most PAHs are hydrophobic, the chemicals released from the mixtures or solid carriers are most likely to easily penetrate lipophilic absorption barriers. Accordingly, determination of the physicochemical properties (octanol/water partition coefficients and dissolution rates under the physiological conditions) of PAHs is useful for predicting their bioavailability [Törnquist *et al.*, 1988; Bevan and Ruggio, 1991]. Octanol/water partition coefficients are known for most common PAHs.

Human exposure to PAH-containing mixtures may occur through all exposure routes. For mixtures, the absorption profiles can be affected by the interactions among individual chemicals. Results from available experiments indicate lower dermal bioavailability following application of the mixtures to animal skin. Dermal bioavailability, in some cases, has been found to be inversely related to the magnitude of the applied dose. Some studies showed that PAHs were bound to DNA on the skin to form high levels of DNA adducts rather than being absorbed into the blood when high levels of PAHs were applied on the skin.

E. CONCLUSIONS

Humans can be exposed to PAHs from the environment through all three exposure routes. Inhalation and dermal contact are the main routes for occupational exposure. Residents living near PAH-contaminated sites may be exposed to airborne PAH particulates. Active or passive inhalation of PAHs from tobacco smoke is an important source of exposure for the general population. Exposure to soil-bound PAHs via ingestion and dermal contact may be significant for high risk groups, particularly for young children with pica behavior. A potential risk for the general population may come from ingestion of contaminated food. Drinking water has only minor exposure potential because of the low solubility of the PAHs.

Quantitative data on respiratory and dermal bioavailability are available for a few PAH compounds. About 55% to 80% of B[a]P is bioavailable following intratracheal administration of the pure chemical in small rodents. These values decline to around 20% to 40% for dosing of B[a]P-containing diesel particulates. Oral bioavailability tends to decrease with an increase of ring number of PAHs in animals: 97% for phenanthrene (three rings), 70% for anthracene (three rings), 60% for chrysene (four rings) and 10% to 30% for B[a]P (five rings). *In vitro* dermal bioavailability for humans is between 24% and 30%. The *in vivo* and *in vitro* values from various animal species are relatively higher than ones from humans, ranging from 40% to 95%. The dermal bioavailability for individual PAHs will markedly decrease for exposure to the PAH-containing mixtures.

Selection of bioavailability values for PAH risk assessment is generally dependent on the mechanisms of their carcinogenesis, specific exposure routes and specific environmental media. Lung cancer, arising from exposure to some PAHs (especially B[a]P) via inhalation, as well as skin cancer via dermal contact with some PAHs appear to be local effects. In such cases, evaluation of deposition fraction for certain PAH aerosols and determination of DNA or protein adduct formation at the applied sites (i.e. within the respiratory tract or on skin surface) may be a preferred starting point for PAH risk assessment. For systemic carcinogenicity (e.g. via ingestion exposure), estimation of the internal dose of PAHs is useful for risk assessment. Meanwhile, characteristics of the bioavailability values which are selected for risk estimation should be close to those characteristics which depict human exposure in the environment, namely the appropriate specific exposure routes, carrier vehicles and ring number of PAHs.

F. RECOMMENDATIONS

Further research on the bioavailability for PAHs should:

- determine the characteristics of PAH-containing aerosols arising at contaminated sites, including: inhalable fraction of aerosols in the air, particle size distribution and dissolution rates under similar physiological conditions.
- determine the profiles of deposition, retention and clearance for PAH-containing aerosols and mixtures.
- determine the respiratory bioavailability for each individual carcinogenic PAH or if possible PAH mixture by either *in vivo* and *in vitro* methods.
- determine the oral bioavailability for each carcinogenic PAH or PAH mixtures.
- identify and quantify the active species of carcinogenic PAHs in the systemic circulation following exposure via ingestion.
- determine the dissolution rates of soil-bound PAHs or the PAH mixtures in the gastrointestinal tract.
- quantify the DNA adduct formation and protein binding at the applied sites.
- establish a correlation between *in vivo* and *in vitro* bioavailability for PAHs via inhalation and ingestion exposure.
- establish a relationship between physicochemical properties (dissolution rates, molecular weight and octanol/water partition coefficients) and bioavailability for PAHs in order to predict their bioavailability by using *in vitro* methods.

IX. BENZENE

Benzene is a highly volatile, relatively nonpolar aromatic hydrocarbon that is widespread in the environment [Sittig, 1991]. Benzene is slightly water soluble and highly miscible with organic solvents, such as alcohol, ether, acetone and chloroform [IARC, 1982a]. This hydrocarbon has low octanol/water partition coefficient (log k_{ow}) of 2.13 and soil adsorption coefficient (log k_{oc}) of 1.8 to 1.9 [ATSDR, 1991a, 1993b; Lide, 1993].

Once released to the environment, generally, over 99% of benzene will partition into the air as a vapor. Based on physicochemical properties, there is exchange between air and water/soil [HWC, 1979]. Benzene in air is readily degraded by reaction with hydroxyl radicals [USEPA, 1985a; ATSDR, 1991a; 1993b]. A very small fraction of benzene will partition into surface water, soil and sediment [IARC, 1982a; Hattemer-Frey *et al.,* 1990; Howard, 1990a]. The biodegradation of benzene in surface water and soil may occur, depending on the presence of an active microbial population, dissolved oxygen and nutrients. Plants can be contaminated by benzene through deposition, root uptake and air-to-leaf transfer [Bedleman, 1988; Hattemer-Frey *et al.*, 1990]. Some foods naturally contain benzene [USEPA, 1978]. Quantitative data for benzene have been reported in several foods: eggs, Jamaican rum, nuts, dairy products, fish, irradiated beef and heat-treated canned beef [USEPA, 1980b; IARC, 1988].

Benzene may produce a series of toxic effects in various human organs. In particular, benzene induces hematological disorders and is also a recognized human carcinogen by inhalation exposure [Ellenhorn and Barceloux, 1988; IARC, 1982a].

Bioavailability of benzene is high through inhalation and ingestion exposure, whereas it is much lower by dermal exposure in humans and animals primarily because of its high volatility.

A. EXPOSURE ROUTES AND RECEPTORS

Humans are primarily exposed to benzene by inhalation of vapor and dermal contact with vapor and liquid. Ingestion of benzene through water, food and soil, or dermal contact with soil-associated benzene are potential routes for the general population.

Occupational exposure to benzene usually occurs via inhalation because of elevated levels in the workplace [OSHA, 1987, 1989]. Workers may be also exposed to benzene through dermal contact [ATSDR, 1991a, 1993b].

The primary exposure route for the general population is inhalation of benzene vapor indoors and outdoors [Brunnemann *et al.,* 1989; Hajimiragha *et al.*, 1989; Wallace, 1986, 1989]. Cigarette smoking is the most significant indoor source. Individual exposures to benzene indoors are about two times higher than those outdoors, accounting for 50% of the total population exposure [Wallace, 1986, 1989]. Sources of atmospheric benzene outdoors come from emissions of automobile exhaust, waste industrial sites and manufacturing chemical sites. However, benzene released to outdoor air only contributes about 20% of the total population exposure [Wallace, 1986, 1989].

Because less than one percent of benzene in the environment partitions into water, soil and biota, the potential hazard through ingestion and dermal exposure is minor for the general population [Hattemer-Frey *et al.,* 1990]. Although benzene occasionally appears in drinking water and is detected in a few foods, the total population exposure is insignificant in comparison with the exposure by inhalation [Wallace, 1986, 1989]. Dermal contact with soil-associated benzene will also be negligible. Only direct skin exposure to liquid solvents containing benzene is likely to pose a serious dermal exposure risk.

B. TOXICOKINETICS (FOLLOWING ABSORPTION)

1. Distribution

Once absorbed following inhalation exposure, benzene is soon distributed to the blood, brain, liver, kidney, stomach and fat in humans [Tauber, 1970; Winek and Collom, 1971]. Animal experiments have shown benzene and its metabolites, through ingestion and dermal exposure, could reach the blood, bone marrow, liver, kidney, lung, brain, untreated skin and fat [ATSDR, 1991a, 1993b]. The parent compound and its metabolites are stored in adipose tissue and bone marrow.

2. Metabolism

The toxicity of benzene has been attributed primarily to its metabolites. Metabolism mainly occurs in the liver. Then, benzene metabolites may be transported to bone marrow via the blood stream for further metabolism or direct toxic action on blood cells.

There are several pathways involving either increased toxicity or detoxification of benzene [IARC, 1982a; Sabourin *et al.*, 1988, 1989; Henderson *et al.*, 1989; Eastmond *et al.*, 1987; Kalf, 1987; Travis *et al.*, 1990; ATSDR, 1991a, 1993b; Andrews and Snyder, 1991; Cox, 1991]. Benzene is initially metabolized to benzene oxide by cytochrome P-450-dependent mixed-function oxidase enzymes in the liver. In the detoxification processes, part of the benzene oxide is conjugated to glutathione to form water soluble phenyl-mercapturic acid which is excreted from the body in the urine. Alternatively, benzene oxide can be rearranged to form phenol, some of which becomes conjugated to glucuronide and sulfate to form water-soluble complexes which are also readily excreted in the urine.

On the other hand, benzene can be bioactivated via other pathways. Benzene oxide may be converted to dihydrodiol by epoxide hydrolase, which is oxidized to catechol by dehydrogenases. Small proportions of catechol may result from the hydroxylation of phenol. Hydroquinone, the most toxic metabolite, is also generated by the hydroxylation of phenol. A minor metabolite, muconaldehyde, comes from breaking the benzene ring. Phenol, catechol and hydroquinone are the principal hepatic metabolites of benzene.

Phenol, catechol and hydroquinone can reach the bone marrow. In the bone marrow, phenol is rapidly oxidized to hydroquinone by the marrow peroxidases, which leads to the formation of benzoquinone by further oxidation. Catechol may convert to trans-trans-muconaldehyde by dioxygenases. These products can interact with cellular DNA, nuclear protein and cell surface membrane protein.

Overall, phenol, catechol and hydroquinone, the so-called transport forms, play a key role in determining the metabolism of benzene in the bone marrow. These toxic hepatic metabolites may directly act at the target sites. Alternatively, they are activated in the bone marrow by other enzyme systems.

3. Excretion

Elimination of benzene and its metabolites depends on the exposure routes. By inhalation, about 12% to 17% of benzene is exhaled unchanged [ATSDR, 1991a, 1993b]. Through other exposure routes, small fractions of benzene absorbed may be eliminated via expired air. The majority of the parent compound and its metabolites are excreted via urine and, to a lesser extent, via feces. Conjugated phenol is a primary product excreted in urine [Sherwood, 1988].

C. BIOAVAILABILITY

1. Inhalation

a. Humans

Benzene vapor is rapidly absorbed into the human body after inhalation. Various factors can influence the respiratory bioavailability of benzene. Because of its limited water solubility and a blood/air partition coefficient of only 7.8 [Sato and Nakajima, 1979], inhaled benzene is not completely the systemic circulation.

In human studies, volunteers inhaled benzene vapor by whole-body or oral-only exposure. The exposure concentration of the vapor in air ranged from about 47 to 110 ppmv, with 1 to 4 hours of exposure duration. The extent of absorption was determined by measuring the exhalation of benzene.

Results from those studies indicated the bioavailability factors ranging from 47% to 60%, following 2 to 4 h post-exposure [Srbova *et al.*, 1950; Hunter, 1968; Nomiyama *et al.,* 1974]. However, a large amount of benzene vapor (70 to 80%) was absorbed during 5 min post-exposure, which then declined sharply [Srbova *et al.*, 1950]. The concentration of benzene in the blood reached a steady state within 30 min.

b. Animals

From animal experiments [Schrenk *et al.,* 1941; Snyder *et al.,* 1981; Sabourin *et al.,* 1987], the absorption of benzene vapor was found to be variable with exposure concentration, exposure duration and species. The degree of absorption for benzene declined through progressive exposure [Snyder *et al.*, 1981; Sabourin *et al.*, 1987].

2. Ingestion

There are a few studies indicating that benzene can be absorbed in the gastrointestinal tract in humans [ATSDR, 1991a, 1993b]. However, little quantitative information on bioavailability by ingestion has been published.

Benzene can be extensively absorbed from the gastrointestinal tract in animals. In rabbits, 84% to 89% of liquid benzene was absorbed following gavage administration at a single massive dose of 0.4 g/kg [Parke and Williams, 1953]. In another study, the bioavailability factor was 97% for rats and mice after oral exposure to benzene at a dose of 0.5 to 150 mg/kg b.w. [Sabourin *et al.,* 1987]. When benzene was adsorbed to different kinds of soils (sandy and clay soil), bioavailability was altered [Turkall *et al.*, 1988; Abdel-Rahman and Turkall, 1989; Travis and Bowers, 1990]. After oral administration of pure benzene (150 μl) and soil-containing benzene (150 μl benzene in 0.5 g soil) in rats, the absorption fraction was 85% for pure benzene, 104% for sandy soil and 63% for clay soil [Turkall *et al.*, 1988]. These three forms of benzene appeared to be absorbed rapidly following ingestion with a half-life time of 4 to 7 min. However, a small portion of benzene would be volatilized in the gastrointestinal tract before it was absorbed.

3. Dermal

a. Humans

Data from either *in vivo* or *in vitro* studies showed a slight absorption of benzene through human skin. Most benzene tended to be rapidly volatilized.

Franz [1984] investigated the dermal absorption of benzene in four volunteers. Pure benzene liquid (0.4 ml) was applied to 80 cm^2 of forearm. The dermal bioavailability was determined by measurement of urinary excretion. The bioavailability factor was 0.05% at 2 h post-exposure. Maibach [1980a, 1980b] reported 0.13% and 0.07% of absorption in human palm and forearm skin, respectively. These exposures involved very high exposure levels because pure benzene was used.

With an *in vitro* human study, the degree of absorption through abdominal skin was proportional to the concentrations (5 to 520 $\mu l/cm^2$) of benzene applied on the skin [Franz, 1984]. The bioavailability factor remained constant and low (0.02%) at each concentration. Furthermore, the permeability of benzene varied with the carrier vehicles [Blank and McAuliffe, 1985]. The mean permeability constant was 111,000 cm/h, 1,400 cm/h, 900 cm/h and 3,700 cm/h for benzene in water, gasoline, hexadecane and isooctane, respectively.

b. Animals

Benzene is also slightly absorbed following dermal exposure in animal experiments. Results from an *in vivo* study revealed that monkeys and mini-pigs had bioavailability factors of 0.14% and 0.09% [Franz, 1984]. Maibach and Anjo [1981] found that 0.17% of pure benzene and 0.08% of a petroleum distillate mixture (0.36% benzene) could penetrate monkey skin. The absorption fraction of benzene in hairless mice was 0.89% [Susten *et al.*, 1985] with a permeability of 0.619 cm/h [Tsuruta *et al.*, 1989]. Using sandy and clay soil-adsorbed benzene, Skowronski *et al.* [1988] and Abdel-Rahman and Turkall [1989] observed that the percutaneous penetration was lower when compared to pure benzene exposure. Pure benzene (300 μl) and soil-adsorbed benzene (300 μl benzene in 1 g soil) were administered to the abdominal skin of hairless mice covered by a glass cap. The extent of absorption was estimated from the degree of excretion in expired air, urine and feces, with a correction for volatilization during the soil treatment. The bioavailability factors were 20%, 16% and 19% for benzene in sandy soil, clay soil and pure benzene, respectively, at 12 h post-exposure, while the bioavailability factors were 70%, 56% and 99% for the above three forms at 48 h post-exposure. This bioavailability was much higher than that reported by Susten *et al.* [1985], possibly because the test substance was applied under a glass cap to avoid volatilization, and bioavailability was determined with a correction for volatilization. Without the correction, the absorption fraction would have only been 6% and 12% for benzene in sandy and clay soil at 12 h post-exposure [Brainard and Beck, 1992]. From an *in vitro* study (diffusion cell technique), the absorption fractions were only 0.20% for monkeys and 0.23% for mini-pigs [Franz, 1984].

4. Summary

Benzene is bioavailable in humans and animals by inhalation, ingestion and dermal contact (Table 5-26, 5-27, 5-28, and 5-29, Figure 5-12). The absorption of benzene through inhalation and ingestion is high, whereas absorption via dermal contact is comparatively low, unless methods to avoid the volatilization of benzene on the tested skin are used in the experiments.

E. UNCERTAINTIES

Benzene is a very volatile compound. When it is released to air, water and soil, it exhibits a short half-life. Atmospheric benzene is readily degraded by photochemical reaction; in water and soil, it is volatilized and biodegraded. Benzene is relatively stable in groundwater. Based on the characteristics of its environmental fate, benzene is unlikely to accumulate in air, surface water and soil. However, benzene may be taken up by plants around waste sites and may contaminate groundwater, particularly because of gasoline from leaking underground storage tanks.

The toxicity of benzene is associated with benzene metabolism. In general, the three hepatic metabolites, phenol, catechol and hydroquinone, are considered to play a key role in benzene carcinogenicity for humans. Using an external dose basis for cancer risk assessment will involve a number of uncertainties. Preferably, the identification and quantification of benzene metabolites should be considered for human health risk assessment. Recently, internal dose which commonly includes the total benzene metabolites formed

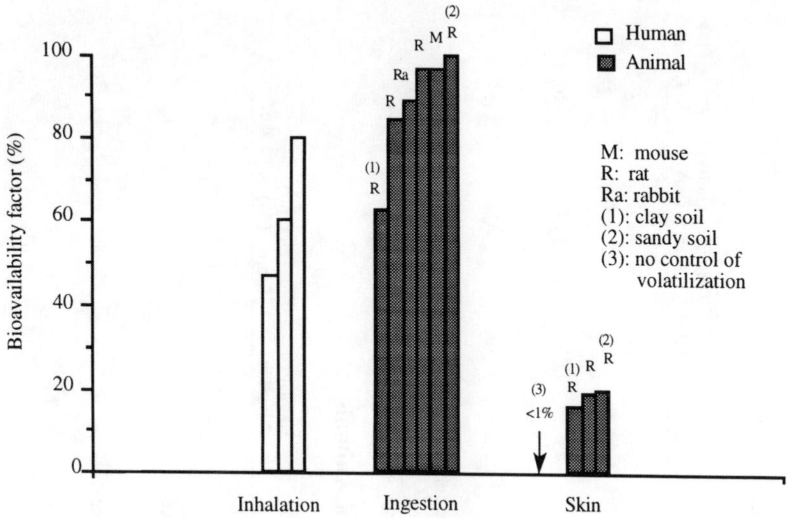

Figure 5-12 Bioavailability factors of benzene in humans and animals by three exposure routes [Adapted from sources in Table 5-26: a, b, c (inhalation); Table 5-27: a, b, c (ingestion); and Table 5-28: a-f (skin)].

has been used for reassessing benzene cancer risk [Cox and Ricci, 1992]. Physiologically-based pharmacokinetic models (PBPK) have been developed for a quantitative analysis of the distribution and metabolism of benzene in various tissues in humans and small rodents [Beliles and Totman, 1989; Paxman and Rappaport, 1990; Travis *et al.*, 1990; Bois *et al.*, 1991a, 1991b; Bois and Paxman, 1992; Woodruff *et al.*, 1992; Watanabe *et al.*, 1994]. The models are used to predict human cancer risk arising from exposure to benzene. However, the models have some limitations, such as the need for a detailed physiological description of benzene kinetics and a lack of understanding of the exact role of each metabolite involved in benzene carcinogenicity.

E. CONCLUSIONS

Humans are extensively exposed to benzene by inhalation in both general and occupational environments. In particular, smoking is a primary source for indoor air contamination by benzene. Nevertheless, exposure to benzene by inhalation for residents living near contaminated sites is likely to be insignificant because of the short half-life for atmospheric benzene. The potential hazard posed by ingestion and dermal exposure is minor for the general population because of the generally low benzene levels in foods, drinking water and soil.

Table 5-26 Respiratory Bioavailability of Benzene in Humans

Form	No. of subjects	Exposure methods	Benzene dosage	Experimental method	Exposure duration	Post-exp duration	TI[1]	Bioavailability factor (%)	Ref.
Vapor	23	Oral-only	47-110 ppmv (150-350 µg/L)	Exhalation measurement	2 - 3 h	5 min 1-2 h	81 50	70-80 50 (20-60)	a
Vapor	4	Whole-body	100 ppmv (300 mg/m^3)	Exhalation measurement	1- 4 h	4 h	-	50-60	b
Vapor	6	Whole-body	52-62 ppmv	Exhalation measurement	4 h	3 h	30	47[2]	c

[1] TI (Total Intake) = (Concentration in inhaled air - concentration in expired air)/ concentration in inhaled air
[2] The result was calculated from total intake plus fraction of air excretion
Adapted from: a. Srbova *et al.*, 1950; b. Hunter, 1968; c. Nomiyama *et al.*, 1974

Table 5-27 Oral Bioavailability of Benzene in Animals

Form	Species	Exposure methods	Benzene carrier	Benzene dosage	Experimental method	Exposure duration	Bioavailability factor (%)	Ref.
Liquid	Rabbit	Gavage	-	Single: 0.4 g/kg	Mass-balance	2 - 3 d	84-89	a
Liquid	Rat Mouse	Gavage	Corn oil	0.5-150 mg/kg	Mass-balance		>97 >97	b
Liquid	Rat	Gavage	-	150 µl	Mass-balance	-	85	c
Liquid	Rat	Gavage	Sandy soil Clay soil	150 µl in 0.5 g soil	Mass-balance	2 h	104 63	c

Adapted from: a. Parke and Williams, 1953; b. Sabourin *et al.*, 1987; c. Turkall *et al.*, 1988

Table 5-28 Dermal Bioavailability of Benzene in Humans and Animals (*in vivo*)

Form	Species	Benzene carrier	Exposure region	Skin area	Benzene concentration	Experimental method	Exposure duration	Bioavailability factor (%)	Ref.
Liquid	Human	Pure	Forearm	80 cm^2	0.4 ml	Urinary excretion	2 - 4 h	0.05	a
Liquid	Human	Pure	Forearm	-	-	-	-	0.07	b
Liquid	Human	Pure	Palm	-	-	-	-	0.13	c
Liquid	Monkey Mini-pig	Pure	Back	55-75 cm^2	0.5 ml	Urinary excretion	2 - 4 h	0.14 0.09	a
Liquid	Monkey	Pure Distillate	Forearm	12.9 cm^2	0.5 ml 50 µl	Urinary and fecal excretion	-	0.17 0.08	d
Liquid	Hairless mouse	Pure	Back	0.8 cm^2	5 µl	Mass-balance	4 h	0.89	e
Liquid	Hairless mouse	Pure	Whole-body	-	200-3,000 ppm	Mass-balance	-	0.619 cm/h[1]	f
Liquid	Rat	Pure	Abdominal	13 cm^2	300 µl	Mass-balance	12 h 48 h	19 99	g
		Sandy soil			300 µl benzene in 1 g soil		12 h 48 h	20 70	
		Clay soil			300 µl benzene in 1 g soil		12 h 48 h	16 59	

[1] Permeability

Adapted from: a. Franz, 1984; b. Maibach, 1980a (cited by Susten *et al.*, 1985); c. Maibach, 1980b (cited by Susten *et al.*, 1985); d. Maibach and Anjo, 1981; e. Susten *et al.*, 1985; f. Tsuruta *et al.*, 1989; g. Skowronski *et al.*, 1988.

Table 5-29 Dermal Bioavailability of Benzene in Humans and Animals (*in vitro*)

Form	Species	Benzene carrier	Exposure region	Type of skin	Benzene concentration	Experimental method	Permeability ($x10^3$ cm/h)	Bioavailability factor (%)	Ref.
Liquid	Human	Water	Abdominal	Full-thickness	2 µl	Diffusion cell	111		a
		Isooctane					3.7		
		Hexane					2.4		
		Gasoline					1.4		
		Hexadecane					0.9		
Liquid	Human	Pure	Abdominal	Split-thickness	5-520 µl/cm^2	Diffusion cell		0.20	b
Liquid	Monkey	Pure	Back	Split-thickness	5-520 µl/cm^2	Diffusion cell		0.19 (5 µl/cm^2)	b
Liquid	Mini-pig	Pure	Back	Split-thickness	5-520 µl/cm^2	Diffusion cell		0.23 (5 µl/cm^2)	b

Adapted from: a. Blank and McAuliffe, 1985; b. Franz, 1984

Benzene can be absorbed via three exposure routes. The bioavailability factor for inhalation ranges from 47% to 60% at two hours post-exposure in humans. Usually, the extent of absorption of benzene inhaled appears to be high during the first few minutes of exposure. Quantitative data on the oral bioavailability of benzene in humans are not available. Animal experiments have shown the oral bioavailability factors to be extremely high in small rodents (between 89% to 97%). When benzene is adsorbed to soils, the bioavailability is altered, depending on the characteristics of soil types. Dermal absorption is very low (less than 1%) by direct application of benzene on human or animal skin. Observed bioavailability factors will be increased if means to control benzene volatilization from the skin are used.

F. RECOMMENDATIONS

Benzene bioavailability via certain exposure routes has been well studied in humans or animals. Direct study on oral bioavailability in humans has been limited. There is a need to establish a relationship for oral bioavailability between animals and humans.

Benzene has been accepted as a cause of human leukemia by inhalation exposure as well as contributing other toxic effects. These effects may be attributed to the action of benzene's metabolites at the target sites. Accordingly, there is a need to consider the quantitative relationship between absorbed dose (parent benzene) and the quantity of its metabolites in the systemic circulation, as well as the relationship between active species and benzene carcinogenicity and toxicity. Physiologically-based pharmacokinetic models will be useful for these purposes.

X. TOLUENE

Toluene is a highly volatile, low water-solubility compound which is readily miscible with most organic solvents and which has an octanol/water partition coefficient (log k_{ow}) of 2.69 [Lide, 1993].

Toluene readily partitions to air in the environment with much smaller fractions partitioning to water and soil [Howard, 1990b]. Toluene rapidly degrades by reacting with hydroxyl radicals in air [WHO, 1985; Howard, 1990b] and toluene is rapidly biodegradable in surface water and soil, depending on microbial acclimation, temperature and mixing conditions. Toluene has been reported in several foods, such as baked potatoes, cheese, fried bacon, fried chicken and fish [WHO, 1985; Howard, 1990b].

Toluene produces a range of toxic response in humans, including narcotic and neurotoxic effects [Ellenhorn and Barceloux, 1988; ATSDR, 1992]. Both general and occupational populations are primarily exposed to toluene through inhalation. The potential hazard via ingestion or dermal exposure is minor.

Bioavailability for toluene through all exposure routes is well studied in humans and animals. Toluene is completely absorbed in the gastrointestinal tract, but there is a relatively low dermal bioavailability.

A. EXPOSURE ROUTES AND RECEPTORS

Humans may be exposed to toluene primarily by inhalation. Ingestion and dermal contact are potential exposure routes of lesser importance. In the occupational environment, workers in the printing industry, chemical manufacturing and solvent operations are subject to toluene exposure by inhalation [Fishbein, 1985; WHO, 1985; ATSDR, 1992]. Dermal contact with toluene-based products (e.g. inks and dyes) can be a significant exposure route for workers.

The general population may be exposed to toluene via inhalation either indoors or outdoors. Toluene vapor may arise from emissions of automobile exhaust, gasoline, hazardous wastes, cigarette smoke and toluene-containing products (paints, glue and cleaner) [Fishbein, 1985; ATSDR, 1992]. Smoking is a major concern for indoor exposure of the general population to toluene [Hjelm *et al.*, 1988; Brunnemann *et al.*, 1989; Inoue *et al.*, 1993]. Because toluene generally occurs at low levels in drinking water, foods and soil, exposure by ingestion is unlikely to be significant [USEPA, 1985b]. People may be potentially exposed to toluene by dermal contact with toluene-based solvent, cosmetic products and, to a limited degree, drinking water [Brown *et al.*, 1984; WHO, 1985].

B. TOXICOKINETICS (FOLLOWING ABSORPTION)

1. Distribution

In humans, toluene absorbed via inhalation is rapidly distributed in the blood, brain, liver, lung and fat tissue [Carlsson and Ljungquist, 1982; Paterson and Sarvesvaran, 1983; Takeichi *et al.,* 1986; Ameno *et al.*, 1989].

The highest concentrations are found in the higher vascular tissue (e.g. brain and fat tissue) [Cohr and Stokholm, 1979].

2. Metabolism and Excretion

About 70% to 80% of toluene absorbed is metabolized by the cytochrome P-450 mixed function oxidase system in the liver of humans [USEPA, 1985b; WHO, 1985; ATSDR, 1992]. The majority of toluene is converted to benzyl alcohol which is subsequently catalyzed to benzoic acid by alcohol and aldehyde dehydrogenase. Benzoic acid is conjugated with glycine to produce hippuric acid and the conjugation of benzoic acid with glucuronic acid forms benzoyl glucuronide. Small portions (less than 1%) of toluene are catalyzed to o-cresol and p-cresol by the cytochrome P-450 system. These ultimately conjugate with sulfate glucuronide, glutathione or cysteine.

Approximately 20% of unchanged toluene is exhaled from the lung. Small amounts of parent compound and all the metabolites are primarily eliminated in urine. Hippuric acid is a major metabolite in urine, accounting for about 80% of the total excreted dose [Antti-Poika *et al.,* 1987]. Fecal excretion accounts for a very small fraction (less than 2%) of total toluene excretion [Ellenhorn and Barceloux, 1988].

C. BIOAVAILABILITY

1. Inhalation

Respiratory bioavailability of toluene in humans is well documented. As a vapor, toluene is readily absorbed following inhalation. In one study, the maximal blood toluene concentration was achieved after a two hour inhalation exposure [Sato and Nakajima, 1978].

Because of its relatively low water solubility, the extent of the inhaled toluene depends on its partition coefficient (percentage of total intake of the administered toluene) rather than exposure concentration [Veulemans and Masschelein, 1978]. In many studies, only total intake of toluene was reported, which is often used to estimate the respiratory bioavailability factor for toluene. Total intake of toluene is that fraction of inhaled toluene which is retained within the lung (see equation 4-8). Part of this will be deposited on the mucous lining of the airways, particularly in the pulmonary region (alveoli). Only deposited toluene readily diffuses to the systemic circulation because of its relatively high lipophilicity. The dose of toluene which deposits on the mucous lining and enters the systemic circulation is only a portion of the total respiratory intake. Thus, the concentration of toluene in alveolar air, reflecting the applied dose, is a more useful measurement for evaluating bioavailability. Bælum [1990] reported that averaged alveolar concentrations of toluene in humans only accounted for 18% and 19% of the total inhaled dose at rest and during exercise, respectively. However, there were large variations in alveolar concentrations of toluene among exposed individuals.

Respiratory bioavailability of toluene varies with the workload conditions in humans. Total intake was inversely proportional to workload during exposure [Carlsson, 1982; Åstrand, 1983]. The bioavailability factors ranged from 39% to 51% at rest, two hour post-exposure, while the bioavailability factor decreased to 29% under extremely heavy work. In other studies, the respiratory bioavailability factor was between 37% and 51% [Nomiyama *et al.*, 1974; Löf *et al.*, 1990, 1993].

2. Ingestion

a. Humans

Human quantitative data on oral bioavailability of toluene are limited. In one human study, four volunteers were orally given a commercial toluene-containing food at a dose rate of 2 mg toluene/min for three hours (total 368 mg toluene) [Bælum *et al.*, 1993]. Hippuric acid in urine accounted for 87% to 100% of the total exposure dose. The hepatic extraction ratio (ratio of the amount of toluene metabolized in the liver to the amount of toluene in the portal blood) was close to one (99% to 100%). This study indicated that almost all the toluene in the gastrointestinal tract was absorbed to the portal blood and metabolized in the liver. The fraction of parent toluene in the systemic circulation was close to zero.

b. Animals

A study with rats showed that the absorption rate of toluene via ingestion was slower than that via inhalation with the maximal concentrations in various tissues occurring after 2 to 3 h oral dosing [Pyykkö *et al.*, 1977]. When toluene was adsorbed to sandy and clay soil, the absorption rate in the gastrointestinal tract differed slightly from that via ingestion of pure toluene in rats [Turkall *et al.*, 1990, 1991]. Toluene was readily absorbed to the portal blood for all three exposure forms (pure, adsorbed to soil or clay). The fractions of urinary metabolites of toluene were 98%, 93% and 97% for pure, sandy and clay exposure forms, respectively. These data reflected the high partition of toluene into the portal blood from the gastrointestinal tract.

3. Dermal

a. Humans

Following dermal exposure to toluene vapor, toluene appeared to be rapidly absorbed to the blood [Sato and Nakajima, 1978; Riihimäki and Pfäffli, 1978; Aitio *et al.,* 1984]. Dutkiewcz and Tyras [1968] investigated the quantitative absorption of toluene via dermal exposure in humans. The absorption rate was derived from indirect measurement, namely, the measurement of urinary metabolites and the direct method, that is, the calculation of differences in the tested solution concentrations before and after dermal exposure by immersion of one hand in toluene solution [Dutkiewcz and Tyras, 1967]. The absorption rate of toluene was relatively high (14 to

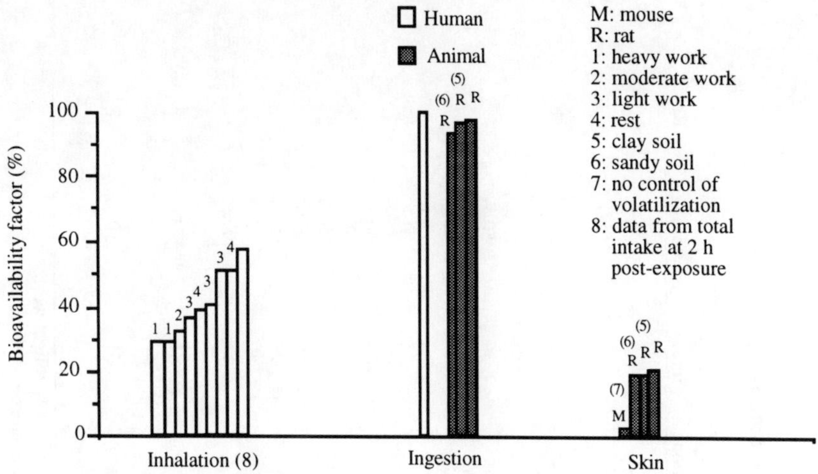

Figure 5-13 Bioavailability factors of toluene in humans and animals by three exposure routes. Adapted from sources in Table 5-30: a-e (inhalation); Table 5-31: a, b (ingestion); and Table 5-32: d, e (skin).

23 mg/cm^2 h). According to Fick's law, the permeability of liquid toluene was 0.9 cm/h for human skin [Brown *et al.*, 1984].

b. Animals

Dermal bioavailability has also been observed in animals. When hairless mice were exposed to toluene vapor at air concentrations of 200, 1,000 and 3,000 ppmv, the permeability was 1.24 cm/h [Tsuruta, 1989]. The bioavailability factor was 2.1% after toluene (5 μl) was directly applied to the back skin of hairless mice [Susten *et al.,* 1990]. Dermal bioavailability in rats tended to be higher if toluene was applied on the skin with a glass cap to avoid volatilization [Turkall *et al.*, 1990]. Absorption fractions estimated from blood, expired air, urine and feces were 20%, 19% and 19% for toluene in pure, sandy and clay exposure forms, respectively.

4. Summary

Toluene is bioavailable through all three exposure routes in humans and animals (Table 5-30, 5-31, 5-32, and Figure 5-13). Toluene in the gastrointestinal tract is almost completely absorbed to the portal blood. Respiratory bioavailability varies with rest and exercise conditions. Absorption via dermal contact is lower than that via the other two routes, partially because of volatilization.

Table 5-30 Respiratory Bioavailability of Toluene in Humans

Form	No. of subjects	Exposure methods	Toluene dosage	Experimental method	Exposure duration	Post-exp duration	Total intake[1]	Workload[2]	Ref.
Vapor	10	Whole-body	98-130 ppmv (430 mg/m^3)	Exhalation measurement	4 h	2 h	37	37 (57)[3]	a
Vapor	12	Whole-body	80 ppmv (300 mg/m^3)	Exhalation and blood measurement	30 min x 4	30 min	55	150 W	b
						2 h	29	150 W	
						2 h	32	100 W	
						2 h	36	50 W	
						2 h	39	Rest	
Vapor	13	Whole-body	300 mg/m^3	Exhalation and blood measurement	30 min x 4	30 min	52	50 W	c
							52-55	Rest	
						1 h	47-49	50 W	
							50	Rest	
						1.5 h	40	100 W	
							50	Rest	
						2 h	29	150 W	
							40	50 W	
							51	Rest	
Vapor	3	Whole-body	3.25 mmol/m^3	Exhalation and blood measurement	4 h	4 h	51	-	d
Vapor	9	Oral-nose	194 mg/m^3	Exhalation and blood measurement	2 h	2 h	51	50 W	e

[1] Total Intake = (Concentration in inhaled air - concentration in expired air)/ concentration in inhaled air; [2] 50W = light work; 100W = moderate work; 150W = heavy work; [3] The result was derived from the sum of total intake and air excretion

Adapted from: a. Nomiyama *et al.*, 1974; b. Carlsson, 1982; c. Åstrand, 1983; d. Löf *et al.*, 1990; e. Löf *et al.*, 1993

Table 5-31 Oral Bioavailability of Toluene in Humans and Animals

Form	Species	Exposure methods	Toluene carrier	Toluene dosage	Experimental method	Exposure duration	Bioavailability factor (%)	Ref.
Liquid	Human	Feed-tube	Tube feeding product	368 mg	Exhalation and urinary excretion	40 min x 4 (5 min break)	100[1]	a
Liquid	Rat	Gavage	-	150µl	Mass-balance and urinary excretion	48 h	98[1]	b
Liquid	Rat	Gavage	Sandy soil Clay soil	150 µl toluene in 0.5 g soil	Mass-balance and urinary excretion	48 h	93[1] 97[1]	b

[1] the values are derived from the fraction of toluene metabolites in urine
Adapted from: a. Bælum *et al.*, 1993; b. Turkall *et al.*, 1991

Table 5-32 Dermal Bioavailability of Toluene in Humans and Animals

Form	Species	Toluene carrier	Exposure region	Toluene concentration	Experimental method	Exposure duration	Permeability (cm/h)	Bioavailability factor (%)	Ref.
Liquid	Human	Pure	One hand	100-600 mg/L	Loss in solution or urinary excretion	1 h	0.9 (rate: 14-23 mg/cm^2 h)		a,b
Vapor	Hairless mouse	Pure	Whole-body	200 ppm 1,000 ppm 3,000 ppm	Mass-balance	6 h	1.24		c
Liquid	Hairless mouse	Pure	Back	5 µl	Mass-balance	-		2.1	d
Liquid	Rat	Pure	Abdominal	225 µl	Exhalation, blood, urinary, and fecal excretion	12 h 48 h		20 79	e
		Sandy soil		750 mg soil		12 h 48 h		19 90	
		Clay soil		750 mg soil		12 h 48 h		19 81	

Adapted from: a. Dutkiewcz and Tyras, 1967, 1968; b. Brown *et al.*, 1984; c. Tsuruta, 1989; d. Susten *et al.*, 1990; e. Turkall *et al.*, 1990

D. UNCERTAINTIES

Almost all the toluene released to the environment partitions to the air as a vapor phase. Toluene has a short half-life in the atmosphere. To determine adverse human health effects, specific information in relation to the atmospheric toluene levels is needed. However, available data at specific contaminated sites are limited. In general, residents living around contaminated sites will likely have a low level of exposure to toluene vapor by inhalation because dispersed air concentrations of toluene are likely to be low. With regard to other environmental media, toluene has a short half-life in surface water or soil, but it can be relatively stable in groundwater or deep soil because of its slow biodegradation rate.

To date, estimation of respiratory bioavailability in humans is based on the total intake of toluene. In effect, this parameter only represents the external dose. An attempt has been made to estimate the alveolar concentration of toluene. The internal dose of toluene in humans or animals is generally not available.

Active species and mechanisms of action for toluene are currently not fully understood. There are three postulated mechanisms: alteration of the lipid structure of cell membranes by toluene, interaction with cell protein by toluene and formation of the o-cresol and p-cresol intermediates (arene oxide intermediates) which may bind to cell protein or RNA [ATSDR, 1992]. Thus, toluene toxicity may be related to either the parent form or its metabolites. Although most absorption data via ingestion in humans and animals are derived from urinary metabolites, these values, as the upper bound of oral bioavailability factors, are useful for toluene risk assessment.

E. CONCLUSIONS

Exposure to toluene via inhalation is a significant route in comparison to the other routes for either the general or occupational population. Dermal exposure may occur in industries using toluene as a solvent (e.g. printing). The potential background risk via ingestion and dermal exposure is minor for the general population.

Toluene is readily absorbed by all three exposure routes in humans. Respiratory bioavailability (based on total intake) is variable with the workload rather than air concentration of toluene. This value is inversely proportional to the workload, ranging from 39% to 51% at rest to 29% after heavy work measured two hours post-exposure. The alveolar concentration (the applied dose) in humans is reported to be 18% to 19%. Toluene in the gastrointestinal tract is completely absorbed to the portal blood in humans and animals. For dermal exposure to toluene liquid or vapor, the permeability is 0.9 cm/h in humans. The bioavailability factor was 2.1% in hairless mice, and 20%, 19% and 19% in rats with dermal contact involving pure toluene, sandy soil and clay soil containing toluene, respectively, when volatilization was controlled.

F. RECOMMENDATIONS

Because evaluation of exposure to toluene vapor via inhalation depends on data relevant to air concentrations, more specific information would be needed for assessing particular contaminated sites. Moreover, data on the toluene levels in groundwater are also needed if drinking water is derived from groundwater near waste sites.

Currently, estimation of respiratory bioavailability is based on a total intake of toluene. This will overestimate the respiratory bioavailability factor for toluene because total intake of toluene reflects the administered dose of toluene within the lung which does not completely enter the alveolar region, but which is deposited on the mucous lining and then absorbed into the systemic circulation. Only one study reported the alveolar concentration of toluene in humans. Thus, research to determine the alveolar concentration or internal dose in the systemic circulation is needed.

XI. POLYCHLORINATED PHENOLS: PENTACHLOROPHENOL, 2,4,5-TRICHLOROPHENOL, 2,4,6-TRICHLOROPHENOL, 2,3,4,5-TETRACHLOROPHENOL

Pentachlorophenol (PCP) is relatively volatile with very low water solubility. PCP is readily soluble in most organic solvents, such as benzene, acetone, ethanol and methanol, and it has an octanol/water partition coefficient (log k_{ow}) of 3.32 at pH 7.2 [Lide, 1993]. Technical-grade PCP products include a variety of impurities, such as tetrachlorophenol, trichlorophenol, hexachlorobenzodioxins, hexachlorobenzofurans and polychlorinated biphenyls [IARC, 1991]. 2,4,5-Trichlorophenol (2,4,5-T_3CP), 2,4,6-trichlorophenol (2,4,6-T_3CP) and 2,3,4,6-tetrachlorophenol (2,3,4,6-T_4CP) are also discussed in this section.

PCP is a widespread environmental pollutant which has been detected in the air, freshwater, fish and different foods. It mainly partitions into soil and the food chain when released to the environment. Commonly, pentachlorophenol occurs in the air as a vapor or aerosol which is directly emitted by evaporation from surface water, surface soil and treated-wood materials [USEPA, 1979b; Geyer *et al*., 1986; WHO, 1987]. Data on ambient air levels for 2,4,5-T_3CP, 2,4,6-T_3CP and 2,3,4,6-T_4CP are limited. High levels of PCP in surface and groundwater have been extensively reported, particularly around wood treatment sites [WHO, 1987; IARC, 1991].

Almost all PCP (96.5%) released to the environment may partition into the soil, arising from direct soil application, volatilization, and wet deposition of atmospheric PCP [Hattemer-Frey and Travis, 1989]. PCP may be readily adsorbed to sediment [WHO, 1987; IARC, 1991]. In principle, biodegradation occurs rapidly in soil under aerobic and anaerobic conditions for PCP and other chlorophenol compounds [ATSDR, 1989c; WHO, 1989a; Lamar and Dietrich, 1990; Seech *et al*., 1990]. Roots of plants, rather than the upper part of plants, may absorb and accumulate PCP [Hattemer-Frey and Travis, 1989]. High PCP levels have been found in certain foodstuffs: grains, vegetables and fruits [Agriculture Canada, 1989]. Low levels of 2,4,5-T_3CP and 2,4,6-T_3CP are typically detected in fish, drinking water and corn or pea seedlings [IARC, 1979; WHO, 1989a; ATSDR, 1990b].

Toxic effects caused by PCP consist of skin injury, liver and renal function alterations, and central nervous disorders [IARC, 1991]. Adverse human health effects caused by 2,4,5-T_3CP, 2,4,6-T_3CP and 2,3,4,6-T_4CP are not well known. Because PCP has been associated with the occurrence of hepatic cancer in mice by ingestion, it has been classified as a possible human carcinogen [McConnell, 1989; IARC, 1991]. 2,4,5-T_3CP and 2,4,6-T_3CP are observed to produce cancer in mice, but the data are currently inadequate to reach a conclusion on human carcinogenicity [IARC, 1979, 1987].

PCP is bioavailable through inhalation, ingestion and dermal exposure in humans and animals. In particular, the absorption of the chemical in the gastrointestinal tract is almost complete. Bioavailability via inhalation is also high. From *in vitro* studies, PCP demonstrated a relatively high permeability through human skin.

A. EXPOSURE ROUTES AND RECEPTORS

Humans are exposed to PCP through inhalation, ingestion and dermal contact. Inhalation and dermal contact are primary exposure routes for workers, while ingestion is a major pathway for the general population. PCP, 2,4,6-T_3CP and 2,3,4,6-T_4CP have been detected at high levels in the urine of exposed workers [Gilbert *et al.*, 1990; Pekari *et al.*, 1991]. The major sources come from wood treatment products and other materials used in the chemical industry [Exon, 1984]. Workers are primarily exposed to PCP by inhalation [Casarett *et al.*, 1969; Williams, 1982; ATSDR, 1990b]. Alternatively, they may have dermal contact with PCP during handling of chemical products [WHO, 1987, 1989a, 1989b; ATSDR, 1990b].

The general population may be exposed to PCP at low levels [Murphy *et al.*, 1983; Cline *et al.*, 1989; Hill *et al.*, 1989]. Ingestion of foods containing PCP is a primary exposure route [Hattemer-Frey and Travis, 1989]. Ingestion of drinking water and soil appears to be less significant because of the relatively low levels of PCP found in water [WHO, 1989b]. Because increased levels of PCP in ambient air have been found around waste sites or urban areas, the general population has a potential exposure to PCP via inhalation indoors and outdoors. However, general population exposure by dermal contact is likely insignificant. General population exposure to 2,4,5-T_3CP, 2,4,6-T_3CP and 2,3,4,6-T_4CP by ingestion of food and drinking water may be significant [NRCC, 1982; NHW, 1988; WHO, 1989a; ATSDR, 1990b].

B. TOXICOKINETICS (FOLLOWING ABSORPTION)

1. Distribution

Once absorbed, PCP is distributed in various organs and tissues. PCP is highly protein-bound in plasma [Braun *et al.*, 1977; Uhl *et al.*, 1986]. Free PCP in plasma will reflect a recent exposure [ACGIH, 1991]. High levels of PCP have been found in the liver, kidney and brain in humans, and small amounts have been found in fat, bone and muscle [Exon, 1984; WHO, 1987; IARC, 1991]. 2,3,4,6-T_4CP has been found in human liver tissue [Mussalo-Rauhamaa *et al.*, 1989].

2. Metabolism

Studies from humans and animals have showed that PCP, 2,4,5-T_3CP, 2,4,6-T_3CP and 2,3,4,6-T_4CP were readily metabolized [WHO, 1989a; Reigner *et al.,* 1992a]. They tended to conjugate with glucuronic acid. To a lesser extent, a small portion of PCP and 2,3,4,6-T_4CP undergoes oxidation to tetrachloro-hydroquinone (TCHQ) [Ahlborg and Larsson, 1978; Edgerton *et al.*, 1979; Ahlborg and Thunberg, 1980; Juhl *et al.,* 1985].

The majority of PCP is protein bound in plasma. Both the unchanged forms and their metabolites are excreted primarily in urine, and to a lesser extent in feces. A review of human studies showed the percentage of total pentachlorophenol-glucuronide (PCPG) recovered in urine ranged from 14% to 98.5% [Reigner *et al.*, 1992a]. This extreme variation has been attributed to the use of different analytical methods for measuring PCP.

C. BIOAVAILABILITY

1. Inhalation

PCP has been observed to be absorbed in humans during occupational studies [Casarett *et al.,* 1969; Jones *et al.*, 1986]. Casarett *et al.* [1969] investigated the extent of absorption of PCP vapor following inhalation by two volunteers. Estimation of the bioavailability factor was based on the air concentration of PCP inhaled, the respiratory rate and the tidal volume with a correction of daily urinary excretion of the chemical. After dosing PCP at an air concentration of 230-432 ng/m^3, the bioavailability factor was 76% at 5 d post-exposure and 88% at 7 d post-exposure in the two subjects, respectively.

Absorption of PCP via inhalation occurred in animals [Hoben *et al.*, 1976]. Rats inhaled sodium-pentachlorophenate (Na-PCP) aerosols at a single dose of 12.5 mg for 20 min. About 70-75% of the chemical was absorbed over a 24 h period of exposure, according to measures of the amount of PCP in the lung, liver, plasma and urine. Even though the rats received a continuous exposure for 5 d, the total absorption was close to that from a single exposure. No information was available with regard to the respiratory bioavailability for 2,4,5-T_3CP, 2,4,6-T_3CP and 2,3,4,6-T_4CP in humans and animals.

2. Ingestion

a. Humans

Evidence indicating the absorption of PCP in humans via ingestion exposure has been reported [Braun *et al.*, 1979; Uhl *et al.*, 1986]. The toxicokinetics of PCP were examined in four human volunteers [Braun *et al.*, 1979]. After oral administration of sodium pentachlorophenate at a single dose of 0.1 mg/kg, absorption took place rapidly with an absorption half-life of 1.3 h and an absorption rate constant of 1.16 h^{-1}. According to the extent of excretion of PCP in urine (74% of PCP and 12% of the conjugated PCP) and in feces (2% of both forms), the bioavailability factor was estimated to range from 74% to 98%.

In another study, the half-life for absorption after oral administration of PCP was 16 d in human volunteers [Uhl *et al.*, 1986]. Data on oral bioavailability in humans for 2,4,5-T_3CP, 2,4,6-T_3CP and 2,3,4,6-T_4CP have not been reported.

b. Animals

A number of studies showed that PCP was well-absorbed in various animals. The results from studies revealed a high degree of absorption for PCP following oral administration, regardless of the species (monkeys, rats and mice), vehicles (e.g. corn oil, diet and water) or different doses [Braun and Sauerhoff, 1976; Braun *et al.*, 1977; Meerman *et al.*, 1983; Reigner *et al.*, 1991, 1992b]. Rhesus monkeys were orally administered a corn oil containing PCP at a single dose of 10 mg/kg [Braun and Sauerhoff, 1976]. About 70% of unchanged PCP was present in urine and 11% of the chemical was detected in tissues and carcass. Approximately 18% of PCP in feces was in the unchanged form. As a result, at least 81% of PCP was estimated to be absorbed. Similarly, the value in rats was 82% at a dose of 10 mg/kg and 64% at a dose of 100 mg/kg, using the same experimental methods as in the monkey [Braun *et al.*, 1977]. In another three studies [Meerman *et al.*, 1983; Reigner *et al.*, 1991; 1992b], bioavailability was determined by comparison of the plasma concentration-time curve (AUC) via oral and intravenous administration. After receiving the PCP-containing water or diet, rats absorbed about 90% of the total administered dose over a 24 h post-exposure period [Meerman *et al.*, 1983]. Reigner *et al.* [1991, 1992b] reported the oral bioavailability factors ranging from 91% to 97% in rats and 106% in mice.

Two studies revealed that over 80% of 2,4,6-T_3CP was absorbed in rats after oral administration [Korte, 1978; Bahig *et al.*, 1981]. The upper bound of values (from fecal excretion) ranged from 88% to 93%, while the lower bound values were between 82% to 94%.

3. Dermal

Few quantitative results of dermal bioavailability for PCP are available. Dermal absorption for PCP varies with the carrier vehicles. For *in vitro* human studies, the bioavailability factors were 16% with an aqueous vehicle, 62% with a diesel vehicle, 8% to 10% with an acetone vehicle and 15% to 19% with a soil vehicle [Horstman *et al.*, 1989; Wester *et al.*, 1993]. The result from an *in vivo* study in monkey revealed high bioavailability [Wester *et al.*, 1993b]. With PCP-containing soil and acetone contact to abdominal skin, 25% and 29% of the compound was absorbed into the systemic circulation over 14 d of exposure.

For *in vitro* studies, the permeability of 2,4,6-T_3CP through human skin was 5,940,000 cm/h and for hairless mouse skin was 174,000 cm/h at pH 5, 870 cm/h at pH 6 and 41,000 cm/h at pH 7.4 [Roberts *et al.*, 1977; Huq *et al.*, 1986].

4. Summary

PCP can be bioavailable through all three exposure routes (Table 5-33, 5-34, 5-35, and Figure 5-14). PCP is readily and rapidly absorbed via inhalation and ingestion in humans. Dermal bioavailability from *in vitro*

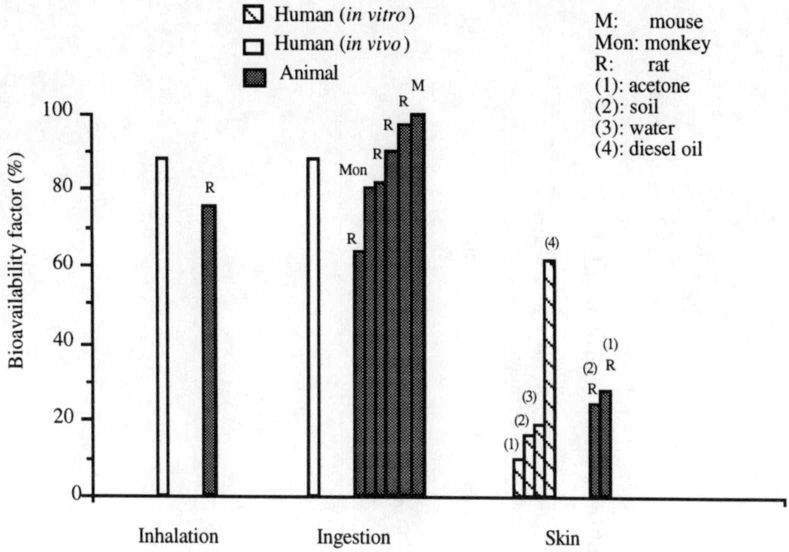

Figure 5-14 Bioavailability factors for pentachlorophenol in humans and animals by three exposure routes [Adapted from sources in Table 5-33: a, b (inhalation); Table 5-34: a-f (ingestion); and Table 5-35: a, b (skin)]

human studies has been shown to be relatively high, depending on the types of vehicles.

2,4,6-T_3CP can be absorbed from the gastrointestinal tract and skin. Data on bioavailability were not available for any exposure routes for 2,4,5-T_3CP and 2,3,4,6-T_4CP.

D. UNCERTAINTIES

Humans may be exposed to PCP vapor or aerosols. Exposure to aerosols may be significant for residents around PCP contaminated sites because of the persistence of PCP in soil. Knowledge of particle size distribution and inhalable fraction for PCP is limited. For respiratory bioavailability, the results were derived from two subjects who were exposed to PCP vapor. Nevertheless, data on deposition, retention and clearance of PCP aerosols are lacking in both humans and animals. Respiratory bioavailability for 2,4,5-T_3CP, 2,4,6-T_3CP and 2,3,4,6-T_4CP is also not clear, although these chemicals are similar to PCP with respect to physicochemical properties.

The general population may be readily exposed to polychlorinated phenol compounds via ingestion of food, soil and drinking water. The oral bioavailability in humans has been estimated from urinary and fecal excretion of PCP in four subjects, a small experimental cohort. Because most values are

Table 5-33 Respiratory Bioavailability of Pentachlorophenol in Humans and Animals

Form	Species	Exposure methods	PCP dosage	Experimental method	Exposure duration	Post-exp duration	Bioavailability factor (%)	Ref.
Vapor	Human	Whole-body	230-432 ng/m^3	Exhalation and urine measurement	45 min	5 d 7 d	76 88	a
Na-PCP Aerosol	Rat	Whole-body	Single: 12.5 mg (0.1 mg/kg)	Mass-balance	20 min	24 h	70-75	b

Adapted from: a. Casarett *et al.*, 1969; b. Hoben *et al.*, 1976

Table 5-34 Oral Bioavailability of PCP and 2,4,6-T_3CP in Humans and Animals

Compound	Species	PCP carrier	Exposure methods	PCP dosage	Experimental method	Exposure duration	Post-exp duration	Bioavailability factor (%)	Ref.
Na-PCP	Human	-	Oral	Single: 0.1 mg/kg	Exhalation, blood, urinary and fecal excretion	-	8 d	98[1] 88[2] 74[3]	a
PCP	Monkey	Corn oil	Oral	Single: 10 mg/kg	Mass-balance	-	15 d	81 (88)[1]	b
PCP	Rat	Corn oil	Oral	10 mg/kg 100 mg/kg	Mass-balance	-	7 d	82 64	c
PCP and Na-PCP	Rat	Diet Water	Oral	350 ppm in diet 288 mg/L in water	Blood measurement	7 d	24 h	90	d
PCP	Rat	-	Oral	Single: 2.5 mg/kg	Blood, urinary and fecal excretion	-	48 h	91-97	e
PCP	Mouse	-	Oral	Single: 15 mg/kg	Blood, urinary and fecal excretion	-	36 h	106[4]	f
2,4,6-T_3CP	Rat	Diet	Oral	1 ppm in diet	Urinary and fecal excretion	15 d	3 d	94[2] 93[3]	g
2,4,6-T_3CP	Rat	Diet	Oral	1 mg/kg in diet	Mass-balance	3 d	5 d	88[2] 82[3]	h

[1] values from fecal excretion data that are considered as the upper bound of the bioavailability factor; [2] values from urinary excretion data, including the unaltered form and its metabolites; [3] values are the fraction of the unaltered form in urine; [4] this value is adapted from original paper.
Adapted from: a. Braun *et al.*, 1979; b. Braun and Sauerhoff, 1976; c. Braun *et al.*, 1977; d. Meerman *et al.*, 1983; e. Reigner *et al.*, 1991; f. Reigner *et al.*, 1992b; g. Bahig *et al.*, 1981; h. Korte, 1978

Table 5-35 Dermal Bioavailability of PCP and 2,4,6-T_3CP in Humans and Animals

Compound	Species	PCP carrier	Exposure region	Skin area	PCP application	Experimental method	Exposure duration	Permeability ($x\ 10^3$ cm/h)	BF (%)[1]	Ref.
PCP	Human	Liquid Diesel oil	Abdominal cadaver	1.0 cm^2	435 μg 1,540 μg	Diffusion cell (*in vitro*)	24 h		16 62	a
PCP	Human	Soil Acetone	Cadaver	1.0 cm^2	0.07 μg/cm^2 0.8 μg/cm^2	Diffusion cell (*in vitro*)	15 h		15-19 8-11	b
PCP	Monkey	Soil Acetone	Abdominal skin	12 cm^2	0.07 μg/cm^2 0.8 μg/cm^2	Urinary excretion (*in vivo*)	14 d		25 29	b
2,4,6-T_3CP	Human	Solution	Abdominal cadaver	2.5 cm^2	-	Diffusion cell (*in vitro*)	-	$9.9\ x\ 10^4$ [cm/min]		c
2,4,6-T_3CP	Hairless mouse	Solution	Abdominal skin	-	0.05 g/ 100 ml	Diffusion cell (*in vitro*)	-	174 (pH 5) 87 (pH 6) 41 (pH7.4)		d

[1] BF = Bioavailability Factor

Adapted from: a. Horstman *et al.*, 1989; b. Wester *et al.*, 1993; c. Roberts *et al.*, 1977; d. Huq *et al.*, 1986

derived from cumulative urinary and fecal excretion data in animals, without quantification of their metabolites, these values are relatively unreliable.

A few quantitative studies evaluated dermal exposure for these compounds. Dermal bioavailability in humans was determined by *in vitro* methods. There is not good agreement of these values between *in vitro* and *in vivo* experiments.

Humans may be generally exposed to mixtures of polychlorinated phenol compounds including PCP rather than pure PCP in the environment. These contaminants may alter the toxic profiles which are produced by pure PCP. For instance, dioxins and dibenzofurans are inducers of hepatic microsomal enzymes, some of which may enhance detoxification of PCP, so that PCP toxicity will be reduced [ATSDR, 1989c]. However, toxic effects arising from exposure to mixtures of polychlorinated phenol compounds including PCP may be attributed to other contaminants (e.g. dioxins and dibenzofurans) besides pure PCP. Bioavailability for technical-grade PCP has not been investigated.

E. CONCLUSIONS

Humans can be exposed to PCP through inhalation, ingestion and dermal contact. Inhalation and dermal contact are primary exposure routes for workers, while exposure via ingestion is the primary route for the general population. Residents living around contaminated sites or in houses which have used PCP-treated wood may be subjected to higher levels of PCP by inhalation. Also, dermal contact with PCP-containing materials is a potential exposure route for the general population.

PCP is readily bioavailable via three exposure routes in humans and animals. The bioavailability factor for inhalation of PCP vapor is 88% over 7 d post-exposure in humans. For PCP aerosols, the bioavailability factor ranges from 70% to 75% in small rodents. Oral bioavailability in humans is estimated to be between 74% to 98% by indirect measurement. However, PCP and 2,4,6-T_3CP in small rodents via ingestion are completely absorbed, being 64% to 100% for PCP and 82% to 94% for 2,4,6-T_3CP. The dermal bioavailability depends on the carrier vehicles. By using *in vitro* methods, about 62% of PCP in diesel oil, 8% to 10% in acetone, 15% to 19% in soil and 16% in liquid form can penetrate human cadaver skin. An *in vivo* study indicates that the dermal bioavailability in monkey is 25% in soil and 29% in acetone. The permeability of 2,4,6-T_3CP through humans and hairless mouse skin is 9.9×10^4 cm/min and 2.9×10^3 cm/min, respectively.

Oral bioavailability for PCP appears to be only slightly affected by species differences, carrier vehicles, exposure levels and exposure duration. Absorption of PCP via ingestion is generally very high. Thus, exploration of methods to reduce absorption of PCP from the gastrointestinal tract can mitigate accidental exposures to PCP. For instance, an agent, cholestryramine, may enhance fecal excretion of PCP in monkeys, and may be recommended for use in cases of human overexposure to PCP [ATSDR, 1989c].

Dermal bioavailability was found to be strongly influenced by different carrier vehicles and pH conditions *in vitro*. Diesel oil and a low pH (acidic condition) will enhance dermal absorption. Therefore, control of these factors may reduce dermal absorption for PCP.

F. RECOMMENDATIONS

For residents living around PCP contaminated sites, inhalation of PCP aerosols may be an important source of exposure. Data relevant to the bioavailability of PCP aerosols in humans and animals are critical for completing this assessment. These include inhalable fraction in air, particle size distribution, dissolution rate of soil-bound PCP, fractions of deposition, retention and mechanical clearance. For 2,4,5-T_3CP, 2,4,6-T_3CP, and 2,3,4,6-T_4CP, a study on respiratory bioavailability is needed.

Information on oral bioavailability for soil-bound PCP and its dissolution rate in the gastrointestinal tract are required. An accurate assessment of oral bioavailability requires the use of blood measurements and the mass-balance technique. A study on the correlation between *in vivo* and *in vitro* dermal bioavailability is needed.

Bioavailability of technical-grade PCP by the main exposure routes and interaction among these contaminants needs to be investigated.

XII. PHENOL

Phenol is a colorless crystalline solid at room temperature which is soluble in water and most organic solvents and which exhibits an aromatic odour [Baich and Davist, 1981; Ellenhorn and Barceloux, 1988; Sittig, 1991]. Phenol may be present in surface water, groundwater, soil and, to a lesser extent, air from natural and anthropogenic sources [Bruce *et al.,* 1987; Howard, 1989a]. Phenol exists in air as a vapor which is rapidly removed by photochemical reaction [Howard, 1989a]. In surface water and soil, phenol is primarily biodegraded by microorganisms quickly. Because phenol has a low adsorptivity to soil and a high mobility through the soil, it tends to leach to groundwater [ATSDR, 1989d]. Phenol has been measured in several foods, such as smoked summer sausage, fried bacon, fried chicken and cheese [Howard, 1989a]. Toxic effects caused by phenol include central nervous system depression with coma and respiratory arrest, diarrhea, sore throat, and chemical burns to skin [Ellenhorn and Barceloux, 1988; ATSDR, 1989d]. No evidence has been found for carcinogenicity of phenol in humans and animals.

A. EXPOSURE ROUTES AND RECEPTORS

Workers and residents living in the vicinity of contaminated areas may be exposed to phenol vapor. However, information on air concentrations of phenol and the importance of this exposure route is limited.

For the general population, ingestion of foods and drinking water which are contaminated by phenol, or phenol products, may be a primary exposure pathway [ATSDR, 1989d]. Also, dermal contact with waterborne phenol and phenol products might be significant if exposure concentrations were high. Because of the characteristics of phenol in soil, inhalation of airborne phenol particulates and ingestion or dermal contact with the soil-bound phenol are unlikely to be significant exposure routes.

B. TOXICOKINETICS (FOLLOWING ABSORPTION)

1. Distribution

Phenol absorbed via ingestion was found to be rapidly distributed in the liver, lung, blood, brain, kidney, heart, spleen, adrenal and thyroid glands in animals [Deichmann, 1944; Liao and Oehme, 1982].

2. Metabolism

Most phenol can be conjugated with various acids to form phenylsulfate, phenylglucuronide, 1,4-dihydroxybenzene sulfate and 1,4-dihydroxybenzene glucuronide in the liver or other organs [ATSDR, 1989d]. A small portion of phenol is oxidized to carbon dioxide, water, quinol (1,4-dihydroxybenzene) and catechol.

Phenol is mainly excreted in urine as sulfate or glucuronide conjugate forms with a small fraction in the unaltered form [Deichmann and Keplinger,

1981]. However, phenol in urine is also likely to arise from the metabolism of common aromatic amino acids (e.g. phenylalanine).

C. BIOAVAILABILITY

1. Inhalation

Phenol was found to be absorbed via inhalation in humans [Piotrowski, 1971; Ohtsuji and Ikeda, 1972]. Human volunteers inhaled phenol vapor at concentrations of 6-20 mg/m^3 for 8 h [Piotrowski, 1971]. The mean retention fraction of phenol within the respiratory tract was 70%. About 84% to 99% of the retained phenol was eliminated in urine. Hence, the respiratory bioavailability factor was estimated to be 60% to 70% for humans. Qualitative and quantitative information on the respiratory bioavailability of phenol is lacking for animals.

2. Ingestion

Phenol can be absorbed in the gastrointestinal tract of humans and various animal species [Capel *et al.*, 1972; Kao and Bridges, 1979; Edwards *et al.*, 1986]. Aqueous phenol was orally administered to humans (0.01 mg/kg), rhesus monkeys (50 mg/kg), and small rodents (25 mg/kg). Using cumulative urinary excretion measurement, the oral bioavailability factor was 90% in humans, 44% in rhesus monkeys, and 50% to 95% in small rodents after 24 h exposure. Kao and Bridges [1979] reported that only 0.5% of phenol which was orally administered to rats at a dose of 25 mg/kg was eliminated in feces, with 97% in urinary excretion during 24 h post-exposure. In another study, an oral bioavailability factor of 80% was determined in rats following oral administration of a dose of 1.2 mg/kg phenol [Edwards *et al.*, 1986].

The foregoing bioavailability factor includes both the fractions of the parent phenol and its metabolites. Commonly, unaltered phenol accounts for a very small part of the urinary excretion [Cassidy and Houston, 1979].

3. Dermal

Piotrowski [1971] studied the dermal absorption of phenol in humans. Following whole-body exposure to phenol vapor, humans were found to absorb the vapor at a constant rate of 0.35 m^3/h (this means that the quantity of phenol contained in 0.35 m^3 of air will penetrate the skin each hour). Thus, the degree of absorption of phenol vapor may be calculated by the product of the absorption rate constant and the phenol concentrations in air. Phenol aqueous solution was applied to human forearm skin at various concentrations [Baranowska-Dutkiewicz, 1981b]. Dermal bioavailability factors ranged from 12% to 14% after 30 min exposure, and increased to 23% at one hour exposure. Feldmann and Maibach [1970] used phenol acetone solution applied on human forearm skin at a dose of 4 $\mu g/cm^2$. The dermal bioavailability factor was low (4.4%). Another study showed dermal bioavailability in humans to be about 24% [Bucks *et al.*, 1990].

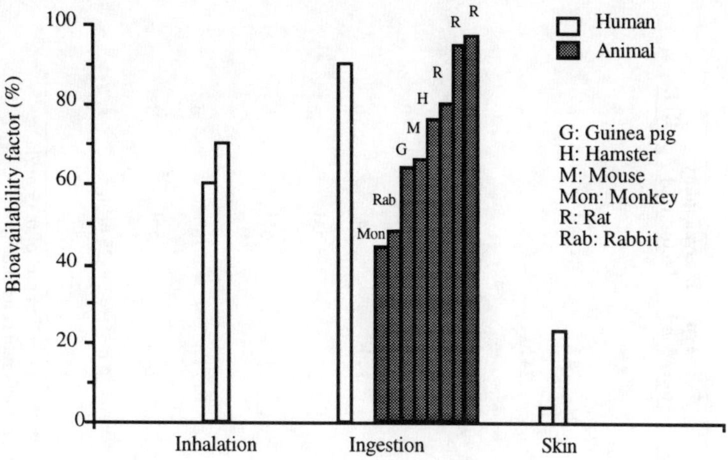

Figure 5-15 Bioavailability factors of phenol (*in vivo*) in humans and animals by three exposure routes (Adapted from sources in Table 5-36, a, b, c, d, f, g)

Using *in vitro* studies, the dermal bioavailability factor for humans was reported to be 11% to 19% [Franz, 1975; Hotchkiss *et al.*, 1992]. The *in vitro* values in rats were between 16% and 27% [Bronaugh and Stewart, 1985a, 1985b; Bronaugh *et al.*, 1982; Hotchkiss *et al.*, 1992].

The dermal permeability through human skin was $0.8x10^3$ cm/h [Roberts *et al.*, 1977]. The permeability in hairless mice and rats was higher than measured for humans, being $20x10^3$ cm/h and $11x10^3$ cm/h, respectively [Roberts and Anderson, 1975; Behl *et al.*, 1983; Jetzer *et al.*, 1986]. Because phenol can cause skin damage, high concentrations of phenol applied to the skin caused high dermal permeability in hairless mice [Behl *et al.*, 1983].

4. Summary

Phenol can be readily and rapidly absorbed through inhalation, ingestion and dermal exposure routes in humans and animals (Table 5-36, 5-37, and Figure 5-15). The dermal permeability is proportional to the applied phenol levels on skin.

D. UNCERTAINTIES

Bioavailability of phenol via three exposure routes has been well studied in humans and animals. Because phenol is primarily eliminated in urine, most *in vivo* values are determined by measurement of the cumulative urinary excretion. However, the majority of the excreted phenol occurs as metabolites in the urine. The significance of determining the parent compound or its metabolites relies on an adequate understanding of the mechanisms of phenol

Table 5-36 Bioavailability of Phenol in Humans and Animals (*in vivo*)

Form	Species	Exposure methods	Phenol dosage	Experimental method	Exposure duration	Post-exp duration	Bioavailability factor (%)	Ref.
Vapor	Human	Inhalation	6-20 mg/m^3	Urinary excretion	8 h	8 h	60-70	a
Aqueous	Human	Oral	0.01 mg/kg	Urinary excretion	-	24 h	90[1]	b
Aqueous	Rhesus monkey	Oral	50 mg/kg	Urinary excretion	-	24 h	44[1]	b
Aqueous	Rabbit	Oral	25 mg/kg	Urinary excretion	-	24 h	48[1]	b
Aqueous	Rat	Oral	25 mg/kg	Urinary excretion	-	24 h	95[1]	b
Aqueous	Mouse	Oral	25 mg/kg	Urinary excretion	-	24 h	64[1]	b
Aqueous	Hamster	Oral	25 mg/kg	Urinary excretion	-	24 h	76[1]	b
Aqueous	Guinea pig	Oral	25 mg/kg	Urinary excretion	-	24 h	64[1]	b
Aqueous	Rat	Oral	1.2 mg/kg	Urinary excretion	-	24 h	80	c
Aqueous	Rat	Oral	25 mg/kg	Urinary excretion	-	24 h	97[1]	d
Vapor	Human	Dermal (whole body)	5-25 mg/m^3	Urinary excretion	6 h	6 h	0.35 m^3/h [2]	a

[1] Data from cumulative urinary excretion including the unaltered form and metabolites.[2] Data from the absorption rate of 0.35 m^3/h. The extent of absorption via dermal per hour is calculated by the product of the air phenol levels and this rate constant.

Adapted from: a. Piotrowski, 1971; b. Capel *et al.*, 1972; c. Edwards *et al.*, 1986; d. Kao and Bridges, 1979

(Continued)

Form	Species	Exposure methods	Phenol dosage	Experimental method	Exposure duration	Post-exp duration	Bioavailability factor (%)	Ref.
Aqueous (water)	Human	Dermal (forearm)	0.3 mg/cm^2	Surface disappearance	-	1 h	23	e
						0.5 h	13	
			0.6 mg/cm^2			0.5 h	14	
			1.25 mg/cm^2			0.5 h	12	
Phenol in liquid acetone	Human	Dermal (forearm)	4 µg/cm^2	Urinary excretion	24 h	5 d	4.4	f
Aqueous	Human	Dermal	-	-	-	-	24	g

Adapted from: e. Baranowska-Dutkiewicz, 1981b; f. Feldmann and Maibach, 1970; g. Bucks *et al.*, 1990

Table 5-37 Dermal Bioavailability of Phenol in Humans and Animals (*in vitro*)

Form	Species	Phenol carrier	Exposure region	Phenol concentration	Experimental method	Exposure duration	Permeability (x10^3 cm/h)	Bioavailability factor (%)	Ref.
Aqueous	Human	Pure	Abdominal cadaver	10 µl/cm^2	Diffusion cell	24 h		11	a
Aqueous	Human	Pure	Breast	1.3-2.7 µg/cm^2	Diffusion cell	72 h		19	b
Aqueous	Rat	Pure	Dorsal	1.3-2.7 µg/cm^2	Diffusion cell	73 h		27	b
Aqueous	Rat	Pure	-	-	Diffusion cell	-		16	c
Aqueous	Human	Pure	Abdominal cadaver	-	Diffusion cell	-	0.8		d
Aqueous	Hairless mouse	Pure	Dorsal and Abdominal	- 0.5 g/l 1.0 g/l 2.0 g/l 4.0 g/l 6.0 g/l	Diffusion cell	-	20.3 22.7 30.7 46.6 165.4 202.7		e
Aqueous	Hairless mouse	Pure	Abdominal	-	Diffusion cell	-	18.8		f
Aqueous	Rat	Pure	Dorsal	-	Diffusion cell	-	11.4		g

Adapted from a. Franz, 1975; b. Hotchkiss *et al.*, 1992; c. Bronaugh and Stewart, 1985a,1985b; Bronaugh *et al.*, 1982; d. Roberts *et al.*, 1977; e. Behl *et al.*, 1983; f. Jetzer *et al.*, 1986; g. Roberts and Anderson, 1975

toxicity which at present remain inadequately understood. Some *in vitro* studies indicate that the reactive metabolites of phenol (catechol, hydroxyquinol, p-benzoquinone and 4,4'-biphenol) may mediate phenol toxicity [ATSDR, 1989d].

E. CONCLUSIONS

Humans may be exposed to phenol through three exposure routes. Ingestion of the phenol-contaminated foods and drinking water could be a significant route for the general population. Exposure via inhalation of phenol vapor may be a concern for workers and to a much lesser extent to residents living near contaminated sites.

Phenol is bioavailable via all three exposure routes. The respiratory bioavailability factor is 60% to 70% for humans. Absorption of phenol from the gastrointestinal tract into the portal blood is almost complete (90%). Dermal bioavailability (*in vivo*) is reported in humans to be about 4.4% to 24% with the dermal permeability through human skin being 0.8×10^{3} cm/h.

F. RECOMMENDATIONS

Bioavailability of phenol has been extensively studied in humans. However, better understanding of the toxicity mechanism of phenol is still needed for reliable quantitative risk assessment.

XIII. ANILINE

Aniline is a volatile and oily liquid, at room temperature, which is very soluble in water and immiscible with most organic solvents [ACGIH 1991]. Aniline has been detected in air, surface water, groundwater, soil, sediment and some foods [IARC, 1982b; Howard, 1989b; ACGIH, 1991]. Aniline in air is degraded by photochemical reaction. In surface water, it is subject to biodegradation and photoxidation. A small part may be adsorbed to sediment. In addition to biodegradation, aniline released to soil tends to be slightly or moderately adsorbed to organic matter. Aniline has been found in fresh fruits, and vegetables.

Aniline is considered to be very acutely toxic to humans with a lethal dose of 50 to 500 mg/kg [Sittig, 1991]. The characteristic of acute aniline toxicity is blue skin arising from the formation of methemoglobin in the blood. Chronic toxic effects occur in the blood and on the skin. Evidence on the carcinogenicity of aniline in humans and animals is not sufficient [IARC, 1982b].

A. EXPOSURE ROUTES AND RECEPTORS

Occupational exposure to aniline via inhalation of vapor and dermal contact with vapor or liquid may be significant [Howard, 1989b; Sittig, 1991]. General populations may be exposed to aniline by ingestion of contaminated foods and drinking water. Potential risk from this exposure route is minor because of the low levels of aniline usually found in these media. People may be exposed to aniline vapor from smoking or from heavily contaminated air [IARC, 1982b]. Ingestion exposure of soil-bound aniline or dermal contact with aniline-containing soil is unlikely to be significant for the general population.

B. TOXICOKINETICS (FOLLOWING ABSORPTION)

1. Distribution

In animals, absorbed aniline is distributed throughout the body, with the highest levels in the liver, blood, kidney and intestine [IARC, 1982b]. The distribution of aniline in humans has not been documented.

2. Metabolism and Excretion

In humans, aniline is primarily metabolized in the liver by hydroxylation in the para-position of the aromatic ring and subsequent conjugation with sulfate and glucuronide to form the p-aminophenol conjugates [IARC, 1982b; ACGIH, 1991]. N-hydroxylation of aniline also occurs to form phenylhydroxylamine, resulting in the formation of methemoglobin.

Aniline is mainly excreted by urine. The extent of the urinary p-aminophenol has been found to be linearly proportional to the degree of absorption of aniline via inhalation and dermal exposure [Piotrowski, 1957; Dutkiewicz and Piotrowski, 1961]. Thus, levels of urinary p-aminophenol are used as an absorption index (35%) of aniline for humans [Piotrowski, 1957]. No phenyl-hydroxylamine has been observed in the urine for either humans or animals.

C. BIOAVAILABILITY

1. Inhalation

The retention and absorption of aniline in humans have been investigated [Dutkiewcz and Piotrowski, 1961; Dutkiewcz, 1962a, 1962b]. Following inhalation of aniline vapors at concentrations of 5-30 mg/m^3, about 90% of the inhaled aniline was retained within the respiratory tract. Because a very small portion of the inhaled aniline was mechanically cleared to the gastrointestinal tract, this value represented the respiratory bioavailability.

2. Ingestion

Aniline has been observed to be absorbed from the gastrointestinal tract in workers experiencing accidental ingestion [ACGIH, 1991]. Quantitative data are not available. Freitage *et al.* [1982] orally administered aniline liquid to rats at a dose of 1 mg/kg for 3 d. The oral bioavailability factor was estimated to be 62% at 7 d post-exposure using the mass-balance technique.

3. Dermal

Aniline vapor or liquid can be absorbed via contact with human skin [Piotrowski, 1957; Dutkiewcz and Piotrowski, 1961; Dutkiewicz, 1961, 1962; Baranowska-Dutkiewicz, 1982]. Because of the existence of a linear relationship between aniline absorption and urinary p-aminophenol, the absorption profiles of aniline were obtained based on the level of urinary p-aminophenol [Piotrowski, 1957]. The absorption rate of aniline liquid through the human skin was reported to be 0.18 to 0.72 mg/cm^2/h from the Piotrowski study [1957] and 2.0-4.6 mg/cm^2/h from the Baranowska-Dutkiewicz study [1982].

Susten *et al.* [1990] investigated the dermal bioavailability of aniline through hairless mouse skin. About 4.7% of aniline was absorbed through the applied skin.

4. Summary

Aniline can be bioavailable via inhalation, ingestion and dermal exposure routes in humans (Table 5-38). The respiratory bioavailability is high in humans. The dermal absorption rate may be used to calculate the extent of dermal bioavailability for humans. Quantitative data on oral bioavailability are limited for humans.

D. UNCERTAINTIES

Aniline is very toxic for humans. Occupational populations may be exposed to high levels of aniline in air. Exposure to aniline vapor would usually be insignificant for residents living near contaminated sites.

Table 5-38 Bioavailability of Aniline in Humans and Animals

Form	Species	Exposure methods	Aniline dosage	Experimental method	Exposure duration	Absorption rate	Bioavailability factor (%)	Ref.
Vapor	Human	Inhalation	5-30 mg/m^3	Exhalation	-		90	a
Liquid	Rat	Oral	1 mg/kg x 3 d	Mass-balance	7 d		62	b
Liquid	Rhesus monkey	Dermal	6.25 µl/cm^2	Mass-balance	4 h		4.7	c
Liquid	Human	Dermal (forearm)	0.25 ml in sponge patches	Surface disappearance	5 h	0.18-0.72 mg/cm^2/h		d
Liquid	Human	Dermal (forearm)	-	Urinary excretion[1]	24 h	2.0-4.6 mg/cm^2/h		e

[1] The value is calculated based on the levels of the urinary p-aminophenol and absorption index of 35% that is derived from Piotrowski's study (1957)
Adapted from: a. Dutkiewcz and Piotrowski,1961; and Dutkiewicz, 1961,1962; b. Freitage *et al.*, 1982; c. Susten *et al.*, 1990; d. Piotrowski, 1957; e. Baranowska-Dutkiewicz, 1982

The adsorptivity of aniline to soil depends on the type of soil and soil pH condition. However, data regarding the levels of soil-bound aniline are generally limited at contaminated sites. Evaluation of exposure routes requires specific information on the environmental profiles of aniline at contaminated sites.

Bioavailability has been studied for human exposure to aniline vapor or liquid through inhalation or skin contact. The quantitative oral bioavailability, respiratory bioavailability of airborne aniline particulates and dermal absorption of soil-bound aniline are essentially unknown.

E. CONCLUSIONS

Humans may be exposed to aniline via inhalation, ingestion and dermal contact. In general, occupational exposure to aniline vapor via inhalation and dermal contact is of concern. Potential exposure routes for the general population may be related to the ingestion of the aniline-contaminated foods and drinking water, dermal contact with aniline liquid and inhalation of aniline vapor from smoking.

Aniline can be absorbed by inhalation, ingestion and dermal contact in humans and animals. Respiratory bioavailability is high (90%) in humans. The absorption rate of aniline through the human skin ranged from 0.18 to 4.6 $mg/cm^2/h$. About 62% and 4.7% of aniline are bioavailable via oral and dermal exposure, respectively, in small rodents.

F. RECOMMENDATIONS

Data are needed for quantitative risk assessment of aniline including: oral bioavailability in humans and bioavailability for aniline aerosols via all three exposure routes.

Chapter 6
BIOAVAILABILITY OF XENOBIOTICS FROM CONTAMINATED SITES

Information on bioavailability for selected xenobiotics from contaminated sites has been compiled and summarized in Chapter 5 from a review of the literature. These bioavailability data result in a relatively high level of uncertainty because of limitations inherent in the way they were derived. These limitations include: different experimental methods, variable analytical capabilities, species differences, inadequate knowledge of the mode action of xenobiotics, alteration of bioavailability caused by exposure to mixtures, variable chemical forms and inadequate knowledge of exposure duration/frequency.

Because the reported values for bioavailability are subject to considerable variability and uncertainty, users of these data are cautioned to obtain the original publications for data they may intend to rely upon. This caution is necessary to be sure that adequate attention can be given to the applicability of the reported data in consideration of the influence of the factors described below.

I. IDENTIFICATION OF EXPOSURE ROUTES

Exposure to volatile compounds (such as benzene, toluene and creosote vapors) by inhalation is least likely to be significant for residents near contaminated sites because of the relatively short half-life of these vapors in air. General populations would face greater inhalation risks from airborne particulates that adsorb high concentrations of persistent xenobiotics such as cadmium, chromium (VI), lead, copper, hydrophobic PAHs, polychlorinated dioxins and, possibly, pentachlorophenol.

Water soluble compounds tend to easily partition into surface and ground waters. A hazard for the general population may arise from exposure to these xenobiotics via ingestion of contaminated drinking water. Xenobiotics which are water soluble to some practical extent include: chromium, arsenic, copper, lead, pentachlorophenol, 2,4,5-trichlorophenol, 2,4,6-trichlorophenol, 2,3,4,6-tetrachlorophenol, phenol, 2,4,6-trimethylphenol, aniline, quinoline and two or three ring PAHs.

Locally grown vegetables may also accumulate xenobiotics (e.g. cadmium, pentachlorophenol, polychlorinated dioxins, phenol, aniline and some PAHs) by deposition on leaves from vapors and aerosols or by absorption via roots from soil. Young children with pica behavior may be exposed to high levels of soil-bound xenobiotics, such as lead, pentachlorophenol, polychlorinated dioxins and some PAHs, because of their habit of ingesting soil.

Dermal exposure at contaminated sites may occur for soil-bound hydrophobic PAHs and to a lesser extent polychlorinated dioxins and pentachlorophenol. Occasionally, humans may be exposed to waterborne xenobiotics during swimming and bathing, but these exposures are relatively rare and unlikely to be significant for most populations.

Overall, an evaluation of the significance of human exposure to xenobiotics through these exposure routes requires specific information on levels of xenobiotics and their stability in environmental media. Also, absorption profiles of xenobiotics are likely to affect the importance of different human exposure routes. For example, chromium species possess a very low oral bioavailability in humans. As a result, although humans could ingest a high level of chromium, the health risk may be relatively low. Likewise, human exposure to lead via dermal contact would usually be insignificant because of the low dermal bioavailability of lead. Potential routes for the general population exposure to xenobiotics from contaminated sites are summarized in Table 6-1.

II. RELIABILITY AND VALIDITY OF THE BIOAVAILABILITY FACTORS

The variability of bioavailability factor estimates for selected xenobiotics are summarized in Table 6-2. A number of factors influence the reliability and validity of these estimates.

A. CHOICE OF EXPERIMENTAL METHODS

A variety of experimental methods are involved in bioavailability studies. Each methodology offers both strengths and limitations, which inevitably result in either overestimation or underestimation of true bioavailability factors for the conditions under assessment.

Although the use of blood measurement is generally reliable and accurate, it is not usually practical for humans. Measurement of cumulative excreta has been common for human studies. In animal studies, use of a mass-balance technique has been common. Data derived from an accurate mass balance are more reliable than are estimates from other experimental methods.

The most common method for determining the respiratory bioavailability in humans is the measurement of exhaled airborne compounds. Such values (e.g. bioavailability values for benzene, toluene and aniline) reflect the total mass of inhaled xenobiotics which are deposited within the respiratory tract. Using these values for estimating the respiratory bioavailability factors assumes complete absorption of the deposited compounds into the systemic circulation. Some bioavailability factors which are directly derived from the deposited dose (e.g. arsenic and lead) require the external dose to be corrected for the deposition fraction. Other bioavailability factors derived using the same methods, including a correction for urinary excretion or the deposition

Table 6-1 Potential Exposure Routes to Xenobiotics for the General Population

Compound	Inhalation aerosol /vapor	Ingestion food	Ingestion water	Ingestion soil	Dermal contact water	Dermal contact soil
Arsenic	++	++	+++	+	+	+
Cadmium	+++	+++	+	+	+	+
Chromium (VI)	+++	+	++	+	++	+
Copper	++	+	+++	+	+	+
Mercury	+	+++	++	+	++	++
Lead	+++	+++	+++	+++	+	+
TCDD	++	+++	+	++	+	++
PAH vapors[1]	+	++	++	+	++	+
Hydrophobic PAH[2]	+++	+++	+	+++	+	+++
Benzene	++	+	+	+	+	+
Toluene	++	+	+	+	+	+
Pentachlorophenol	++	+++	+	++	+	+
2,4,5-trichlorophenol	++	+++	++	++	+	+
2,4,6-trichlorophenol	++	+++	++	++	+	+
2,3,4,6-tetrachlorophenol	++	+++	+	++	+	+
Phenol	-	++	++	+	+	+
Aniline	+	+	+	+	+	+

[1] PAH vapors include naphthalene, 2-methylnaphthalene, acenaphthene and fluorene.
[2] Hydrophobic PAHs include anthracene, phenanthrene, fluoranthene, pyrene, chrysene, benzo[a]pyrene, benzo[a]anthracene, benzo[b]fluoranthene, benzo[k]fluoranthene, benzo[g,h,i]perylene, dibenz[a,h]anthracene, and indeno[1,2,3-c,d]pyrene.
Significance of exposure:
- data are not available; + low; ++ moderate potential; +++ significant potential

Table 6-2 Summary of Bioavailability Factors (%), or Permeability in the Case of Dermal Contact, for Selected Xenobiotics

Compound	Inhalation		Ingestion		Dermal contact	
	human	animal	human	animal	human	animal
Arsenic	85-90[a]	70-80 [b,c] 50-75 [c,d] 90 [b,e]	50-100[b,c] 100 [b,e] 25-30 [d,e]	30-100[b,c] 15-25 [c,d] 65-90 [b,e]	2.71×10^{-6} (cm/h) 0.8-1.9[b,f]	2-6 [b]
Cadmium	30-95	25-70	2-9	1-7	0.01-0.6	0.2-2
Chromium (VI)	+	20-50 [b] 2 [d]	10 [b,g]	2 [b] 10 [b,g]	1×10^{3}- 3.5×10^{3} (cm/h)	1-4
Chromium (III)	+	4-10	<1-2	<1-3 10 [g]	+	2
Copper	+	+	15-70	+	+	+
Mercury vapor	75-80	50-90	-	-	2	-
Mercury (inorganic)	-	-	7	2-25	0.26×10^{-5} (cm/h)	2-4.5
Mercury (organic)	-	-	95	100	-	2-6
Lead	50-80 [a,c] 55-75 [e]	+	40-55 [h,i] 4-13 [i,j] 30-70[g,j] 16-20 [k] 3-10 [l]	+	6.3 [e] 0.05-29 [c]	6.5 [e] 4.4 [c]
TCDD	-	100	90	50-85 45 [m]	40--50 [f] 10-15 [f,m]	20-40 1.3 [m]
Naphthalene	+	-	+	-	-	88-95
2-methylnaphthalene	-	-	-	-	-	8 [f,m]
Acenaphthene	-	-	-	70	+	-

for notes see page 220

Table 6-2 (Continued)

Compound	Inhalation		Ingestion		Dermal contact	
	human	animal	human	animal	human	animal
Carbazole	-	-	-	-	+	50 ^{m} 45 ^{f,m}
Fluorene	-	-	-	-	+	15 ^{f,m}
Phenanthrene	-	-	-	100	+	80 70-80 ^{f} 20 ^{m} 15 ^{f,m}
Anthracene	-	-	-	-	+	50 55 ^{f}
Fluoranthene	-	-	-	-	+	-
Pyrene	-	-	-	-	+	95 40-70 ^{f} 8 ^{m}
Chrysene	+	40-55	-	60-90	+	15 ^{m}
Benzo[a]pyrene	+	20-80	-	10-25	25-30 ^{f}	35-85 45-70 ^{f} 5-16 ^{m} 3 ^{f,m}
Benzo[a]anthracene	+	-	-	-	+	20 ^{m}
Benzo[b]-fluoranthene	+	-	-	-	-	20 ^{m}
Benzo[k]-fluoranthene	-	-	-	-	+	-
Benzo[g,h,i]perylene	-	-	-	-	+	-
Dibenz[a,h]-anthracene	+	-	-	-	+	8-80

for notes see page 220

Table 6-2 (Continued)

Compound	Inhalation human	Inhalation animal	Ingestion human	Ingestion animal	Dermal contact human	Dermal contact animal
Indeno[1,2,3-c,d]-pyrene	-	-	-	-	+	-
Benzene	50-60 [n]	+	+	90-100	<1	<1 20 [p]
Toluene	30-50 [n]	+	100	90-100	0.9 cm/h	+
Pentachlorophenol	90 [n]	70-75	75-100	65-100	10-60 [f]	25-30
2,4,5-trichlorophenol	-	-	-	-	-	-
2,4,6-trichlorophenol	-	-	-	80-95	$9.9x10^4$ (cm/h)	$4.1x10^3$ (cm/h)
2,3,4,6-tetrachlorophenol	-	-	-	-	-	-
Phenol	60-70 [n]	-	90	45-100	4-25	10-30
Aniline	90 [n]	-	+	60	+	5

- Data are not available; + Qualitative data available
[a] Data from the compounds deposited within the respiratory tract;
[b] Data from water soluble compounds;
[c] Inorganic compound;
[d] Data from slightly water soluble or water insoluble compounds;
[e] Organic compound;
[f] *in vitro* data;
[g] In fasting status;
[h] Data for children;
[i] In feeding status;
[j] Data for adults;
[k] With drinking water;
[l] With lead in diet;
[m] Data from individual compounds in the mixtures;
[n] Vapor;
[p] Application sites are covered by a glass cover to avoid volatilization of the compounds

fraction (e.g. for pentachlorophenol and phenol), are more accurate and reliable.

For oral bioavailability studies, a wide range of values has been observed. The upper bound values (by the measurement of cumulative fecal excretion data) represent the total mass of xenobiotics in the portal blood. The use of such values as the oral bioavailability factors makes an assumption that all the compounds in the portal blood will enter the systemic circulation without undergoing hepatic presystemic metabolism. Lower bound estimates (based on the measurement of cumulative urinary excretion) account for only a portion of the absorbed mass of the xenobiotics.

Some bioavailability estimates are based on data including both the parent xenobiotic and its metabolites. Other studies are incomplete because they fail to use a reference dose to account for the effects of metabolism and distribution of xenobiotics. In general, the oral bioavailability factors based strictly on the measurement of excreta data are unlikely to be reliable and accurate. True values are expected between the lower and upper bound of estimates. Consequently, this range of estimates only provides some insight into the absorption profiles for most xenobiotics.

In dermal studies, in addition to the bioavailability factor, the dermal permeability (occasionally, the absorption rate) is often determined. This parameter is suitable for assessing the internal dose from dermal exposure to vapors and waterborne xenobiotics. Estimates from surface disappearance techniques are generally higher than estimates from cumulative urinary data.

Both *in vivo* and *in vitro* techniques are commonly employed for dermal bioavailability studies and, to a lesser extent, for respiratory bioavailability studies. *In vitro* data may differ from *in vivo* data for most xenobiotics. In some cases, *in vitro* data may be used to predict *in vivo* data if a correlation has been established. For example, a good correlation of the dermal bioavailability factors between *in vivo* and *in vitro* estimates has been found for some PAHs (carbazole, phenanthrene, anthracene and benzo[a]pyrene).

Human *in vivo* values may be estimated from *in vitro* human data and from animal data based on *in vivo* and *in vitro* experiments. A relationship for dermal bioavailability has been proposed [USEPA, 1992b]:

$$\text{BF (Human } \textit{in vivo} \text{ estimate)} = \frac{\text{Human } \textit{in vitro} \text{ BF x Species A } \textit{in vivo} \text{ BF}}{\text{Species A } \textit{in vitro} \text{ BF}}$$

Application of this equation assumes that the ratio between *in vivo* and *in vitro* rates is the same for human and animal skin. For example, *in vitro* values of B[a]P in humans ranged from 24% to 31% [Storm *et al.*, 1990; Wester *et al.*, 1990] and the data in rats were 46% *in vivo* and 50% *in vitro* [Yang *et al.*, 1986b]. Thus, the *in vivo* dermal bioavailability factor for B[a]P in humans may be estimated to be from 22% to 29%.

B. ANALYTICAL CAPABILITIES

The reliability of bioavailability estimates is often restricted by limitations of the analytical methods. Current analytical techniques may not detect all relevant compounds (the unaltered form and the metabolites) in the biological fluids. Also, the treated biological samples may lose or change their original form. For instance, a wide variety of methylated arsenic compounds appear in urine samples depending upon different treatment and analytical methods for studies of arsenic exposure [Hopenhayn-Rich *et al.*, 1993].

C. SPECIES DIFFERENCES

Bioavailability studies of xenobiotics are now rarely carried out in humans for ethical reasons. Most bioavailability factors must be obtained from animal studies, particularly from small rodents. Extrapolation of data from animal to humans is necessary but cannot be absolutely reliable. For some xenobiotics (chromium, arsenic, toluene, pentachloro-phenol and phenol), the oral bioavailability in animals is close to estimates for humans. Dermal permeability for selected compounds in small rodents is often higher than that in humans. Unfortunately, reliable information is limited on the correlation of respiratory bioavailability between humans and animals for a wide range of xenobiotics.

D. MIXTURES AND VEHICLES

Most bioavailability studies are conducted in animals and humans by administration of a single pure xenobiotic. In reality, individuals are exposed to complex mixtures rather than pure chemicals. Exposure to these mixtures results in interaction effects which usually alter the bioavailability of each individual xenobiotic. Currently, there are no practical approaches for evaluating the effects of mixtures on bioavailability using bioavailability factors for each pure chemical. Direct determination of bioavailability for chemical mixtures, in reality, is necessary to properly understand the health risks posed by a mixture. Based on a few bioavailability studies for PAH-containing mixtures, each component in the mixture was found to exhibit a marked reduction of bioavailability in dermal exposure compared with studies of the pure PAH alone. Bioavailability will also be affected by environmental vehicles (soil, food, water and solvents).

E. MECHANISM OF TOXICITY

Understanding the mechanisms of toxicity for xenobiotics is necessary for the development of quantitative approaches for estimating health risk. A critical determinant of toxicity is the quantity of the active species which reacts at the target sites. The active species refers to either the parent xenobiotic or its reactive metabolites. Target sites can be the applied sites or remote sites where the active species may be delivered by the systemic circulation.

Bioavailability focuses on the quantity of parent xenobiotics present in the systemic circulation. When the active species are related to the parent compounds (e.g. inorganic arsenic), determination of bioavailability may provide good estimates for quantitative risk assessment. In some cases, the active species are associated with reactive metabolites (e.g. reactive metabolites of benzene and benzo[a]pyrene). In these cases, identification and quantification of the reactive metabolites or of the rate at which they are produced from the parent compound will be essential for reliable quantitative risk assessment. For these cases, the definition of bioavailability in toxicokinetics should be extended to include the total quantity of both the parent compounds and the biologically active metabolites in the systemic circulation, based on knowledge of the specific mechanism of toxicity for these selected xenobiotics.

Results from carcinogenicity testing of some xenobiotics suggest that tumors may be induced at the sites they were applied (such as skin and lung cancer produced by carcinogenic PAHs). In this context, data on the systemic bioavailability of such xenobiotics will not be useful for risk assessment of cancer at these sites.

Some quantitative risk assessment methods for evaluating skin cancers posed by dermal exposure to PAHs have inappropriately extrapolated dermal dose-response relationships from an oral dose-response relationship. This approach incorrectly assumes that dermal cancer risk can be primarily evaluated on the basis of the risk posed by reactive metabolites in the systemic circulation rather than at the site of application. However, dermal absorption may contribute to PAH loading in the systemic circulation and should not be neglected when evaluating total PAH exposure and non-inhalation cancer risk.

F. CHEMICAL FORMS

Bioavailability is often affected by the administration of different chemical forms, particularly metals. Generally, various forms of a xenobiotic (elemental form, oxidation states and their salts) will result in different absorption profiles. For example, none of the bioavailability values reported for metals in this book with exception of two oral bioavailability studies relevant to As_2O_5 [Charbonneau *et al.,* 1980; Hollins *et al.,* 1979] are related to the chemical forms known to be in the commonly used CCA-based wood preserving mixtures (CrO_3, As_2O_5, CuO). As a result, application of these values to quantitative risk assessment for CCA-based wood preserving mixtures should be limited.

G. EXPOSURE DURATION AND FREQUENCY

Exposure duration and frequency in bioavailability experiments are commonly shorter than those of humans exposed to actual environmental xenobiotics. This is certainly true where risk assessments attempt to evaluate cancer risks posed by exposure to chemicals over a lifetime. Some results are experimentally obtained for very short exposure periods, without determining whether the absorption has reached steady-state. In such cases, the bioavailability factors could be underestimated.

III. PRIORITY NEEDS FOR FURTHER RESEARCH

Currently, there is a lack of reliable information about bioavailability for most xenobiotics. This situation creates substantial uncertainty for quantitative risk assessment. Studies of bioavailability for individual xenobiotics would provide better understanding of the health risks posed by environmental exposure.

The general population is exposed to complex mixtures of xenobiotics within soil, food and water rather than individual pure xenobiotics. In most cases, bioavailability estimates of complex mixtures and xenobiotics within carriers differ from those of individual pure xenobiotics. Consequently, bioavailability studies of complex mixtures and xenobiotics within carriers should be pursued for high priority xenobiotics. Determination of priority needs for further research and the feasibility of bioavailability studies for xenobiotics depend on their physicochemical, environmental, biological and toxicological properties, primary exposure routes for the general population, feasible experimental designs and economic considerations (cost and time).

Respiratory bioavailability of chromium trioxide aerosols, hydrophobic PAHs aerosols, PAH-containing mixtures and PCP aerosols may be important and require further study. The reasons for singling out these compounds include: (1) chromium trioxide and hydrophobic PAHs aerosols are associated with an increased incidence of lung cancer in animals via inhalation; (2) carcinogenicity of chromium (VI) and PAH appears to be a localized rather than a systemic effect, so that characterization of the profiles of deposition and retention within the respiratory tract is important for these compounds; (3) PCP can produce systemic toxicity to humans via inhalation, and (4) the respiratory bioavailability of PAH-based mixtures may vary considerably from that of the individual components in these mixtures.

In vivo methods for studying respiratory bioavailability in animals are complex, expensive and time-consuming. Alternatively, there are some simple *in vitro* methods for determining the physicochemical and environmental properties of xenobiotic-containing aerosols. These data may be used to predict the respiratory bioavailability of xenobiotics. The first step is to determine the inhalable fraction (aerosol size less than 10 μm MMAD) and the particle size distribution in air. These parameters can be used to predict the deposition of aerosols within the respiratory tract. The second step is to estimate the fractions of mechanical clearance and retention of aerosols based on available experimental data or mathematical models. The third step is to determine the dissolution fraction or rate of aerosols under conditions that simulate the respiratory environment (pH 7.4). Finally, the absorption fraction may be roughly estimated by the product of the retention fraction and dissolution fraction, by assuming absorption of all the dissolved constituents into the systemic circulation. Also, it is necessary to use the octanol/water partition coefficients of the chemicals to estimate the extent that dissolved compounds may penetrate the air-blood barrier into the systemic circulation.

Oral bioavailability studies are needed for xenobiotics which are likely to show large differences in bioavailability for complex mixtures compared to individual xenobiotics. These xenobiotics consist of TCDD, soil-bound PCP, PCP-based mixtures, soil-bound PAHs, PAH-containing mixtures and soil-bound lead at specific contaminated sites. For example, exposure via ingestion of locally grown vegetables may be not important for local residents unless these locally grown vegetables accumulate significant levels of xenobiotics from contaminated sites and they also comprise a significant portion of individual diets.

Measurement of biological fluids and the mass-balance technique are the most common approaches for studying oral bioavailability. Compared to the mass-balance technique, measurement of biological fluids is simple, less expensive and less time-consuming. The experimental design for measuring biological fluids must use the reference dose (intravenous dose) to account for the effects of distribution and metabolism.

Soil-bound xenobiotics which reach absorption sites must dissolve in the biological fluid from the soil matrix prior to entering the systemic circulation. Thus, determination of the dissolution fraction of soil-bound xenobiotics under conditions simulating the gastrointestinal tract (e.g. pH 1-3 in the stomach and pH 7-8 in the intestine) may aid in better understanding and estimating oral bioavailability for soil-bound xenobiotics. Dissolution tests, either *in vivo* or *in vitro*, are very simple, inexpensive and less time-consuming as compared to full bioavailability experiments.

Dermal bioavailability studies which focus on PAH-containing mixtures from specific contaminated sites are needed because some PAHs are dermal carcinogens in animals and mixtures of these show marked differences in their dermal bioavailability compared to individual pure PAHs. Both *in vivo* and *in vitro* methods could be used for these types of studies.

Chapter 7

CONSIDERATIONS FOR APPLYING BIOAVAILABILITY TO HEALTH RISK ASSESSMENT

Bioavailability estimates are available in the literature for some individual contaminants. These data can be used for quantitative risk assessment to improve its relevance, reliability and accuracy. These bioavailability estimates are often dependent upon physicochemical and environmental properties of xenobiotics (specific chemical form, water solubility, physical states and inorganic or organic form), human or animal physiological conditions (age), nutritional conditions (deficiencies of some trace-element and vitamins and fasting or normal feeding status) and disease states (genetic diseases). Thus, selection of bioavailability estimates for calculation of internal doses or adjustment of toxicity values must consider these variables. The characteristics of xenobiotics which are employed for bioavailability studies should be similar and related to those found in the environment where humans may be exposed.

I. METALS

Bioavailability of toxic metals (arsenic, cadmium, chromium, copper, lead and mercury) via some exposure routes has been extensively investigated for humans and animals. Metal bioavailability is strongly affected by their chemical form, water solubility, age of humans or animals and nutritional conditions. The following generalizations can be made:

- Chromium (VI) compounds (anionic form) are more readily and rapidly absorbed than chromium (III) compounds (cationic form) via the main exposure routes.
- Inorganic arsenic and inorganic mercury compounds are less readily absorbed in the respiratory tract and gastrointestinal tract than organic arsenic and mercury compounds.
- Cadmium compounds are less readily absorbed in the gastrointestinal tract and through the skin than in the respiratory tract.
- Organic lead compounds will more easily penetrate human or animal skin than inorganic lead compounds.
- Water soluble cadmium and chromium or arsenic salts have higher bioavailability in the respiratory tract and gastrointestinal tract than the slightly water soluble or water insoluble compounds.
- Chromium (VI) in gastric fluid (acidic environment) is readily reduced to chromium (III) which will be less readily absorbed from the gastrointestinal tract.

- Children have a higher capacity for lead absorption from the gastrointestinal tract than adults.
- Fasting can enhance chromium (III) or (VI) and lead absorption from the gastrointestinal tract.
- Deficiencies of essential elements and some vitamins can increase cadmium, chromium and lead bioavailability in the gastrointestinal tract.
- Daily absorption of copper (an essential element for humans) in the gastrointestinal tract is regulated based on daily requirements for human metabolism.
- Some genetic diseases may enhance lead and copper absorption in humans.

II. ORGANIC XENOBIOTICS

Generally, bioavailability of organic compounds is strongly influenced by their physical states, water solubility, lipophilicity and form of exposure (pure compound, mixtures or within carriers). The following generalizations can be made:

- Organic vapors (benzene, toluene, PCP, phenol, aniline and PAH vapors) are readily absorbed after they are deposited within the respiratory tract.
- Pure organic liquids and solids (benzene, toluene, PCP, phenol, aniline and relatively water-soluble PAHs) are readily absorbed in the gastrointestinal tract and easily penetrate animal skin.
- Hydrophobic PAHs (e.g. B[a]P) have lower oral bioavailability than the more water-soluble PAHs.
- Dermal bioavailability of PAH-containing mixtures or soil-bound PAHs has been found to be dramatically lower than bioavailability for pure PAH compounds in animals.
- TCDD in acetone-corn oil or acetone is more readily absorbed in the gastrointestinal tract and through the skin than soil-bound TCDD exposure.

III. OTHER APPLICATION CONSIDERATIONS

In general, the internal dose is a refined parameter for assessing exposure and health effects. Based on its inherent properties, the application of internal dose for quantitative risk assessment has certain limitations. General considerations for using the bioavailability factor to adjust an external dose or a toxicity value to an internal dose basis include:

- Internal dose represents the quantity of a xenobiotic in the systemic circulation which may be transported to various organs and tissues to exert various toxic effects. By incorporating the bioavailability factor, the internal dose is applicable for xenobiotics that produce systemic toxicity or carcinogenicity rather than those that cause local toxic effects (i.e. at the external site of application).
- Use of a bioavailability factor for the adjustment of external dose is only essential when the toxicity data, such as reference doses or slope factors, are based on internal dose or will be better understood by seeking internal dose estimates for the toxicity data.
- The route-specific bioavailability factor for each exposure route is useful because there may be large differences in bioavailability by different exposure routes. For instance, the respiratory bioavailability factor must only be applied for assessing the internal dose via inhalation exposure.
- When assessing human health risks posed by exposure to xenobiotics in the environment, the xenobiotic characteristics modelled in the risk assessment should be similar to those from bioavailability studies and toxicity studies in order to be relevant. These characteristics include chemical form and environmental vehicles (media). For instance, when the risk assessment is conducted for a general population exposure to inorganic arsenic from drinking water via ingestion, the best choice of parameters for assessing risk are the toxicity value (RfD or slope factor) and bioavailability factor from the administration of inorganic arsenic in drinking water.
- When the xenobiotic characteristics modelled in the risk assessment are not similar with those from bioavailability studies and toxicity studies, the relative bioavailability factor can be a useful parameter for the adjustment of external dose. For example, when the risk assessment is conducted for general population exposure to lead from soil via ingestion, but the RfD is available for lead acetate in the diet, relative bioavailability of 20%, which is the ratio of the extent of absorption of lead in mining soil to that of lead acetate (a standard form) in the diet, could be used for adjustment.
- When toxicity values are not available for a xenobiotic for one of the three main exposure routes, route-to-route extrapolation may be necessary for quantitative risk assessment. For instance, dermal toxicity data are often unavailable for most xenobiotics. In this case, dermal

toxicity values may be estimated from oral or inhalation toxicity values. If the oral or inhalation values are based on external doses, these values need to be converted into internal doses by using their bioavailability factors (oral bioavailability or respiratory bioavailability). Thus, the internal dose can be related back to an external dermal dose by considering dermal bioavailability of the absolute bioavailability.

- The distribution of reported bioavailability factors is affected by a variety of factors, such as different designs and analytical methods in the bioavailability experiments, chemical forms, types of vehicles, exposure duration and frequency, species, age, gender and nutritional conditions.

Available information on bioavailability studies for the selected xenobiotics is summarized in Table 7-1. A checklist approach for application of bioavailability factors in health risk assessment is provided in Figure 7-1. This logic diagram seeks to provide a framework for organizing the information which will be encountered when trying to determine the applicability of bioavailability to health risk assessment.

IV. RISK MANAGEMENT CONSIDERATIONS

For prevention of toxicity caused by exposure to environmental contaminants, there are several options to reduce and, if possible, block absorption of xenobiotics by humans. Major risk management options, in order of decreasing utility and practicality, include:

- Reduction of xenobiotic levels in the environment. Reduce exposure to xenobiotics in the environment by blocking or eliminating exposure routes (e.g. removal or capping of buried contaminants, encasing of buried contaminants to protect groundwater, etc.).
- Identification of high risk groups. Individuals in high risk groups, such as young children, pregnant women, the elderly, heavy smokers and people with nutritional deficiencies and some genetic diseases, are more vulnerable to exposure to environmental contaminants. Thus, correction of nutritional deficiencies, improved hygienic behavior for young children, reduction of smoking and treatment of relevant diseases are important approaches for reducing absorption of certain xenobiotics into the human body.
- Use of antidotes. Using an antidote to block absorption or other steps relevant to toxicity can be an effective method for treating severe acute poisoning. This method is not feasible

for large populations who are chronically exposed to environmental contaminants at relatively low levels. In some cases, a specific antidote may be applied for very small groups of people at high risk to reduce absorption of specific xenobiotics by the human body, as a strictly reactive measure.

- Reduction of xenobiotic body burden. After absorption, xenobiotics are distributed to various organs and tissues. One way to minimize toxic effects to humans is to reduce the body burden. Chelation therapy for metal intoxication provides a means for reducing the body burden of toxic metals. Chelators can bind free metal ions to remove these toxic metals from the body. The most common chelating agents include dimercaprol (BAL) for lead and arsenic, EDTA for lead and penicillamine for copper and lead [Goyer, 1991]. However, chelation therapy is only suitable for acute metal intoxication and it must be practiced carefully by consulting appropriately specialized physicians; otherwise, chelation therapy itself can pose a direct toxic risk.

Table 7-1 Available Information on the Bioavailability Factors for Selected Xenobiotics

	Inhalation		Ingestion		Dermal contact	
Compound	**human**	**animal**	**human**	**animal**	**human**	**animal**
Arsenic (inorganic)	√	√	√	√	-	√
Arsenic (organic)	-	√	√	√	-	-
Cadmium	√	√	√	√	√	√
Chromium (VI)	-	√	√	√	-	√
Chromium (III)	-	√	√	√	-	√
Copper	-	-	√	-	-	-
Lead (inorganic)	√	-	√	-	√	√
Lead (organic)	√	-	-	-	√	√
Mercury (inorganic)	-	-	√	√	-	√
Mercury (organic)	-	-	√	√	-	√
TCDD	-	√	√	√	√	√
Naphthalene	-	-	-	-	-	√
2-methylnaphthalene	-	-	-	-	-	√
Acenaphthene	-	-	-	√	-	-
Carbazole	-	-	-	-	-	√
Fluorene	-	-	-	-	-	√
Phenanthrene	-	-	-	√	-	√
Anthracene	-	-	-	-	-	√
Fluoranthene	-	-	-	-	-	-

(Continued)

Compound	Inhalation		Ingestion		Dermal contact	
	human	animal	human	animal	human	animal
Pyrene	-	-	-	-	-	√
Chrysene	-	√	-	√	-	√
Benzo[a]pyrene	-	√	-	√	√	√
Benzo[a]-anthracene	-	-	-	-	-	√
Benzo[b]-fluoranthene	-	-	-	-	-	√
Benzo[k]-fluoranthene	-	-	-	-	-	-
Benzo[g,h,i]-perylene	-	-	-	-	-	-
Dibenz[a,h]-anthracene	-	-	-	-	-	√
Indeno[1,2,3-c,d]-pyrene	-	-	-	-	-	-
Benzene	√	-	-	√	√	√
Toluene	√	-	√	√	-	-
Pentachlorophenol	√	√	√	√	√	√
2,4,5-trichlorophenol	-	-	-	-	-	-
2,4,6-trichlorophenol	-	-	-	√	√	√
2,3,4,6-tetrachlorophenol	-	-	-	-	-	-
Phenol	√	-	√	√	√	√
Aniline	√	-	-	√	-	√

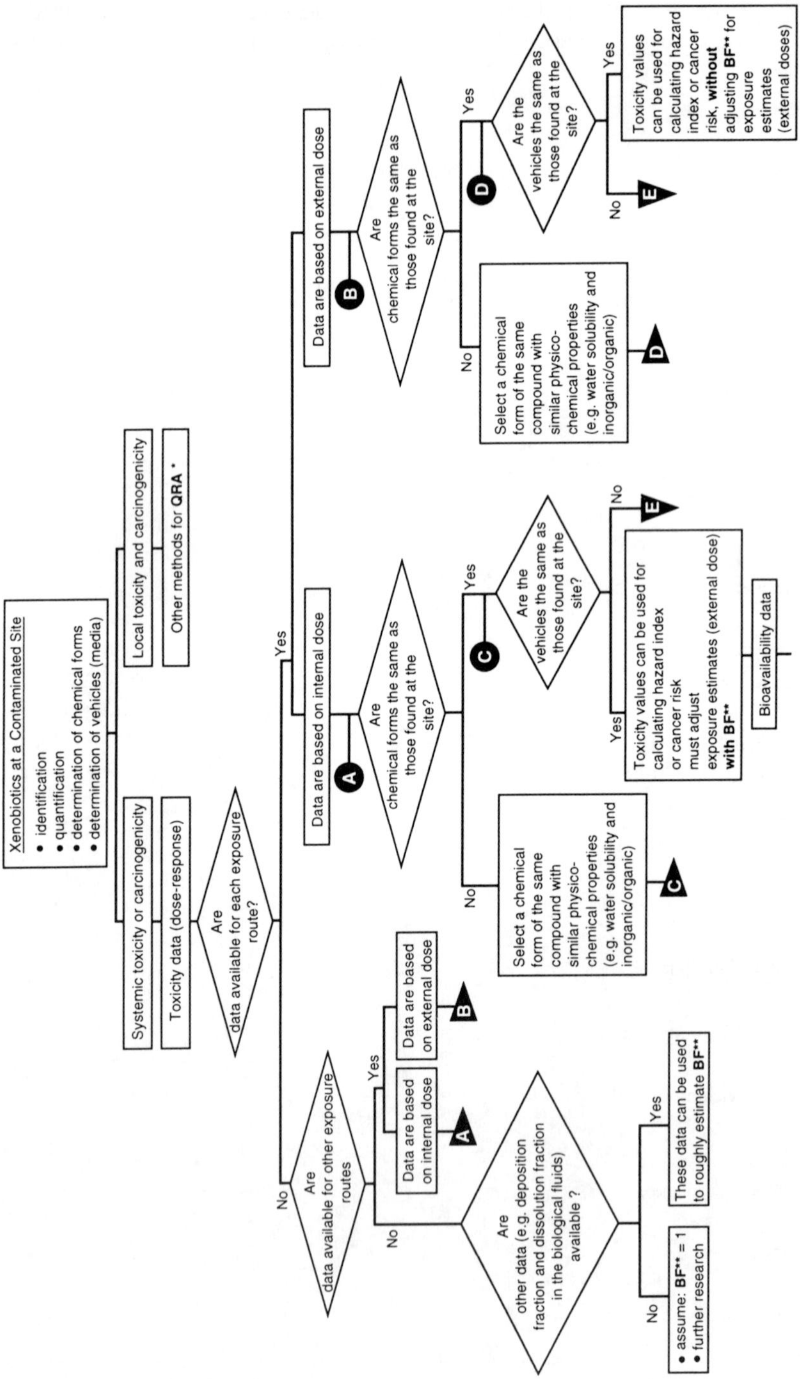
Xenobiotics at a Contaminated Site
• identification
• quantification
• determination of chemical forms
• determination of vehicles (media)
Systemic toxicity or carcinogenicity
Toxicity data (dose-response)
Local toxicity and carcinogenicity
Other methods for QRA *
Are data available for each exposure route?
Yes
No
Data are based on internal dose
Data are based on external dose
Are chemical forms the same as those found at the site?
Are the vehicles the same as those found at the site?
Select a chemical form of the same compound with similar physico-chemical properties (e.g. water solubility and inorganic/organic)
Toxicity values can be used for calculating hazard index or cancer risk must adjust exposure estimates (external dose) with BF**
Bioavailability data
Toxicity values can be used for calculating hazard index or cancer risk, without adjusting BF** for exposure estimates (external doses)
Are data available for other exposure routes
Data are based on internal dose
Data are based on external dose
Are other data (e.g. deposition fraction and dissolution fraction in the biological fluids) available ?
• assume: BF** = 1
• further research
These data can be used to roughly estimate BF**
A
B
C
D
E

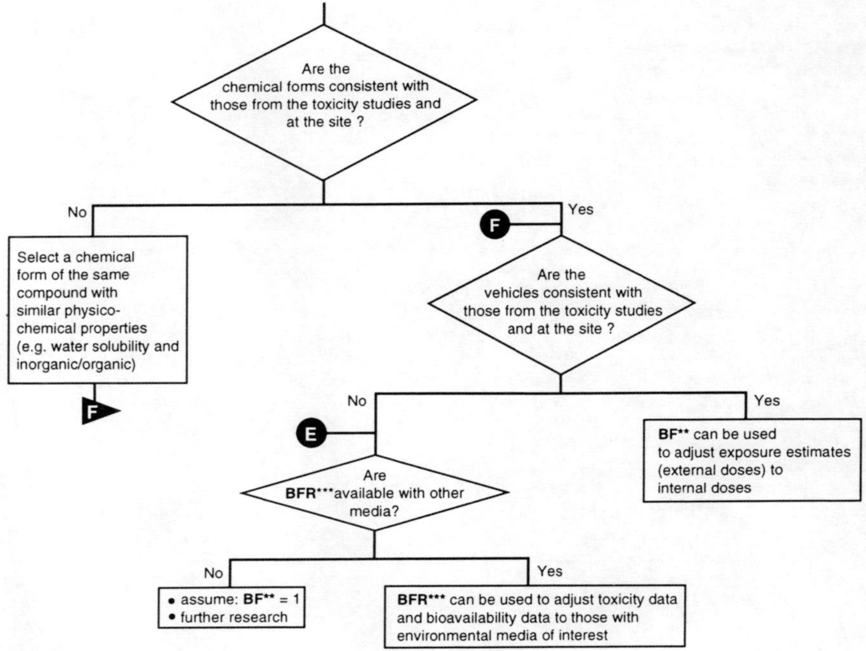

* **QRA** = Quantitative Risk Assessment
** **BF** = Bioavailability Factor
*** **BFR** = Relative Bioavailability factor

Figure 7-1 Framework for application of bioavailability factors in quantitative risk assessment

REFERENCES

Abdel-Rahman, M. S. and Turkall, R. M., Determination of exposure of oral and dermal benzene from contaminated soils, in *Petroleum Contaminated Soils*, Vol. I, Kostecki, P. T. and Calabrese, E. J., Eds., Lewis Publishers, Chelsea, Michigan, 1989, 301.

Abdou, H. M., *Dissolution, Bioavailability and Bioequivalence,* MACK Publishing Co., Easton, Pennsylvania, 1989, 297.

Aberg, B., Ekman, L., Falk, R., Persson, G., and Snihs, O., Metabolism of methyl mercury (^{203}Hg) compounds in man, *Arch. Environ. Health,* 19, 478, 1969.

Abernathy, C. O. and Ohanian, E. V., Non-carcinogenic effects of inorganic arsenic, *Environ. Geochem. Health*, 14, 35, 1992.

ACGIH (American Conference of Governmental Industrial Hygienists), Documentation of the Threshold Limit Values and Biological Exposure Indices, 6th ed., American Conference of Governmental Industrial Hygienists, Cincinnati, Ohio, 1991.

Ackermann, C., Flynn, G. L., and Smith, W. M., Ether-water partitioning and permeability through hairless mouse skin *in vitro*. II. Hydrocortisone 21-n-alkyl esters, alkanols and hydrophilic compounds, *Int. J. Pharma.*, 36, 67, 1987.

Adamsson, E., Piscator, M., and Nogawa, K., Pulmonary and gastrointestinal exposure to cadmium oxide dust in a battery factory, *Environ. Health Persp.*, 28, 219, 1979.

Agriculture Canada, Annual Report on Residue and Formula Compliance Monitoring for Fiscal Year 1988-1989, Agri-Food Safety Division, Food Inspection Directorate, Ontario, 1989.

Aguiar, A. J. and Weiner, M., Percutaneous absorption studies of chlorophenicol solutions, *J. Pharma. Sci.,* 58, 210, 1969.

Ahlborg, U. G. and Larsson, K., Metabolism of tetrachlorophenols in rat, *Arch. Toxicol.*, 40, 63, 1978.

Ahlborg, U. G. and Thunberg, T. M., Chlorinated phenols: occurrence, toxicity, metabolism, and environmental impact, *CRC Crit. Rev. Toxicol.*, 7(1), 1, 1980.

Aitio, A., Pekari, K., and Jävisalo, J., Skin absorption as a source of error in biological monitoring, *Scand. J. Work Environ. Health,* 10, 317, 1984.

Albery, W. J., Burke, J. F., Leffler, E. B., and Hadgraft, J., Interfacial transfer studies with a rotating diffusion cell, *J. Chem. Soc. (Faraday Trans. I)*, 72, 1618, 1976.

Albery, W. J. and Hadgraft, J., Percutaneous absorption: interfacial transfer kinetics, *J. Pharm. Pharmacol.*, 31, 65, 1979.

Alexander, Y. W., Clayton, B. E., and Delves, H. T., Mineral and trace-metal balances in children receiving normal and synthetic diets, *Quart. J. Med.*, 43, 89, 1974.

Ameno, K., Fuke, C., Ameno, S., Kiriu, T., Sogo, K., and Ijiri, I., A fatal case of oral ingestion of toluene, *Forensic. Sci. Int.*, 41, 255, 1989.

Andersen, O., Oral cadmium exposure in mice: toxicokinetics and efficiency of chelating agents, *Crit. Rev. Toxicol.*, 20(2), 83, 1989.

Andersen, O., Factors affecting the intestinal uptake of cadmium from the diet, in *Cadmium in the Human Environment,* Nordberg, G.F., Herber, R. F. M., and Alessio, L., Eds., International Agency for Research on Cancer, Lyon, 1992, 173.

Andersen, O., Nielsen, J. B., and Svendsen, P., Oral cadmium chloride intoxication in mice: effects of dose on tissue damage, intestinal absorption and relative organ distribution, *Toxicology*, 48, 225, 1988.

Andersen, O. and Nielsen, J. B., Effects of diethyldithiocarbamate on the toxicokinetics of cadmium chloride in mice, *Toxicology*, 55, 1, 1989.

Andersen, Y. S., Jackson, J. A., and Birnbaum, L. S., Maturational changes in dermal absorption of 2,3,7,8-tetrachlorodibenzo-p-dioxin (TCDD) in Fischer 344 Rats, *Toxiocol. Appl. Pharmacol.*, 119, 214, 1993.

Anderson, R. A., Nutritional role of chromium, *Sci. Total Environ.*, 17, 13, 1981.

Anderson, R. A., Essentiality of chromium in humans, *Sci. Total Environ.*, 86, 75, 1989.

Anderson, R. A. and Kozlovsky, A. S., Chromium intake, absorption and excretion of subjects consuming self-selected diets, *Am. J. Clin. Nutr.*, 41, 1177, 1985.

Anderson, R. A., Polansky, M. M., Bryden, N. A., Patterson, K. Y., Veillon, C., and Glinsmann, W. H., Effects of chromium supplementation on urinary Cr excretion of human subjects and correlation of Cr excretion with selected clinical parameter, *J. Nutr.*, 113, 276, 1983.

Andrews, L. S. and Snyder, R., Toxic effects of solvents and vapors, in *Casarett and Doull's Toxicology, The Basic Science of Poisons,* 4th ed., Amdur, M. O., Doull, J., and Klaassen, C. D., Eds., Pergamon Press, New York, 1991, 681.

Antti-Poika, M., Kalliokoski, P., and Hänninen, O., Toluene, in *Ethel Browning's Toxicity and Metabolism of Industrial Solvents,* Vol. I, Snyder, R., Ed., Elsevier, New York, 1987, 38.

Ariyoshi, T. and Ikeda, T., On the tissue distribution and the excretion of arsenic in rats and rabbits of administration with arsenical compounds, *J. Hyg. Chem.*, 20, 290, 1974.

Aschner, M. and Aschner, J. L., Mercury neurotoxicity: mechanisms of blood-brain barrier transport, *Neurosci. Biobehav. Rev.*, 14, 169-176, 1990.

Åstrand, I., Effect of physical exercise on uptake, distribution and elimination of vapors in man, in *Modeling of Inhalation Exposure to Vapor: Uptake, Distribution, and Elimination,* Vol. II, Fiserova-Bergerova, V., Ed., CRC Press, Boca Raton, Florida, 1983, 108.

Atkinson, R., Atmospheric lifetimes of dibenzo-p-dioxins and dibenzofurans, *Sci. Total Environ.*, 104, 17, 1991.

ATSDR (Agency for Toxic Substances and Disease Registry), Toxicological Profile for Chromium, Agency for Toxic Substances and Disease Registry, U.S. Public Health Service, Atlanta, GA, 1989a.

ATSDR (Agency for Toxic Substances and Disease Registry), Toxicological Profile for Arsenic, Agency for Toxic Substances and Disease Registry, U.S. Public Health Service, Atlanta, GA, 1989b.

ATSDR (Agency for Toxic Substances and Disease Registry), Toxicological Profile for Pentacholorophenol, Agency for Toxic Substances and Disease Registry, U.S. Public Health Service, Atlanta, GA, 1989c.

ATSDR (Agency for Toxic Substances and Disease Registry), Toxicological Profile for Phenol, Agency for Toxic Substances and Disease Registry, U.S. Public Health Service, Atlanta, GA, 1989d.

ATSDR (Agency for Toxic Substances and Disease Registry), 2,3,7,8-Tetrachlorodibenzo-p-dioxin, Agency for Toxic Substances and Disease Registry, U.S. Public Health Service, Atlanta, GA, 1989e.

ATSDR (Agency for Toxic Substances and Disease Registry), Toxicological Profile for Copper, Agency for Toxic Substances and Disease Registry, U.S. Public Health Service, Atlanta, GA, 1990a.

ATSDR (Agency for Toxic Substances and Disease Registry), Toxicological Profile for 2,4,6-Trichlorophenol, Agency for Toxic Substances and Disease Registry, U.S. Public Health Service, Atlanta, GA, 1990b.

ATSDR (Agency for Toxic Substances and Disease Registry), Toxicological Profile for Creosote, Agency for Toxic Substances and Disease Registry, U.S. Public Health Service, Atlanta, GA,1990c.

ATSDR (Agency for Toxic Substances and Disease Registry), Toxicological Profile for Naphthalene. 2-methylnaphthalene, Agency for Toxic Substances and Disease Registry, U.S. Public Health Service, Atlanta, GA, 1990d.

ATSDR (Agency for Toxic Substances and Disease Registry), Toxicological Profile for Polycyclic Aromatic Hydrocarbons, Agency for Toxic Substances and Disease Registry, U.S. Public Health Service, Atlanta, GA, 1990e.

ATSDR (Agency for Toxic Substances and Disease Registry), Toxicological Profile for Benzo[a]pyrene, Agency for Toxic Substances and Disease Registry, U.S. Public Health Service, Atlanta, GA, 1990f.

ATSDR (Agency for Toxic Substances and Disease Registry), Toxicological Profile for Benzo[a]anthracene, Agency for Toxic Substances and Disease Registry, U.S. Public Health Service, Atlanta, GA, 1990g.

ATSDR (Agency for Toxic Substances and Disease Registry), Toxicological Profile for Benzo[b]fluoranthene, Agency for Toxic Substances and Disease Registry, U.S. Public Health Service, Atlanta, GA, 1990h.

ATSDR (Agency for Toxic Substances and Disease Registry), Toxicological Profile for Dibenz[a,h]anthracene, Agency for Toxic Substances and Disease Registry, U.S. Public Health Service, Atlanta, GA, 1990i.

ATSDR (Agency for Toxic Substances and Disease Registry), Toxicological Profile for Chrysene, Agency for Toxic Substances and Disease Registry, U.S. Public Health Service, Atlanta, GA, 1990j.

ATSDR (Agency for Toxic Substances and Disease Registry), Toxicological Profile for Benzene, Agency for Toxic Substances and Disease Registry, U.S. Public Health Service, Atlanta, GA, 1991a.

ATSDR (Agency for Toxic Substances and Disease Registry), Toxicological Profile for Lead, Agency for Toxic Substances and Disease Registry, U.S. Public Health Service, Atlanta, GA, 1991b.

ATSDR (Agency for Toxic Substances and Disease Registry), Toxicological Profile for Toluene, Agency for Toxic Substances and Disease Registry, U.S. Public Health Service, Atlanta, GA, 1992.

ATSDR (Agency for Toxic Substances and Disease Registry), Toxicological Profile for Arsenic (Update), Agency for Toxic Substances and Disease Registry, U.S. Public Health Service, Atlanta, GA, 1993a.

ATSDR (Agency for Toxic Substances and Disease Registry), Toxicological Profile for Benzene (Update), Agency for Toxic Substances and Disease Registry, U.S. Public Health Service, Atlanta, GA, 1993b.

ATSDR (Agency for Toxic Substances and Disease Registry), Toxicological Profile for Lead (Update), Agency for Toxic Substances and Disease Registry, U.S. Public Health Service, Atlanta, GA, 1993c.

ATSDR (Agency for Toxic Substances and Disease Registry), Toxicological Profile for Cadmium (Update), Agency for Toxic Substances and Disease Registry, U.S. Public Health Service, Atlanta, GA, 1993d.

ATSDR (Agency for Toxic Substances and Disease Registry), Toxicological Profile for Mercury, Agency for Toxic Substances and Disease Registry, U.S. Public Health Service, Atlanta, GA, 1993e.

Aungst, B. J., Dolce, J. A., and Fung, H., The effect of dose on the disposition of lead in rats after intravenous and oral administration, *Toxicol. Appl. Pharmacol.*, 61, 48, 1981.

Aungst, B. and Shen, D. D., Gastrointestinal absorption of toxic agents, in *Gastrointestinal Toxicology,* Rozman, K. and Hänninen, O., Eds., Elsevier, New York, 1986, 29.

Bælum, J., Toluene in alveolar air during controlled exposure to constant and to varying concentrations, *Int. Arch. Occup. Environ. Health,* 62, 59, 1990.

Bælum, J., Mølhave, L., Hansen, S. H., and Døssing, M., Hepatic metabolism of toluene after gastrointestinal uptake in humans, *Scand. J. Work Environ. Health,* 19, 55, 1993.

Baetjer, A. M., Damron, C., and Budacz, V., The distribution and retention of chromium in men and animals, *Arch. Ind. Health,* 20, 136, 1959.

Baich, H. and Davist, D. L., Phenol: a review of environmental and health risks, *Reg. Toxicol. Pharmacol.*, 1, 90, 1981.

Bailey, M. R., Fry, F. A., and James, A. C., Long-term retention of particles in the human respiratory tract, *J. Aerosol Sci.*, 16, 295, 1985.

Bair, W. J., Human respiratory tract model for radiological protection: a revision of the ICRP dosimetric model for the respiratory system, *Health Phys.*, 57, Sup.1, 249, 1989.

Bahig, M. E., Kraus, A., and Klein, W., Excretion and metabolism of 2,4,6-T_3CP-^{14}C in rats, *Chemosphere*, 10, 323, 1981.

Bagdon, R. E. and Hazen, R. E., Skin permeation and cutaneous hypersensitivity as a basis for making risk assessments of chromium as a soil contaminant, *Environ. Health Persp.*, 92, 111, 1991.

Baker, H. and Kligman, A. M., A simple method *in vivo* for studying the permeability of the human stratum corneum, *J. Invest. Dermatol.*, 48, 273, 1967.

Baldini, M., Coni, E., Mantovani, A., Stacchini, A., and Zanasi, F., Effect of unbalanced diets on the long-term metabolism of a toxicant. 1. Lead in rats: preliminary note, *Food Addi. Contm.*, 6, 117, 1989.

Banks, Y. B., Brewster, D. W., and Birnbaum, L. S., Age-related changes in dermal absorption of 2,3,7,8-tetrachlorodibenzo-p-dioxin and 2,3,4,7,8-pentachlorodibenzofuran, *Fund. Appl. Toxicol.*, 15, 163, 1990.

Banks, Y. B. and Birnbaum, L.S., Absorption of 2,3,7,8-tetrachlorodibenzo-p-dioxin (TCDD) after low dose dermal exposure, *Toxicol. Appl. Pharmacol.*, 107, 302, 1991.

Baranowska-Dutkiewicz, B., Absorption of hexavalent chromium by skin in man, *Arch. Toxicol.*, 47, 47, 1981a.

Baranowska-Dutkiewicz, B., Absorption of phenol from aqueous solution in men, *Int. Arch. Occup. Enrivon. Health,* 49, 99, 1981b.

Baranowska-Duttiewicz, B., Skin absorption of aniline from aqueous solutions in man, *Toxicol. Lett.*, 10, 367, 1982.

Barltrop, D. and Khoo, H. E., The influence of nutritional factors on lead absorption, *Postgra. Med. J.*, 51, 795, 1975.

Barltrop, D. and Meek, F., Absorption of different lead compounds, *Postgra. Med. J.*, 51, 805, 1975.

Barltrop, D., Strehlow, C. D., Thornton, I., and Webb, J. S., Absorption of lead from dust and soil, *Postgra. Med. J.*, 51, 801, 1975.

Barr, W. H., The use of physical and animal models to assess bioavailability, *Pharmacology*, 8, 55, 1972.

Barregard, L., Sälleten, G., Schütz, A., Attewell, R., Skerfving, S., and Järvholm, B., Kinetics of mercury in blood and urine after brief occupational exposure, *Arch. Environ. Health*, 47, 176, 1992.

Barrett, H. M., Irwin, D. A., and Semmons, E., Studies on the toxicity of inhaled cadmium. I. The acute toxicity of cadmium oxide by inhalation, *J. Ind. Hyg. Toxicol.*, 29, 249, 1947.

Barry, B. W. and El Eini, D. I. D., Influence of non-ionic surfactants on permeation of hydrocortisone, dexamethasone, testosterone and progesterone across cellulose acetate membrane, *J. Pharm. Pharmacol.*, 28, 219, 1976.

Barry, B. W., *Dermatological Formulations. Percutaneous Absorption,* Marcel Dekker, Inc., New York, 1983.

Barry, P. S. I., A comparison of concentrations of lead in human tissue, *Br. J. Ind. Med.*, 38, 61, 1975.

Bartlett, R. J. and James, B. R., Mobility and bioavailability of chromium in soil, in *Chromium in the Natural and Human Environments*, Nriagu, J. O. and Nieboer, E., Eds., John Wiley & Sons, New York, 1988, 267.

Bartlett, R. J., Chromium cycling in soil and water: links, gaps, and methods, *Environ. Health Persp.*, 92, 17, 1991.

Baskin, L. B. and Falco, J. W., Assessment of human exposure to gaseous pollutants, *Risk Anal.*, 9, 365, 1989.

Bates, M. N., Smith, A. H., and Hopenhayn-Rich, C., Arsenic ingestion and internal cancers: a review, *Am. J. Epidemiol.*, 135, 462, 1992.

Baxter, P. J., Samuel, A. M., and Holkham, M. P. E., Lead hazard in British standing glass workers, *Br. Med. J.*, 291, 383, 1985.

Bearn, A. G. and Kunkel, H. G., Metabolic studies in Wilson's disease using Cu^{64}, *J. Lab. Clin. Med.*, 45, 623, 1955.

Beck, B. D., Brain, J. D., and Bohannon, D. E., The pulmonary toxicity of an ash sample from the Mt. St. Helens volcano, *Exp. Lung Res.*, 2, 289, 1981.

Beck, H., Eckart, K., Mathar, W., and Wittkowski, E., Occurrence of PCDD and PCDF in different kinds of paper, *Chemosphere*, 17, 51, 1988.

Behl, C. R., Linn, E. E., Flynn, G. L., Pierson, C. L., Higuchi, W. I., and Ho, N. F. H., Permeation of skin and eschar by antiseptic. I. Baseline studies with phenol, *J. Pharm. Sci.*, 72, 391, 1983.

Behl, C. R., Bellantone, N. H., Dethlefsen, L., and Goldberg, A. H., Design of Transdermal Delivery Systems. I. Characterization of Permeability Behaviors of Rate Controlling Synthetic Membranes. Evaluation of the Alkyl Chain Length and Vehicle Influences, paper presented at the 131st Annual Meeting of the American Pharmaceutical Association, Montreal, May 5-10, 1984, Basic Pharmaceutics Abstract #1.

Behl, C. R., Malick, A. W, and Goldberg, A. H., Transdermal Delivery of Ro 22-1327. I. Partitioning and *in vitro* Skin Permeability Studies. Correlation between Human and Rat Skins,

Silastic® Permeability Studies, paper presented at the 39th National Meeting of the Academy Pharmaceutical Science, Minneapolis, MN, Oct. 20-24, 1985, Basic Pharmaceutics Abstract #25.

Behl, C. R., Pei, J. Y., and Malick, A. W., Skin Permeability Profile of 14C Diazepam, paper presented at the 133rd Annual Meeting of the American Pharmaceutical Association, San Francisco, CA, March 16-20, 1986, Basic Pharmaceutics Abstract #S 814-2.

Behl, C. R., Kumar, S., Malick, A. W., Patel, S. P., and Piemontese, D., Choice of membranes for *in vitro* skin uptake studies and general experimental technics, in *Methods for Skin Absorption*, Kemppainen, B. W. and Reifenrath, W. G., Eds., CRC Press, Boca Raton, Florida, 1990, 1.

Beliles, R. P. and Totman, L. C., Pharmacokinetically based risk assessment of workplace exposure to benzene, *Regul. Toxicol. Pharmacol.*, 9, 186, 1989.

Bend, J. R., Serabjit-Singh, C. J., and Philpot, R. M., The pulmonary uptake, accumulation, and metabolism of xenobiotics, *Ann. Rev. Pharmacol. Toxicol.*, 25, 97, 1985.

Benford, D. J. and Bridges, J. W., Xenobiotic metabolism in lung, in *Progress in Drug Metabolism,* Vol. 9, Bridges, J. W. and Chasseaud, L. F., Eds., Taylor & Francis, Ltd., New York, 1986, 53.

Bentur, Y., Koren, G., McGuigan, M., and Spielberg, S. P., An unusual skin exposure to copper; clinical and pharmacokinetics evaluation, *Clin. Toxicol.*, 26, 371, 1988.

Berlin, M., Mercury, in *Handbook on the Toxicology of Metals*, Friberg, L., Nordberg, G. F., and Vouk, V. B., Eds., Elsevier, Amsterdam, 1986, 387.

Berlin, M., Nordberg, G., and Hellberg, J., The uptake and distribution of methylmercury in the brain or Saimiri Sciureus in relation to behavioral and morphological changes, in *Mercury, Mercurials and Mercaptans*, Miller, M. W. and Clarkson, T. W., Eds., Charles C Thomas Publisher, Springfield, Illlinois, 1973, 187.

Bertazzi, P. A., Zocchetti, C., Pesatori, A. C., Guercilena, S., Sanarico, M., and Radice, L., Ten-year mortality study of the population involved in the Seveso incident in 1976, *Am. J. Epidemiol.*, 129, 1187, 1989.

Bertazzi, P. A., Pesatori, A. C., Consonni, D., Tironi, A., Landi, M. T., and Zocchetti, C., Cancer incidence in a population accidentally exposed to 2,3,7,8-tetrachlorodibenzo-para-dioxin, *Epidemiology*, 4, 398, 1993.

Bettley, F. R. and O'Shea, J. A., The absorption of arsenic and its relation to carcinoma, *Br. J. Dermatol.*, 92, 563, 1975.

Bevan, D. R. and Yonda, N. T., *In vitro* technique to study elution of benzo[a]pyrene from particulates into biomembranes with application to woodstove particulates, *Anal. Biochem.*, 150, 105, 1985.

Bevan, D. R. and Worrell, W. J., Elution of benzo[a]pyrene from carbon blacks into biomembranes *in vitro*, *J. Toxicol. Environ. Health*, 15, 697, 1985.

Bevan, D. R. and Ruggio, D. M., Bioavailability *in vivo* of benzo[a]pyrene adsorbed to diesel particulate, *Toxicol. Ind. Health*, 7, 125, 1991.

Beveridge, S. J., Garrett, I. R., Whitehouse, M. W., Vernon-Roberts, B., and Brooks, P. M., Biodistribution of ^{64}Cu in inflamed rats following administration of two anti-inflammatory copper complexes, *Agents Actions*, 14, 291, 1984.

Bhattacharyya, M. H., Sellers, D. A., and Peterson, D. P., Postlactational changes in cadmium during pregnancy and lactation, *Environ. Res.*, 40, 145, 1986.

Bidleman, T. F., Atmospheric Processes, *Environ. Sci. Technol.*, 22, 361, 1988.

Birnbaum, L. S., EPA's reassessment of dioxin risk: directed health research, *Chemosphere*, 27, 469, 1993.

Black, A. and Hounam, R. F., Penetration of iodine vapor through the nose and mouth and clearance and metabolism of the deposited iodine, *Ann. Occup. Hyg.*, 11, 209, 1968.

Blake, K. C. H., Absorption of ^{203}Pb from gastrointestinal tract of man, *Environ. Res.*, 11, 1, 1976.

Blake, K. C. H., Barbezat, G. O., and Mann, M., Effect of dietary constituents on the gastrointestinal absorption of ^{203}Pb in man, *Environ. Res.*, 30, 182, 1983.

Blake, K. C. H. and Mann, M., Effect of calcium and phosphorus on the gastrointestinal absorption of ^{203}Pb in man, *Environ. Res.*, 30, 188, 1983.

Blank, I. H., Scheuplein, R. J., and MacFarlane, D. J., Mechanism of percutaneous absorption, *J. Invest. Dermatol.*, 49, 582, 1967.

Blank, I. H. and Scheuplein, R. J., Transport into and within the skin, *Br. J. Dermatol.*, 81 (supp l4), 4, 1969.

Blank, I. H. and McAuliffe, D. J., Penetration of benzene through human skin, *J. Inves. Dermatol.*, 85, 522, 1985.

Bois, F., Smith, M. T., and Spear, R. C., Mechanisms of benzene carcinogenesis: application of a physiological model of benzene pharmacokinetics and metabolism, *Toxicol. Lett.*, 56, 283, 1991a.

Bois, F., Woodruff, T. J., and Spear, R. C., Comparison of three physiologically based pharmacokinetic models of benzene disposition, *Toxicol. Appl. Pharmacol.*, 110, 79, 1991b.

Bois, F. and Paxman, D. G., An analysis of exposure rate effects for benzene using a physiologically based pharmacokinetic model, *Regul. Toxicol. Pharmacol.*, 15, 122, 1992.

Boisset, M., Girard, F., Godin, J., and Boudene, Cl., Cadmium content of lung, liver and kidney in rats exposed to cadmium oxide fumes, *Int. Arch. Occup. Environ. Health*, 41, 41, 1978.

Bonaccorsi, A., di Domenico, A., Fanelli, R., Merli, F., Motta, R., Vanzati, R., and Zapponi, G. A., The influence of soil particle adsorption on 2,3,7,8-tetrachlorodibenzo-p-dioxin biological uptake in the rabbit, *Arch. Toxicol. Suppl.*, 7, 431, 1984.

Booker, D. V., Chamberlain, A. C., Newton, D., and Stott, A. N. B., Uptake of radioactive lead following inhalation and injection, *Br. J. Radiol.*, 42, 457, 1969.

Boroujerdi, M., Kung, H., Wilson, A. G., and Anderson, M. W., Metabolism and DNA binding of benzo[a]pyrene *in vivo* in the rat, *Cancer Res.*, 41, 951, 1981.

Bottari, F., DiColo, G., Nannipieri, E., Saettone, M. F., and Serafini, M. F., Release of drug from ointment bases. II. *In vitro* release of benzocaine from suspension type aqueous gels, *J. Pharm. Sci.*, 66, 926, 1977.

Bragt, P. C. and van Dura, E. A., Toxicokinetics of hexavalent chromium in the rat after intratracheal administration of chromates of different solubilities, *Ann. Occup. Hyg.*, 27, 315, 1983.

Bragt, P. C., Zwennis, W. C. M., Kouwenberg, F. J. A., van Pijeren, M. E., and Franssen, A. C., Relevance of the Solubility of Chromium (VI) Salts for Biological Monitoring with Special Reference to Lead Chromate, paper presented at ACS Meeting, Miami, FL, April 29-May 2, 1985.

Brain, J. D. and Mosier, M. J., Deposition and clearance of grain dusts in the human lung, in *Occupational Pulmonary Disease: Focus on Grain Dust and Health,* Dossman, J. A. and Cotten, D. J., Eds., Academic Press, New York, 1977, 77.

Brain, J. D., Macrophage damage in relation to the pathogenesis of lung disease, *Environ. Health Persp.*, 35, 21, 1980.

Brain, J. D., Mechanism, measurement, and significance of lung macrophage function, *Environ. Health Persp.*, 97, 5, 1992.

Brainard, J. and Beck, B. D., A review of the bioavailability of petroleum constituents, *J. Soil Contam.*, 1, 273, 1992.

Braun, W. H. and Sauerhoff, M. W., The pharmacokinetic profile of pentachlorophenol in monkeys, *Toxicol. Appl. Pharmacol.*, 38, 525, 1976.

Braun, W. H., Young, J. D., Blau, G. E., and Gehring, P. J., The pharmacokinetics and metabolism of pentachlorophenol in rats, *Toxicol. Appl. Pharmacol.*, 41, 395, 1977.

Braun, W. H., Blau, G. E., and Chenoweth, M. B., The metabolism/pharmacokinetics of pentachlorophenol in man and a comparison with the rat and monkey, *Dev. Toxicol. Environ.*, 4, 289, 1979.

Bress, W. C., Percutaneous *in vivo* and *in vitro* absorption of lead, *Vet. Hum. Toxicol.*, 33, 212, 1991.

Brewster, D. W., Banks, Y. B., Clark, A., and Birnbaum, L. S., Comparative dermal absorption of 2,3,7,8-tetrachlorodibenzo-p-dioxin and three polychlorinated dibenzofurans, *Toxicol. Appl. Pharmacol.*, 97, 156, 1989.

Bronaugh, R. L., Stewart, R. F., Congdon, E. R., and Giles, A. L., Methods for *in vitro* percutaneous absorption studies. I. Comparison with *in vivo* results, *Toxicol. Appl. Pharmacol.*, 62, 474, 1982.

Bronaugh, R. L. and Stewart, R. F., Methods for *in vitro* percutaneous absorption studies. III. Hydrophobic compounds, *J. Pharm. Sci.*, 73, 1255, 1984.

Bronaugh, R. L. and Stewart, R. F., Methods for *in vitro* percutaneous absorption studies. IV. The flowthrough diffusion cell, *J. Pharm. Sci.*, 74, 64, 1985a.

Bronaugh, R. L. and Stewart, R. F., Methods for *in vitro* percutaneous absorption studies. V. Permeation through damaged skin, *J. Pharm. Sci.*, 74, 1062, 1985b.

Bronaugh, R. L., Stewart, R. F., and Simon, M., Methods for *in vitro* percutaneous absorption studies. VI. Use of excised human man skin, *J. Pharm. Sci.*, 75, 1094, 1986.

Bronaugh, R. L., Skin permeation: *in vitro* techniques, in *Methods for Cutaneous Investigation,* Rietschel, R. L. and Spencer, T. S., Eds., Marcel Dekker, Inc., New York, 1990, 171.

Brown, H. S., Bishop, D. R., and Rowan, C. A., The role of skin absorption as a route of exposure for volatile organic compounds (VOCs) in drinking water, *Am. J. Public Health,* 74, 479, 1984.

Bruce, R. M., Santodonato, J., and Neal, M. W., Summary review of the health effects associated with phenol, *Toxicol. Ind. Health,* 3, 535, 1987.

Brune, D. and Evje, D. M., Man's mercury loading from a dental amalgam, *Sci. Total Environ,* 44, 51, 1985.

Brunnemann, K. D., Kagan, M. R., Cox, J. E., and Hoffmann, D., Determination of benzene, toluene and 1,3-butadiene in cigarette smoke by GC-MSD, *Exp. Pathol.*, 37, 118, 1989.

Buchet, J. P., Lauwerys, R., and Roles, H., Urinary excretion of inorganic arsenic and its metabolites after repeated ingestion of sodium metaarsenite by volunteers, *Int. Arch. Occup. Environ. Health,* 48, 111, 1981a.

Buchet, J. P., Lauwerys, R., and Roles, H., Comparison of the urinary excretion of arsenic metabolites after a single oral dose of sodium arsenic, monomethylarsenate, or dimethylarsenate, *Int. Arch. Occup. Environ. Health*, 48, 71, 1981b.

Buck, D. A. W., McMaster, J. R., Maibach, H. I., and Guy, R. H., Bioavailability of topically administered steroids: a 'mass balance' technique, *J. Invest. Dermatol.*, 91, 29, 1988.

Buck, D. A., Guy, R. H., and Maibach, H. I., Percutaneous penetration and mass balance accountability: technique and implications for dermatology, *J. Toxicol. Cutan. Ocular Toxicol.*, 9, 439, 1990.

Bulman, T. L., Lesage, S., Fowlie, P. J. A., and Webber, M. D., The Persistence of Polynuclear Aromatic Hydrocarbons in Soil, PACE Report No. 85-2, Environment Canada, Ontario, 1985.

Burg, R. V. and Liu, D., Toxicology update: chromium and hexavalent chromium, *J. Appl. Toxicol.*, 13, 225, 1993.

Cai, X., Wang, N., Wei, J., and Zheng, G., Experimental investigation of the light extinction method for measuring aerosol size distributions, *J. Aerosol Sci.*, 23, 749, 1992.

Calesnick, B., Wase, A., and Overby, L. R., Availability during human consumption of the arsenic in tissues of chicks fed arsanilic-^{74}As acid, *Toxicol. Appl. Pharmacol.*, 9, 27, 1966.

Capel, I. D., French, M. R., Millborn, P., Smith, R. L., and Williams, R. T., The fate of [^{14}C]phenol in various species, *Xenobiotica,* 2, 25, 1972.

Capraro, V., Permeability and related phenomena: basic concepts, in *Pharmacology of Intestinal Permeation I,* Csáky, T. Z., Ed., Springer-Verlag, Berlin, 1984, 61.

Carlsson, A., Exposure to toluene. Uptake, distribution and elimination in man, *Scand. J. Work Environ. Health,* 8, 43, 1982.

Carlsson, A. and Ljungquist, E., Exposure to toluene. Concentration in subcutaneous adipose tissue, *Scand. J. Work Environ. Health*, 8, 56, 1982.

Carson, B. L., Ellis III, H. V., and McCann, J. L., Eds., *Toxicology and Biological Monitoring of Metals in Humans*, Lewis Publishers, Inc., Chelsea, Michigan, 1986.

Cartwright, G. E. and Wintrobe, M. M., Copper metabolism in normal subjects, *Am. J. Clin. Nutr.*, 14, 224, 1964.

Casarett, J., Bevenue, A., Yauger, W. L., and Whalen, S. A., Observations on pentachlorophenol in human blood and urine, *Am. Ind. Hyg. Assoc. J.*, 30, 360, 1969.

Cassidy, M. K. and Houston, J. B., Protective role of intestinal and pulmonary enzymes against environmental phenol, *Br. J. Pharmacol.*, 69, 316, 1979.

Cavalleri, A. and Minoia, C., Serum and erythrocyte chromium distribution and urinary elimination in person occupationally exposed to chromium (VI) and chromium (III), *F. Ital. Med Lav.*, 7, 35, 1985.

CEC (Commission of the European Communities), CEC Criteria for Cadmium, Commission of the European Communities, Pergamon, Oxford, 1978.

Chamberlain, A. C., Prediction of response of blood lead to airborne and dietary lead from volunteer experiments with lead isotopes, *Proc. R. Soc. Lond.*, B224, 149, 1985.

Chamberlain, A. C., Clough, W. S., Heard, M. J., Newton, D., Stott, A. N. B., and Wells, A. C., Uptake of lead by inhalation of motor exhaust, *Proc. R. Soc. Lond.*, B192, 77, 1975.

Charbonneau, S. M., Spencer, K., Bryce, F., and Sandi, E., Arsenic excretion by monkeys dosed with arsenic-containing fish or with inorganic arsenic, *Bull. Environ. Contam. Toxicol.*, 20, 470, 1978.

Charbonneau, S. M., Hollins, J. G., Tam, G. K. H., Bryce, F., Ridgeway, J. M., and Willes, R.F., Whole-body retention, excretion and metabolism of [^{74}As] arsenic acid in the hamster, *Toxicol. Lett.*, 5, 175, 1980.

Cheng, Y. L., Preslan, J. E., Anderson, M. B., and George, W. J., Solubility and bioavailability of lead following oral ingestion of vitrified slagged aggregate, *J. Hazardous Materials,* 27, 137, 1991.

Cheng, Y. S., Su, Y. F., Newton, G. T., and Yen, H. C., Use of a graded diffusion battery in measuring the activity size distributions of thoron progeny, *J. Aerosol Sci.*, 23, 361, 1992.

Cherian, M. G., Hursh, J. B., Clarkson, T. W., and Allen, J., Radioactive mercury distribution in biological fluids and excretion in human subjects after inhalation of mercury vapor, *Arch. Environ. Health*, 33, 109, 1978.

Chowdhury, B. A. and Chandra, R. K., Biological and health implications of toxic heavy metal and essential trace element interactions, *Prog. Food Nutr. Sci.*, 11, 55, 1987.

Clapp, T. C., Umbreit, T. H., Meeker, R. J., Kosson, D. S., Gray, D., and Gallo, M. A., Bioavailability of lead and chromium from encapsulated pigment materials, *Bull. Environ. Contm. Toxicol.*, 46, 271, 1991.

Clarke, S. W. and Pavia, D., Lung mucus production and mucociliary clearance: methods and assessment, *Br. J. Clin. Pharmacol.*, 9, 537, 1980.

Clarkson, T.W., Epidemiological and experimental aspects of lead and mercury contamination of food, *Food Cosmet Toxicol.*, 9, 229, 1971.

Clarkson, T. E., Mercury, *J. Am. Coll. Toxicol.*, 8, 1291, 1989.

Clarkson, T. E., Mercury: major issues in environmental health, *Environ. Health Persp.*, 100, 31, 1993.

Cline, R. E., Hill, R. H., Phillips, D., and Needham, L. L., Pentachlorophenol measurements in body fluids of people in log homes and workplace, *Arch. Environ. Contam. Toxicol.*, 18, 475, 1989.

Cohen, D., Ferromagnetic contamination in the lungs and other organs of the human body, *Science,* 180, 745, 1973.

Cohen, D., Arai, S. F., and Brain, J. D., Smoking impairs long-term dust clearance from the lung, *Science,* 204, 514, 1979.

Cohen, A. J. and Rose, F. J., Review of lead toxicology relevant to the safety assessment of lead acetate as a hair coloring, *Food Chem. Toxicol.*, 29, 485, 1991.

Cohen, M. D., Kargacin, B., Klein, C. B., and Costa, M., Mechanism of chromium carcinogenicity and toxicity, *Crit. Rew. Toxicol.*, 23(3), 255, 1993.

Cohr, K. H. and Stokholm, J., Toluene: a toxicological review, *Scand. J. Work Environ. Health,* 5, 71, 1979.

Collins, J. F., Brown, J. P., Dawson, S. V., and Marty, M. A., Risk assessment for benzo[a]pyrene, *Reg. Toxicol. Pharmacol.*, 13, 170, 1991.

Cotzias, G. C., Borg, D. C., and Selleck, B., Virtual absence of turnover in cadmium metabolism: 109-Cd studies in the mouse, *Am. J. Physiol.*, 201, 927, 1961.

Cox, Jr., L. A., Biological basis of chemical carcinogenesis: insights from benzene, *Risk Anal.*, 11, 453, 1991.

Cox, Jr., L. A. and Ricci, P. F., Reassessing benzene cancer risk using internal dose, *Risk Anal.,* 12, 401, 1992.

Cresia, D. A., Poggenburg, J. K., and Nettesheim, P., Elution of benzo[a]pyrene from carbon particles in the respiratory tract of mice, *J. Toxicol. Environ. Health,* 1, 967, 1976.

Cruzan, G., Low, L. K., Cox, G. E., Meeks, J. R., Mackerer, C. R., Craig, P. H., Singer, E. J., and Mehlman, M. A., Systemic toxicity from subchronic dermal exposure, chemical characterization, and dermal penetration of catalytically cracked clarified slurry oils, *Toxicol. Ind. Health,* 2, 426, 1986.

Csáky, T. Z., Intestinal absorption of xenobiotics, in *Pharmacology of Intestinal Permeation II,* Vol. 70, Csáky, T. Z., Ed., Springer-Verlag, Berlin, 1984, 1.

Cuddihy, R. G. and Yeh, H. C., Respiratory tract clearance of particles and substances dissociated from particles, in *Inhalation Toxicology. The Design and Interpretation of Inhalation Studies and Their Use in Risk Assessment,* Mohr, U., Ed., Springer-Verlag, Berlin, 1988, 169.

Cuddihy, R. G. and McClellan, R. O., Risk assessment for inhaled toxicants, in *Concepts in Inhalation Toxicology,* McClellan, R. O. and Henderson, R. F., Eds., Hemisphere Publishing, Co., New York, 1989, 517.

Cupo, D. Y. and Wetterhahn, K. E., Binding of chromium to chromatin and DNA from liver and kidney of rats treated with sodium dichromate and chromium (III) chloride *in vivo, Cancer Res.,* 45, 1146, 1985.

Cutler, D., Assessment of rate and extent of drug absorption, in *Pharmacokinetics: Theory and Methodology,* Rowland, M. and Tucer, G., Eds., Pergamon Press, New York, 1986, 369.

Czernielewski, A., Brykalski, D. and Depczyk, D., Experimental investigations on penetration of radioactive chromium (Cr^{51}) through the skin, *Dermatologica,* 131, 384, 1965.

Dankovic, D. A., Wright, C. W., Zangar, R. C., and Springer, D.L., Complex mixture effects on the dermal absorption of benzo[a]pyrene and other polycyclic aromatic hydrocarbons from mouse skin, *J. Appl. Toxicol.,* 9, 239, 1989.

Davis, A., Ruby, M. V., and Bergstron, P. D., Bioavailability of arsenic and lead in soil from the Butte, Montana, mining district, *Environ. Sci. Technol.* 26, 461, 1992.

Dawson, D. C., Principle of membrane transport, in *The Gastrointestinal System. Intestinal Absorption and Secretion,* Vol. IV, Rauner, B. B., Ed., American Physiological Society, Bethesda, 1991, 1.

Decker, C. F., Byerrum, R. U., and Hoppert, C. A., A study of the distribution and retention of cadmium-115 in the Albino rat, *Arch. Biochem. Biophys.,* 66, 140, 1957.

Decker, L. E., Byerrum, R. U., Decker, C. F., Hoppert, C. A., and Langhm, R.F., Chronic toxicity studies, *Arch. Ind. Health,* 18, 228, 1958.

Deichmann, W. B., Phenol studies. V. The distribution, detoxification, and excretion of phenol in the mammalian body, *Arch. Biochem.,* 3, 345, 1944.

Deichmann, W. B. and Keplinger, M. L., Phenols and phenolic compounds, in *Patty's Industrial Hygiene and Toxicology, Vol.2A, Toxicology,* Clayton, B. D. and Clayton, F. E., Eds., John Wiley & Sons, New York, 1981, 2567.

Dekaban, A. S., O'Reilly, S., Aamodt, R., and Rumble, W. F., Study of copper metabolism in Kinky hair disease (Monkes' Disease) and in hepatolenticular degeneration (Wilson's Disease) utilizing ^{67}Cu and radioactivity counting in the total body and various tissues, *Transac. Am. Neurol. Assoc.,* 99, 106, 1974.

Demers, R., Overview of radon, lead and asbestos exposure, *Am. Fam. Physician*, 44(suppl), 51s, 1988.

Dickson, L. C. and Buzik, S. C., Health risks of dioxin: a review of environmental and toxicological considerations, *Vet. Hum. Toxicol.*, 35, 68, 1993.

Dieter, M. P. and Matthews, H. B., Comparison of lead bioavailability in F344 rats fed lead acetate, lead oxide, lead sulfide, or lead ore concentrate from Skagway, Alaska, *J. Toxicol. Environ. Health,* 39, 79, 1993.

Dinman, B. D., The mode of absorption, distribution, and elimination of toxic materials, in *Patty's Industrial Hygiene and Toxicology,* 4th ed., Vol. 1, Clayton, G. D., and Clayton, F. E., Eds., John Wiley & Sons, New York, 1991, 211.

DiPolo, R., Sha'afi, R. I., and Solomon, A. K., Transport parameter in a porous cellulose acetate membrane, *J. Gen. Physiol,* 55, 63, 1970.

Doisy, R. J., Streeten, D. H. P, Levine, R. A., and Chodos, R. B., Effects and metabolism of chromium in normal, elderly subjects, and diabetics, in *Trace Substances in Environmental Health* - II, Hemphill, D. D., Ed., University of Missouri, 1968.

Donaldson, R. M. and Barreras, R. F., Intestinal absorption of trace quantities of chromium, *J. Lab. Clin. Med.,* 68, 483, 1966.

Dorato, M. A., Overview of inhalation toxicology, *Environ. Health Persp.*, 85, 163, 1990.

Dowling, H. J., Offenbacher, E. G., and Pi-Sunyer, F. X., Absorption of inorganic, trivalent chromium from the vascularly perfused rat small intestine, *J. Nutr.*, 119, 1138, 1989.

Drew, R. R., The design and operation of systems for inhalation exposure animals, in *Toxicology of Inhaled Materials*, Witschi, H. P. and Srain, J. D., Eds., Springer-Verlag, Berlin, 1988, 3.

Dunn, J. D. and Clarkson, T. W., Dose mercury exhalation signal demethylation of methylmercury? *Health Phys.,* 38, 411, 1980.

Dunnick, J. K. and Fowler, B. A., Cadmium, in *Handbook on Toxicity of Inorganic Compounds*, Seiler, H. G. and Sigel, H., Eds., Marcel Dekker, Inc., New York, 1988, 155.

Dutkiewicz, T. and Piotrowski, J. K., Experimental investigation on the quantitative estimation of aniline absorption in man, *Pure Appl. Chem.*, 3, 319, 1961.

Dutkiewicz, T., Absorption of aniline vapor in man, *Med. Pracy.,* 12, 1, 1961.

Dutkiewicz, T., Absorption and metabolism of aniline in man, *Lodz. Towarz. Nauk. Wydzial,* III, 175, 83, 1962.

Dutkiewcz, T. and Tyras, H., A study of skin absorption of ethylbenzene in man, *Br. J. Ind. Med.,* 24, 330, 1967.

Dutkiewcz, T. and Tyras, H., Skin absorption of toluene, styrene, and xylene by man, *Br. J. Ind. Med.*, 25, 243, 1968.

Dutkiewicz, T., Experimental studies on arsenic absorption routes in rats, *Environ. Health Persp.*, 19, 173, 1977.

Dyer, A., Hayes, G. G., Wilson, J. G., and Catterall, R., Diffusion through skin and model systems, *Int. J. Cosmet. Sci.*, 1, 91, 1979.

Eary, L. E. and Rai, D., Kinetics of chromium (III) oxidation to chromium (VI) by reaction with manganese dioxide, *Environ. Sci. Technol.*, 21, 1187, 1987.

Eary, L. E. and Rai, D., Chromate removal for aqueous wastes by reduction with ferrous iron, *Environ. Sci. Technol.,* 22, 927, 1988.

Eastmond, D., Smith, M., and Iron, R., An interaction of benzene metabolites reproduces the myelotoxicity observed with benzene exposure, *Toxicol. Appl. Pharmacol.*, 91, 85, 1987.

Edgerton, T. R., Moseman, R. F., Linder, R. E., and Wright, L. H., Multi-residue method for the determination of chlorinated phenol metabolites in urine, *J. Chromatog.*, 170, 331, 1979.

Edwards, D. A., Smith, M. T., and Ruzo, L. O., A comparison of the metabolic fate of phenol, phenyl glucoside and phenyl 6-o-malonyl-glucoside in the rat, *Xenobiotica*, 16, 801, 1986.

Eitzer, B. D. and Hites, R. A., Vapor pressures of chlorinated dioxins and dibenzofurans, *Sci. Total Environ.*, 104,9, 1991.

Elias, Z., Poirot, O., Schneider, O., Daniere, M. C., Terzetti, F., Guedenet, J. C., and Cavelier, C., Cellular uptake, cytotoxic and mutagenic effects of insoluble chromic oxide in V78 Chinese hamster Cells, *Mutat. Res.*, 196, 159, 1986.

Elinder, C-G., Cadmium: uses, occurrence, and intake, in *Cadmium and Health: a Toxicological and Epidemiological Appraisal. Exposure, Dose, and Metabolism,* Vol. I, Friberg, L., Elinder, C-G., Kjellström, T., and Nordberg, G. F., Eds., CRC Press, Boca Raton, Florida, 1985, 23.

Elinder, C-G., Cadmium as an environmental hazard, in *Cadmium in the Human Environment*, Nordberg, G. F., Herber, R. F. M., and Alessio, L., Eds., International Agency for Research on Cancer, Lyon, 1992, 123.

Elinder, C-G, Kjellström, T., Friberg, L. F., Lind, B., and Linnman, L., Cadmium in kidney cortex, liver, and pancreas from Swedish autopsies, *Arch. Environ. Health,* 31, 1976, 292.

Ellenhorn, M. J. and Barceloux, D. G., Eds., *Medical Toxicology. Diagnosis and Treatment of Human Poisoning,* Elsevier, New York, 1988.

Eller, M .G., Walker, B. J., Westmark, P. A., Rubera, S. J., Anteny, K. K., and McNutt, B. E., Pharmacokinetics of terfenadine in healthy elderly subjects, *J. Clin. Pharmacol.*, 32, 267, 1992.

Ellis, K. J., Vartsky, D., Zanzi, I., Cohn, S. H., and Yasumura, S., Cadmium: *in vivo* measurement in smokers and nonsmokers, *Science*, 205, 323, 1979.

Endo, T., Nakaya, S., and Kimura, R., Mechanism of absorption of inorganic mercury from rat small intestine. III. Comparative absorption studies of inorganic mercuric compounds *in vitro*, *Pharmacol. Toxicol.*, 66, 347, 1990.

Engström, B. and Nordberg, G. F., Dose dependence of gastrointestinal absorption and biological half-time of cadmium in mice, *Toxicology,* 13, 215, 1979.

Enterline, P. E., Henderson, V. L., and Marsh, G. M., Exposure to arsenic and respiratory cancer. A reanalysis, *Am. J. Epidemiol.*, 125, 929, 1987.

Enterline, P. E., Day, R., and Marsh, G.M., Cancers related to exposure to arsenic at a copper smelter. *Occup. Environ. Med.*, 52, 28, 1995.

Environment Canada, Chromated copper arsenate (CCA) wood preservation facilities: recommendation for design and operation, EPS 2/WP/1, Environmental Protection Service, Environment Canada, Ottawa, 1988.

Esposito, G., Polarity of intestinal epithelial cells: permeability of the brush border and basolateral membranes, in *Pharmacology of Intestinal Permeation I*, Csáky, T. Z., Ed., Springer-Verlag, Berlin, 1984a, 283.

Esposito, G., Intestinal permeability of water-soluble nonelectrolytes: sugar, amino acids, peptides, in *Pharmacology of Intestinal Permeation I*, Csáky, T. Z., Ed., Springer-Verlag, Berlin, 1984b, 567.

Exon, J. H., A review of chlorinated phenol, *Vet Hum. Toxicol.*, 26, 508, 1984.

Fang, S. C., Comparative study of uptake and tissue distribution of methylmercury in female rats by inhalation and oral routes of administration, *Bull. Environ. Contam. Toxicol.*, 24, 65, 1980.

Farmer, J. G. and Johnson, L. R., Assessment of occupational exposure to inorganic arsenic based on urinary concentrations and speciation of arsenic, *Br. J. Ind. Med.*, 47, 342, 1990.

Farris, F., Dedrick, R. L., Allen, P. V., and Smith, J. C., Physiological model for the pharmacokinetics of methyl mercury in the growing rat, *Toxicol. Appl. Pharmacol.*, 119, 74, 1993.

Feldmann, R. J. and Maibach, H. I., Absorption of some organic compounds through the skin in man, *J. Invest. Dermatol.*, 54, 399, 1970.

Ferron, G. A., The size of soluble aerosol particles as a function of the humidity of the air. Application to the human respiratory tract, *J. Aerosol. Sci.*, 8, 251, 1977.

Fiedler, H. and Schramm, K-W., QSAR generated octanol-water partition coefficients of selected mixed halogenated dibenzodioxins and dibenzofurans, *Chemosphere*, 20,1597, 1990.

Fingerhut, M. A., Sweeney, M. H., Halperin, W. E., and Schnorr, T. M., The epidemiology of populations exposure to dioxin, in *Environmental Carcinogens Methods of Analysis and Exposure Measurement. Polychrinated Dioxins and Dibenzofurans*, Vol. 11, Rappe, C., Buser, H. R., Dodet, B., and O'Neill, I. K., Eds., IARC Scientific Publications No. 108, International Agency for Research on Cancer, Lyon, 1991, 31.

Fishbein, L., An overview of environmental and toxicological aspects of aromatic hydrocarbons. II. toluene, *Sci. Total Environ.*, 42, 267, 1985.

Fisher, L. B., Studies on the permeability of infant skin, in *Percutaneous Absorption. Mechanisms-Methodology-Drug Delivery*, 2nd Ed., Bronaugh, R. L. and Maibach, H. I., Eds., Marcel Dekker, Inc., New York, 1989, 135.

Flanagan, P. R., McLellan, J. S., Haist, J., Cherian, G., Chamberlain, M. J., and Valberg, L.S., Increased dietary cadmium absorption in mice and human subjects with iron deficiency, *Gastroenterology*, 74, 841, 1978.

Flanagan, P. R., Chamberlain, A. C., and Valberg, L. S., The relationship between iron and lead absorption in humans, *Am. J. Clin. Nutr.*, 36, 823, 1982.

Flynn, G. L. and Smith, R. W., Membrane diffusion III: influence of solvent composition and permeant solubility on membrane transport, *J. Pharm. Sci.*, 61, 61, 1972.

Flynn, G. L., Yalkowsky, S. H., and Roseman, T. J., Mass transfer phenomena and models: theoretical concepts, *J. Pharm. Sci.*, 63, 479, 1974.

Flynn, G. L., Physicochemical determinants of skin absorption, in *Principles of Route-to-route Extrapolation for Risk Assessment*, Gerrity, T. R. and Henry, C. J., Eds., Elsevier, New York, 93, 1990.

Flynn, G. L., Behl, C. R., Kurihara, T., Smith, W. M, Fox, J. L., Durrheim, H. H., and Higuchi, W. I., Correlation and Prediction of Mass Transport across Membranes. III. Boundary Layer and Membrane Resistances to Diffusion of Water and n-alkanols through Polydimethylsiloxane (Silastic) Membranes, paper presented at the 26th National Meeting of the Academy Pharmaceutical Science, Anaheim, CA, April 21-26, 1979, Basic Pharmaceutics Abstract #30.

Foa, V., Colombi, A., Maroni, M., Buratti, M., and Calzaferri, G., The speciation of the chemical forms of arsenic in the biological monitoring of exposure to inorganic arsenic, *Sci. Total Environ.*, 34, 241, 1984.

Forbes, G. B. and Reina, J. C., Effect of age on gastrointestinal absorption (Fe, Sr, Pb) in the rat, J. *Nutr.*, 102, 647, 1972.

Foth, H, Mollière, M., Kahl, R., Jähnchen, E., and Kahl, G., Covalent binding of benzo[a]pyrene in perfused rat lung following systemic and intratracheal administration, *Drug Metab. Dispos.*, 12, 760, 1984.

Foth, H., Kahl, R. and Kahl, G. F., Pharmacokinetics of low doses of benzo[a]pyrene in the rat, *Food. Cosmet. Toxicol.*, 26, 45, 1988.

Foulkes, E. C., Some determinants of intestinal cadmium transport in the rat, *J. Environ. Pathol. Toxicol.*, 3, 471, 1980.

Foulkes, E. C., Intestinal absorption of heavy metals, in *Pharmacology of Intestinal Permeation I*, Csáky, T. Z., Ed., Springer-Verlag, Berlin, 1984, 543

Foulkes, E. C., On the mechanism of transfer of heavy metals across cell membranes, *Toxicology*, 52, 263, 1988.

Foulkes, E. C., On the mechanism of cellular cadmium uptake, *Bio. Trace Elem. Res.*, 21, 195, 1989.

Foulkes, E. C. and Bergman, D., Inorganic mercury absorption in mature and immature rat jejunum: transcellular and intercellular pathways *in vivo* and in everted sacs, *Toxicol. Appl. Phamacol.*, 120, 89, 1993.

Franchini, R., Mutti, A., Cavatorta, E., Pedroni, C., and Borghetti, A., Chromium, in *Biological Indicators for the Assessment of Human Exposure to Industrial Chemicals*, Alessio, L., Berlin, A. and Roi, M. R., Eds., Directorate-General for Employment and Social Affairs, Health and Safety Directorate, Directorate-General for Science Research and Development, Commission of the European Communities, EUP 8903 EN, Brussels-Luxembourg, 1984.

Frank, N. R., The effects of inhaled pollutants on nasal and pulmonary flow-resistance, *Ann. Otol. Rhinol. Laryngol.*, 79, 540, 1970.

Frank, N. R., Yoder, R. E., Brain, J. D., and Yokoyama, E., SO_2 (^{35}S labeled) absorption by the nose and mouth under conditions of varying concentration and flow, *Arch. Environ. Health*, 18, 315, 1969.

Frantz, S. W., Instrumentation and methodology for *in vitro* skin diffusion cells in methodology for skin absorption, in *Methods for Skin Absorption*, Kemppainen, B. W. and Reifenrath, W. G., Eds., CRC Press, Boca Raton, Florida, 1990, 34.

Franz, T. J., Percutaneous absorption of benzene, in *Applied Toxicology of Petroleum Hydrocarbons*, Vol. VI, Mehlman, M. A., Ed., Princeton Scientific Publishers, Inc., Princeton, New Jersey, 1984, 61.

Franz, T. J., Percutaneous absorption. On the relevance of *in vitro* data, *J. Invest. Dermatol.*, 64, 190, 1985.

Freeman, G. B., Johnson, J. D., Killinger, J. M., Liao, S. C., Davis, A. O., Ruby, M. V., Schoof, R. A., Chaney, R. L., and Bergstrom, P. D., Absolute bioavailability of lead acetate and mining waste lead in rats, *Toxicology*, 91, 151, 1994.

Freeman, G. B., Johnson, J. D., Killinger, J. M., Liao, S. C., Davis, A. O., Ruby, M. V., Chaney, R. L., Lover, S. C. and Bergstrom, P. D., Bioavailability of arsenic in soil impacted by smelter activities following oral administration in rabbits, *Fund. Appl. Toxicol.*, 21, 83, 1993.

Freitage, D., Geyer, H., Kraus, A., Viswanathan, R., Kotzias, D., Attar, A., Klein, W. and Korte, F., Ecotoxicological profile analysis, *Ecotoxicol. Environ. Saf.*, 6, 60, 1982.

Friberg, L., Skog, E., and Wahlberg, J. E., Resorption of mercuric chloride and methyl mercury dicyandiamide in guinea-pigs through normal skin and through skin pre-treated with acetone, alkylaryl-sulphonate and soap, *Acta Dermato. Venereol.*, 41, 40, 1961.

Friberg, L. F. and Nordberg, G. F., Inorganic mercury -- a toxicological and epidemiological appraisal, in *Mercury, Mercurials and Mercaptans*, Miller, M. W. and Clarkson, T. W., Eds., Charles C. Thomas, Springfield, IL, 1973.

Friberg, L. F., Kjellström, T., and Nordberg, G. F., Cadmium, in *Handbook on the Toxicology of Metals,* Vol. II, Friberg, L., Nordberg, G., and Vouk, V. B., Eds., Elsevier, Amsterdam, 1986, 130.

Friberg, L. F., Piscator, M., Nordberg, G. F., and Kjellström, T., *Cadmium in the Environment*, CRC Press, Boca Raton, Florida, 1971.

Friberg, L. F., Piscator, M., Nordberg, G. F., and Kjellström, T., *Cadmium in the Environment*, 2nd Ed., CRC Press, Boca Raton, Florida, 1974.

Fries, G. F. and Marrow, G. S., Retention and excretion of 2,3,7,8-tetrachlorodibenzo-p-dioxin by rats, *Food Chem.*, 23, 265, 1975.

Fry, F. A. and Black, A., Regional particle deposition and clearance of particle in the human nose, *J. Aerosol Sci.*, 4, 113, 1973.

Furst, P., Furst, C., and Wilmers, K., Body burden with PCDD and PCDF from food, in *Biological Basis for Risk Assessment of Dioxins and Related Compounds*, Gallo, M. A., Scheuplein, R. J., and Van Der Heijden, K. A., Eds., Cold Spring Harbor Laboratory Press, Cold Spring Harbor, New York, 1991, 133.

Gaines, T. B., The acute toxicity of pesticides to rats, *Toxicol. Appl. Pharmacol.*, 2, 88, 1960.

Gaines, T. B., The acute toxicity of pesticides, *Toxicol. Appl. Pharmacol.*, 14, 515, 1969.

Gaines, T. B. and Linder, R. E., The acute toxicity of pesticides in adult and weanling rats, *Fund. Appl. Toxicol.*, 7, 299, 1968.

Gallo, M. A. and Doull, J., History and scope of toxicology, in *Casarett and Doull's Toxicology. The Basic Science of Poisons,* Fourth Edition, Amdur, M. O., Doull, J., and Klaassen, C. D., Eds., Pergamon Press, New York, 1991, 3.

Gammelgaard, B., Fullerton, A., Avenstorp, C., and Menné, T., Permeation of chromium salts through human skin *in vitro*, *Contact Dermatitis,* 27, 302, 1992.

Garber, B. and Wei, E., Influence of dietary factors on the gastrointestinal absorption of lead, *Toxicol. Appl. Pharmacol.,* 27, 685, 1974.

Garrett, E. R. and Chemburkar, P. B., Evaluation, control, and prediction of drug diffusion through polymeric membranes I. Methods and reproducibility of steady-state diffusion studies, *J. Pharm. Sci.*, 57, 944, 1968a.

Garrett, E. R. and Chemburkar, P. B., Evaluation, control, and prediction of drug diffusion through polymeric membranes II. Diffusion of aminophenones through silastic membranes: a test of the pH-partition hypothesis, *J. Pharm. Sci.*, 57, 949, 1968b.

Garrett, E. R. and Chemburkar, P. B., Evaluation, control, and prediction of drug diffusion through polymeric membranes III. Diffusion of barbiturates, phenylalkylamines, dextromethorphan, progesterone, and other drugs, *J. Pharm. Sci.*, 57, 1401, 1968c.

Gary-Bobo, C. M., DiPolo, R., and Solomon, A. K., Role of hydrogen-bonding in nonelectrolyte diffusion through dense artificial membranes, *J. Gen. Physiol.*, 54, 369, 1969.

Gerson, B., Lead, *Clin. Lab. Med.*, 10, 441, 1990.

Geyer, H., Scieunert, I., and Korte, F., Bioconcentration potential of organic environmental chemicals in humans, *Reg. Toxicol. Pharmacol.*, 6, 313, 1986.

Gibaldi, M. and Perrier, D., *Pharmacokinetics,* Marcel Dekker, Inc., New York, 1982.

Gilbert, F. I., Minn, C. E., Duncan, R. C., and Wilkinson, J., Effects of pentachlorophenol and other chemical preservatives on the health of wood-treating workers in Hawaii, *Arch. Environ. Contam. Toxicol.,* 19, 603, 1990.

Glaser, U., Klöppel, H., and Hochrainer, D., Bioavailability indicators of inhaled cadmium compounds, *Ecotoxicol. Environ. Safety,* 11, 261, 1986.

Glaser, U., Hochrainer, D., Otto, F. J., and Oldiges, H., Carcinogenicity and toxicity of four cadmium compounds inhaled by rats, *Toxicol. Environ. Chem.,* 27, 153, 1990.

Goode, C. A., Dinh, C. T., and Linder, M. C., Mechanism of copper transport and delivery in mammals: review and recent findings, *Adv. Exp. Med. Biol.*, 258, 131, 1989.

Gordon, T. and Amdur, M. O., Responses of the respiratory system to toxic agents, in *Casarett and Doull's Toxicology. The Basic Science of Poisons,* Fourth Edition, Amdur, M. O., Doull, J., and Klaassen, C. D., Eds., Pergamon Press, New York, 1991, 383.

Gore, D. J. and Patrick, G., A quantitative study of the penetration of insoluble particles into the tissue of the conduction airways, *Ann. Occup. Hyg.*, 26, 146, 1982.

Gorsline, J., Okerholm, R. A., Rolf, C. N., Moos, C. D., and Hwang, S. S., Comparison of plasma nicotine concentration after application of nicotine (nicotine transdermal system) to different skin sites, *J. Clin. Pharmacol.*, 32, 576, 1992.

Gough, M., Human exposure from dioxin in soil - a meeting report, *J. Toxicol. Environ. Health,* 32, 205, 1991.

Goyer, R. A., Toxic effects of metals, in *Casarett and Doull's Toxicology. The Basic Science of Poisons,* Fourth Edition, Amdur, M. O., Doull, J., and Klaassen, C. D., Eds., Pergamon Press, New York, 1991, 643.

Graham, M. J., Lucier, G. W., Linko, P., Maronpot, R. R., and Goldstein, J. A., Increases in cytochrome P-450 mediated 17-A-estradio 2-hydroxylase activity in rat liver microcromes after both acute administration and subchronic administration 2,3,7,8-tetrachlorodibenzo-p-dioxin in a two-stage hepatocarcinogenesis model, *Carcinogenesis,* 9, 1935, 1988.

Grayson, S. and Elias, P. M., Isolation and lipid biochemical characterization of stratum corneum membrane complexes: implication for the cutaneous permeability barrier, *J. Inves. Dermatol.*, 78, 128, 1982.

Greenblatt, D., Sellers, E. and Shader, R., Drug disposition in old age, *N. Engl. J. Med.*, 306, 1081, 1982.

Greger, J. L., Using animals to assess bioavailability of minerals: implications for human nutrition, *J. Nutr.*, 122, 2047, 1992.

Gregus, Z. and Klaassen, C. D., Enterohepatic circulation of toxicants, in *Gastrointestinal Toxicology,* Rozman, K. and Hänninen, O., Eds., Elsevier, New York, 1986, 57.

Grimmer, G., Brune, H., Dettbarn, G., Heinrich, U., Jacob, J., Mohtashamipur, E., Norpoth, K., and Pott, F., Urinary and fecal excretion of chrysene metabolites by rats after oral, intraperitoneal, intratracheal or intrapulmonary application, *Arch. Toxicol.*, 62, 401, 1988.

Groen, K., Vaessen, H., Kliest, J., de Boer, J., van Ooik, T., Timmerman, A., and Vlug, R. F., Bioavailability of inorganic arsenic from bog ore-containing soil in the dog, *Environ. Health Persp.*, 102, 182, 1994.

Gross, S. B., Human oral and inhalation exposure to lead: summary of the kehoe experiments, *J. Toxicol. Environ. Health,* 8, 333, 1981.

Gummer, C. L. and Maibach, H. I., Diffusion cell design, in *In Vitro Percutaneous Absorption: Principles, Fundamentals, and Applications,* Bronaugh, R. L. and Maibach, H. I., Eds., CRC Press, Boca Raton, Florida, 1991, 7.

Guy, R. H. and Fleming, R., The estimation of diffusion coefficients using the rotating diffusion cell, *Int. J. Pharm.*, 3, 143, 1979.

Guyton, A. C., *Textbook of Medical Physiology,* 8th ed., W.B. Saunders Company, Philadelphia, 1991, 422.

Gylseth, B., Gunderson, N., and Langård, S., Evaluation of chromium exposure based on a simplified method for urinary chromium determination, *Scand. J. Work Environ. Health,* 3, 28, 1977.

Hadgraft, J. and Ridout, G., Development of model membranes for percutaneous absorption measurement. I Isopropyl myristate, *Int. J. Pharm.,* 39, 149, 1987.

Hadgraft, J. and Ridout, G., Development of model membranes for percutaneous absorption measurement. II Dipalmitoyl phosphatidycholine, linoleic acid and tetradecane, *Int. J. Pharm.*, 42, 97,1988.

Hadley, J. G., Conklin, A.M., and Sanders, C. M., Rapid solubilization and translocation of 109-CdO following pulmonary deposition, *Toxicol. Appl. Pharmacol.,* 54, 156, 1980.

Hajimiragha, H., Ewers, U., Brockaum, A., and Boettger, A., Levels of benzene and other volatile aromatic compounds in the blood of non-smokers and smokers, *Int. Arch. Environ. Health,* 61, 513, 1989.

Hall, L. L. and Shah, P. V., *In vivo* methods for determining percutaneous absorption and metabolism of xenobiotics: indirect methods, in *Methods for Skin Absorption*, Kemppainen, B. W. and Reifenrath, W. G., Eds., CRC Press, Boca Raton, Florida, 1990, 81.

Hall, L. I., Fisher, H. L., Sumler, M. R., Hughes, M. F., and Shah, P. V., Aged-related percutaneous penetration of 2-sec-butyl-4,6-dinitrophenol (Dinoseb) in rats, *Fund. Appl. Toxicol.*, 19, 258, 1992.

Halpern, M., Williamson, S. J., Spektor, D. M., Schlesinger, R. B., and Lippmann, M., Remanent magnetic field for measuring particle retention and distribution in the lung, *Exp. Lung Res.*, 2, 27, 1981.

Hänninen, O., Biochemistry of the gastrointestinal tract, in *Gastrointestinal Toxicology,* Rozman, K. and Hänninen, O., Eds., Elsevier, New York, 1986, 146.

Hanke, J., Dutkiewicz, T., and Piotrowski, J., The absorption of benzene through human skin, *Med. Pr.,* 12, 413, 1961. (Cited by USEPA 1992d).

Harada, K., Murakami, T., Yata, N., and Yamamoto, S., Role of intercellular lipids in stratum corneum in the percutaneous permeation of drugs, *J. Invest. Dermatol.,* 99, 278, 1992.

Harris, D. R., Papa, C. M., and Stanton, R., Percutaneous absorption and the surface area of occluded skin, *Br. J. Dermatol.*, 91, 27, 1974.

Harris, E. D., Copper transport: an overview, *Proc. Soc. Exp. Biol. Med.*, 196, 130, 1991.

Harrison, G. E., Carr, T. E. F., Sutton, A., and Humphreys, E. R., Effect of alginate on the absorption of lead in man, *Nature*, 224, 1115, 1969.

Harrison, H. E., Bunting, H., Ordway, N., and Albrink, W. S., The effects and treatment of inhalation of cadmium chloride aerosols in the dog, *J. Ind. Hyg. Toxicol.*, 29, 302, 1947.

Hartford, W. H., Chromium compounds, in *Kirk-Othmer Encyclopedia of Chemical Technology,* 3rd ed., Vol. 6, Grayson, M., Ed., John Wiley & Sons, New York, 82, 1979.

Hatanaka, T., Inuma, M., Sugibayashi, K., and Morimoto, Y., Prediction of skin permeability of drugs I. Comparison with artificial membrane, *Chem. Pharm. Bull.,* 38, 3452, 1990.

Hatanaka, T., Inuma, M., Sugibayashi, K., and Morimoto, Y., Prediction of skin permeability of drugs. II. Development of composite membrane as a skin alternative, *Int. J. Pharm.,* 79, 21, 1992.

Hatch, T. E. and Gross, P., *Pulmonary Deposition and Retention of Inhaled Aerosols,* Academic Press, New York, 1964.

Hathaway, G. J., Proctor, N. H., Hughes, J. P., and Fischman, M. L., *Proctor and Hughes' Chemical Hazards of the Workplace,* 3rd ed., Van Nostrand Reinhold, New York, 1991.

Hattemer-Frey, H. A. and Travis, C. C., Pentachlorophenol: environmental partitioning and human exposure, *Arch. Environ. Contam. Toxicol.,* 18, 482, 1989.

Hattemer-Frey, H. A., Travis, C. C. and Land, M. L., Benzene: environmental partitioning and human exposure, *Environ. Res.,* 53, 221, 1990.

Hayashi, M., Yamamoto, K., Yoshimura, M., Kishimoto, T., and Shitara, A., Effects of fasting on distribution and excretion of lead following long-term lead exposure in rats, *Arch. Environ. Contam. Toxicol.,* 24, 201, 1993.

Hayes, A. D. and Rothstein, A., The metabolism of inhalation mercury vapor in the rat studies by isotope techniques, *J. Pharmacol. Exp. Ther.,* 138, 1, 1962.

Heard, M. J. and Chamberlain, A. C., Effects of minerals and food on uptake of lead from the gastrointestinal tract in humans, *Hum. Toxicol.,* 1, 411, 1982.

Heard, M. J. and Chamberlain, A. C., Uptake of lead by humans and effect of minerals and food, *Sci. Total Environ.,* 30, 245, 1983.

Hecht, S. S., Grabowski, W., and Groth, K., Analysis of feces for B[a]P after consumption of charcoal-broiled beef by rats and humans, *Food. Cosmet. Toxicol.,* 17, 223, 1979.

Heidelberger, C. and Weiss, S. M., The distribution of radioactivity in mice following administration of 3,4-Benzpyrene-5-C^{14} and 1,2,5,6-Dibenzanthracene, *Cancer Res.,* 11, 885, 1951.

Heikkila, P. R., Hameila, M., Pyy, L., and Raune, P., Exposure to creosote in the impregnating and handing of impregnated wood, *Scan. J. Work Enrivon. Health,* 13, 431, 1987.

Henderson, R. F., Rebar, A. H., Pickrell, J. A., and Newton, G. J., Early damage indicators in the lung. III. Biochemical and cytological response of the lung to inhaled metal salts, *Toxicol. Appl. Pharmacol.,* 50, 123, 1979.

Henderson, R. F., Sabourin, P. J., Bechtold, W. E., Griffith, W. C., Medinsky, M. A., Birnbaum, L. S., and Lucier, G. W., The effect of dose, dose rate, route of administration, and species on tissue and blood levels of benzene metabolites, *Environ. Health Persp.,* 82, 9, 1989.

Henry, C. J. and Kouri, R. E., Specialized test article administration: nose-only exposure and intratracheal inoculation, in *Inhalation Toxicology. Research Methods, Applications and Evaluation,* Salem, H., Ed., Marcel Dekker, Inc., 1987, 121.

Henry, S., Cramer, G., Bolger, M., and Springer, J., Exposure and risks of dioxin in the U.S. food supply, *Chemosphere,* 25, 235, 1992.

Hertz-Picciotto, I. and Smith, A. H., Observation on the dose-response curve for arsenic exposure and lung cancer, *Scand. J. Work Environ. Health,* 19, 217, 1993.

Hill, R. H., To, T., Holler, J. S., Fast, D. M., Smith, S. J., Needham, L. L., and Binder, S., Residues of chlorinated phenols and phenoxy acid herbicides in the urine of Arkansas children, *Arch. Environ. Contam. Toxicol.*, 18, 469, 1989.

Hjelm, E. W., Naslund, P. H., and Wallen, M., Influence of cigarette smoking on the toxicokinetics of toluene in humans, *J. Toxicol. Environ. Health,* 25, 155, 1988.

Hoben, H. J., Ching, S. A., and Casarett, L. J., A study of inhalation of pentachlorophenol by rats IV. Distribution and excretion of inhaled pentachlorophenol, *Bull. Environ. Contam. Toxicol.,* 15, 466, 1976.

Hoffmann, R. T. and Hrudey, S.E., Evaluation of the reclamation of decommissioned wood preserving plant sites in Alberta, Final Report, University of Alberta, 1990.

Holland, R. H., McCall, M. S., and Lanz, H. C., A study of inhaled arsenic-74 in man, *Cancer Res.*, 19, 1154, 1959.

Holländer, W., Exposure facilities and aerosol generation and characterization for inhalation experiments, in *Inhalation Toxicology. The Design and Interpretation of Inhalation Studies and Their Use in Risk Assessment,* Mohr, U., Ed., Springer-Verlag, Berlin, 1988, 67.

Hollins, J. G., Charbonneau, S. M., Bryce, F., Ridgeway, J. M., Tam, G. K. H., and Willes, R. E., Whole body retention and excretion of [^{74}As] arsenic acid in the adult beagle dog, *Toxicol. Lett.*, 4, 7, 1979.

Hopenhayn-Rich, C., Smith, A. H., and Goeden, H. M., Human studies do not support the methylation threshold hypothesis for the toxicity of inorganic arsenic, *Environ. Res.,* 60, 161, 1993.

Horstman, S. W., Rossner, A., Kalman, D. A., and Morgan, M. S., Penetration of pentachlorophenol and tetrachlorophenol through human skin, *J. Environ. Sci. Health*, A24, 229, 1989.

Hostynek, J. J., Hinz, R. S., Lorence, C. R., Price, M., and Guy, R.H., Metals and the skin, *Crit. Rew. Toxicol.*, 23(2), 171, 1993.

Hotchkiss, S. A. M., Hewitt, P., Caldwell, J., Chen, W. L., and Rowe, R. R., Percutaneous absorption of nicotinic acid, phenol, benzoic acid and triclopyr, butoxyethyl ester through rat and human skin *in vitro:* further validation of an *in vitro* model by comparison with *in vivo* data, *Food. Chem. Toxiol.*, 30, 891, 1992.

Houk, J. and Guy, R.H., Membrane models for skin penetration studies, *Chem. Rev.,* 88, 455, 1988.

Houston, J. B. and Wood, S. G., Gastrointestinal absorption of drugs and other xenobiotics, in *Progress in Drug Metabolism,* Vol. 4, Bridges, J. W. and Chasseaud, L. F., Eds., John Wiley & Sons, New York, 1980, 57.

Howard, P. O., Ed., *Handbook of Environmental Fate and Exposure Data for Organic Chemicals,* Vol. I, Lewis Publishers, Chelsea, Michigan, 1989a, 468.

Howard, P. O., Ed., *Handbook of Environmental Fate and Exposure Data for Organic Chemicals,* Vol. I, Lewis Publishers, Chelsea, Michigan, 1989b, 44.

Howard, P. O., Ed., *Handbook of Environmental Fate and Exposure Data for Organic Chemicals,* Vol. II, Lewis Publishers, Chelsea, Michigan, 1990a, 29.

Howard, P. O., Ed., *Handbook of Environmental Fate and Exposure Data for Organic Chemicals,* Vol. II, Lewis Publishers, Chelsea, Michigan, 1990b, 435-444.

Hughes, M. F., Shrivastava, S. P., Sumler, M. R., Edwards, B. C., Goodwin, J. H., and Shah, P. V., Dermal absorption of chemicals: effect of application of chemicals as a solid, aqueous paste, suspension, or in a volatile vehicle, *J. Toxicol. Environ. Health,* 37, 57, 1992.

Hunke, W. A. and Matheson, Jr., L. E., Mass transport properties of Co(polyether)polyurethane membranes II: permeability and sorption characteristics, *J. Pharm. Sci.,* 70, 1313, 1981.

Hunt, A., Johnson, D. V., Watt, J. M., and Thornton, I., Characterizing the sources of particulate lead in house dust by automated scanning electron microsocopy, *Environ. Sci. Technol.,* 26, 1513, 1992.

Hunter, C. G., Solvents with reference to studies on the pharmacodynamics of benzene, *Proc. R. Soc. Med.,* 61, 913, 1968.

Huq, A. S., Ho, N. F. H., Husari, N., Flynn, G. L., Jetzer, W. E. and Condie, Jr., L., Permeation of water contaminative phenols through hairless mouse skin, *Arch. Environ. Contam. Toxicol.,* 15, 557, 1986.

Hursh, J. B., Particle coefficients of mercury (^{203}Hg) vapor between air and biological fluids, *J. Appl. Toxicol.,* 5, 327, 1985.

Hursh, J. B. and Suomela, J., Absorption of ^{212}Pb from the gastrointestinal tract of man, *Acta Radio. (Therapy Physics Biology),* 7, 108, 1968.

Hursh, J. B. and Mercer, T. T., Measurement of ^{212}Pb loss rate from human lungs, *J. Appl. Physiol.,* 28, 268, 1970.

Hursh, J. B., Clarkson, T., Cherian, M. G., Vostal, J. J., and Vander Mallie, R., Clearance of mercury (Hg-197, Hg-203) vapor inhaled by human subjects, *Arch. Environ. Health,* 31, 302, 1976.

Hursh, J. B., Clarkson, T., Miles, E. F. and Goldsmith, L. A., Percutaneous absorption of mercury vapor by man, *Arch. Environ. Health,* 44, 120, 1989.

Hwang, C-C. and Danti, A. G., Percutaneous absorption of flufenamic acid in rabbits: effect of dimethyl sulfoxide and various nonionic surface-active agent, *J. Pharm. Sci.,* 72, 857, 1983.

HWC (Health and Welfare Canada), Benzene. Human Health Implications of Benzene at Levels Found in the Canadian Environment and Workplace, 79-EHD-40, Environmental Health Directorate Health Protection Branch, Health and Welfare Canada, Ottawa, 1979.

IARC (International Agency for Research on Cancer), *Evaluation of the Carcinogenic Risk of Chemicals to Humans: Cadmium, Nickel, Some Epoxied, Miscellaneous Industrial Chemicals, and General Considerations on Volatile Anaesthetics,* Vol. 11, 39-75, International Agency for Research on Cancer, Lyon, 1976, 95.

IARC (International Agency for Research on Cancer), *Evaluation of the Carcinogenic Risk of Chemicals to Humans: Some Halogenated Hydrocarbons,* Vol. 20, International Agency for Research on Cancer, Lyons, 1979, 349.

IARC (International Agency for Research on Cancer), *Evaluation of the Carcinogenic Risk of Chemicals to Humans: Some Metals and Metallic Compounds,* Vol. 23, International Agency for Research on Cancer, Lyons, 1980, 205.

IARC (International Agency for Research on Cancer), *Evaluation of the Carcinogenic Risk of Chemicals to Humans: Some Industrial Chemicals and Dyestuffs*, Vol. 29, International Agency for Research on Cancer, Lyons, 1982a, 95.

IARC (International Agency for Research on Cancer), *Evaluation of the Carcinogenic Risk of Chemicals to Humans: Some Aromatic Amines, Anthraquinones and Nitroso Compounds, and*

Inorganic Fluorides Used in Drinking-water and Dental Preparations, Vol. 27, International Agency for Research on Cancer, Lyons, 1982b, 39.

IARC (International Agency for Research on Cancer), *Evaluation of the Carcinogenic Risk of Chemicals to Humans: Chemicals, Industrial Processes, and Industries Associated with Cancer in Humans*, Vol. 1 to 29, International Agency for Research on Cancer, Lyon, 1982c, 71.

IARC (International Agency for Research on Cancer), *Evaluation of the Carcinogenic Risk of Chemicals to Humans: Polynuclear Aromatic Compounds, Part 1, Chemical, Environmental and Experimental Data,* Vol. 31, International Agency for Research on Cancer, Lyons, 1983.

IARC (International Agency for Research on Cancer), *Evaluation of the Carcinogenic Risk of Chemicals to Humans: Overall Evaluation of Carcinogenicity: An Updating of IARC Monographs,* Vol. 1 to 42, Suppl. 7, International Agency for Research on Cancer, Lyons, 1987, 165.

IARC (International Agency for Research on Cancer), *Environmental Carcinogens - Methods of Analysis and Exposure Measurement. Benzene and Alkylated Benzenes*, Vol. 10, IARC Scientific Publications No. 85, International Agency for Research on Cancer, Lyons, 1988.

IARC (International Agency for Research on Cancer), *Evaluation of Carcinogenic Risk to Humans: Chromium and Chromium Compounds: Biological Data Relevant to the Evaluation of Carcinogenic Risk to Humans, IARC Monographs,* Vol. 49, International Agency for Research on Cancer, Lyons, 1990, 115.

IARC (International Agency for Research on Cancer), *Evaluation of Carcinogenic Risks to Humans: Occupational Exposures in Insecticide Application, and Some Pesticides, IARC Monographs*, Vol. 53, International Agency for Research on Cancer, Lyons, 1991, 371.

ICRP (International Commission on Radiological Protection), *Report of Committee IV on Evaluation of Radiation Dose to Body Tissues from Internal Contamination due to Occupational Exposure,* International Commission on Radiological Protection Publication 10, Pergamon Press, New York, 1968.

Idson, B. H. and Behl, C. R., Drug structure vs. skin permeation, in *Transdermal Delivery of Drugs,* Vol. 3, Kydonieus, A. F. and Berner, B., Eds., CRC Press, Boca Raton, Florida, 1987.

Illel, B., Schaefer, H., Wepierre, J., and Doucet, O., Follicles play an important role in percutaneous absorption, *J. Pharm. Sci.,* 80, 424, 1991.

Inamasu, T., Hisanaga, A., and Ishinishi, N., Comparison of arsenic trioxide and calcium arsenate retention in the rat lung after intratracheal instillation, *Toxicol. Lett.,* 12, 1, 1982.

Ingram, A. J. and Phillips, J. C., The dermal bioavailability of radiolabelled benzo[a]pyrene from acetone or oils of differing viscosity, assessed by DNA and protein binding, *J. Appl. Toxicol.,* 13, 25, 1993.

Ings, R. M. J., Animal studies and the bioavailability testing of drug products, in *Controlled Drug Bioavailability,* Smolen, V. F. and Ball, L., Eds., John Wiley & Sons, New York, 1984, 43.

Inoue, O., Seiji, K., Watanabe, T., Nakatsuka, H., Jin, C., Liu, S., and Ikeda, M., Effects of smoking and drinking on excretion of hippuric acid among toluene-exposed workers, *Int. Arch. Occup. Environ. Health,* 64, 425, 1993.

IRIS (Integrated Risk Information System), Cadmium, #0141, U.S. Environmental Protection Agency, Washington, D.C., 1993.

Itoh, T., Xia, J., Magavi, R., Nishihata, T., and Rytting, J. H., Use of shed snake skin as a model membrane for *in vitro* percutaneous penetration studies: comparison with human skin, *Pharm. Res.,* 7, 1042, 1990a.

Itoh, T., Magavi, R., Casady, R. L., Nishihata, T., and Rytting, J. H., A method to predict the percutaneous permeability of various compounds: shed snake skin as a model membrane, *Pharm. Res.*, 7, 1302, 1990b.

Iverson, F., and Grant, D. L., Toxicology of the polychlorinated biphenyls, dibenzofurans and dibenzodioxins, in *Environmental Carcinogens Methods of Analysis and Exposure Measurement. Polychrinated Dioxins and Dibenzofurans*, Vol. 11, Rappe, C., Buser, H.R., Dodet, B., and O'Neill, I. K., Eds., IARC Scientific Publications No. 108, International Agency for Research on Cancer, Lyon, 1991, 5.

Iyengar, G. V., Nutritional chemistry of chromium, *Sci. Total Environ.*, 86, 69, 1989.

Jacob, R. A., Skala, J. H., Omaye, S. T., and Turnlund, J. R., Effect of varying ascorbic acid intakes on copper absorption and ceruloplasmin levels in young men, *J. Nutr.*, 117, 2109, 1987.

James, A. C., Lung deposition of submicron aerosols calculated as a function of age and breathing rate, in *Annual Research and Development Report,* National Radiological Protection Board, Harwell, Oxfordshire, 1978.

James, A. C., Birchall, A., Cross, F. T., Cuddihy, R. G., and Johnson, J. R., The current approach of the ICRP Task Group for modeling doses to respiratory tract tissues, *Health Phys.*, 57, Sup.1, 271, 1989.

Jelinek, C. F. and Corneliussen, P. E., Levels of arsenic in the United States food supply, *Environ. Health Persp.*, 19, 83, 1977.

Jetzer, W. E., Huq, A. S., Ho, N. F. H., Flynn, G. L., Duraiswamy, N., and Condie, Jr., L., Permeation of mouse skin and silicone rubber membranes by phenols: relationship to *in vitro* partitioning, *J. Pharm. Pharmacol.*, 75, 1098, 1986.

Jobb, B., Uza, M., Hunsinger, R., Roberts, K., Tosine, H., Clement, R., Bobbie, B., LeBel, G., Williams, D., and Lau. B., A survey of drinking water supplies in the province of Ontario for dioxins and furans, *Chemosphere*, 20, 1553, 1990.

Johnson, C. P., Cowles, V. E., Bonham, L., Buchann, E., Telford, G. L., Roza, A. M., and Adams, M. B., Changes in motility, transit time, and absorption following surgical transection of the jejunum, *Transplant Proc.*, 24, 1110, 1992.

Johnson, P. E., Stuart, M. A., Hunt, J. R., Mullen, L., and Starks, T. L., ^{65}Copper absorption by women fed intrinsically and extrinsically labeled goose meat, goose liver, peanut butter and sunflower butter, *J. Nutr.*, 118, 1522, 1988.

Johnson, P. E., Factors affecting copper absorption in humans and animals, *Adv. Exp. Med. Biol.*, 258, 71, 1989.

Johnson, P. E., Milne, D. B., and Lykken, G., Effects of age and sex on copper absorption, biological half-life, and status in humans, *Am. J. Clin. Nutr.*, 56, 917, 1992.

Jones, R. D., Winter, D. P., and Cooper, A. J., Absorption study of pentachlorophenol in persons working with wood preservatives, *Hum. Toxicol.*, 5, 189, 1986.

Juhl, U., Witte, I., and Butte, W., Metabolism of pentachlorophenol to tetrachlorohydroquinone by human liver homogenate, *Bull. Environ. Contam. Toxicol.*, 35, 596, 1985.

Kalf, G., Recent advances in the metabolism and toxicity of benzene, *CRC Crit. Rev. Toxicol.*, 18, 141, 1987.

Kaminsky, L. S., DeCaprio, A. P., Gierthy, J. F., Silkworth, J. B., and Tumasonis, C., The role of environmental materials and experimental vehicles in chlorinated dibenzodioxin and dibenzofuran toxicity, *Chemosphere*, 14, 685, 1985.

Kao, J. and Bridges, J. M., Metabolism of [14C]phenol by sheep, pig and rat, *Xenobiotica,* 9, 141, 1979.

Kao, J., Patterson, F. K., and Hall, J., Skin penetration and metabolism of topically applied chemicals in six mammalian species, including man: an *in vitro* study with benzo[a]pyrene and testosterone, *Toxicol. Appl. Pharmacol.,* 81, 502, 1985.

Kapoor, S. C., Wieloposki, L., Graziano, J. H., and LoIacono, N. J., Influence of 2,3-Dimercaptosuccinic acid on gastrointestinal lead absorption and whole-body lead retention, *Toxicol. Appl. Pharmacol.,* 97, 525, 1989.

Karg, E., Tuch, T., Ferron, G. A., Haider, B., Kreyling, W. G., Peter, J., Ruprecht, L., and Heyder, J., Design, operation and performance of whole body chambers for long-term aerosol animals, *J. Aerosol Sci.,* 23, 279, 1992.

Kelly, W. F., Ackrill, P., Day, J. P., O'Hara, M., Tye, C. T., Burton, I., Orton, C., and Harris, M., Cutaneous absorption of trivalent chromium: tissue levels and treatment by exchange transfusion, *Br. J. Ind. Med.,* 39, 397, 1982.

Kenoyer, J. L., Phalen, R. F., and Davies, J. R., Particle clearance from the respiratory tract as a test of toxicity: effect of ozone on short and long term clearance, *Exp. Lung Res.,* 2, 111, 1981.

Kies, C. and Harms, J. M., Copper absorption as affected by supplemental calcium, magnesium, manganese, selenium and potassium, *Adv. Exp. Med. Biol.,* 258, 45, 1989.

Kimbrough, R., Falk, H., Stehr, P., and Fries, G., Health implications of 2,3,7,8-tetrachlorodibenzo-p-dioxin (TCDD) contamination of residential soil, *J. Toxicol. Environ. Health,* 14, 47, 1984.

Kimura, M. and Otaki, N., Percutaneous absorption of cadmium in rabbit and hairless mouse, *Ind. Health,* 10, 7, 1972.

Kincl, F. A., Benagiano, G., and Angee, I., Sustained release hormonal preparations I. Diffusion of various steroid through polymer membranes, *Steroid,* 11, 673, 1968.

King, J. C., Raynolds, W. L., and Margen, S., Absorption of stable isotopes of iron, copper, and zinc during oral contraceptive use, *Am. J. Clin. Nutr.,* 31, 1198, 1978.

Klaassen, C. D. and Rozman, K., Absorption, distribution, and excretion of toxicants, in *Casarett and Doull's Toxicology. The Basic Science of Poisons,* Fourth Edition, Amdur, M. O., Doul, J., and Klaassen, C. D., Eds., Pergamon Press, New York, 1991, 50.

Klain, G. J. and Black, K. E., Specialized techniques -- congenitally athymic (nude) animal models, in *Methods for Skin Absorption,* Kemppainen, B. W. and Reifenrath, W. G., Eds., CRC Press, Boca Raton, Florida, 1990, 165.

Kleinman, M. T., Sulfur dioxide and exercise: relationship between response and absorption in upper airways, *Air Poll. Contr. Assoc. J.,* 34, 32, 1984.

Klimisch, H. J., Lung deposition, lung clearance and renal accumulation of inhaled cadmium chloride and cadmium sulphide in rats, *Toxicol.,* 84, 103, 1993.

König, H. P, Heinrich, U., Kock, H., and Peters, L., Effect of photocorrosion on cadmium sulfide suspensions applied in animal inhalation studies with CdS particles, *Arch. Environ. Contam. Toxicol.,* 22, 30, 1992.

Korte, F., Ecotoxicologic profile analysis, *Chemosphere,* 7, 79, 1978.

Kostial, K., Kello, D., Jugo, S., Rabar, I., and Maljkovic, T., Influence of age on metal metabolism and toxicity, *Environ. Health Persp.,* 25, 81, 1978.

Kostial, K., Kello, D., Blanusa, M., Maljkovic, T., and Rabar, I., Influence of some factors on cadmium pharmacokinetics and toxicity, *Environ. Health Persp.*, 28, 89, 1979.

Kostial, K., Simonovic, I., Rabar, I., Blanusa, M., and Landeka, M., Age and intestinal retention of mercury and cadmium in rats, *Environ. Res.*, 31, 111, 1983.

Kostial, K., Blanusa, M., Maljkovic, T., Kello, D., and Rabar, I., Effect of a metal mixture in diet on the toxicokinetics and toxicity of cadmium, mercury and manganese in rats, *Toxicol. Ind. Health*, 5, 685, 1989.

Kutchai, H. C., Gastrointestinal motility, in *Physiology*, 2nd ed., Berne, R. M. and Levy, M. N., Eds., C.V. Mosby, St. Louis, 1988, 649.

Labaune, J. P., *Handbook of Pharmacokinetics. Toxicity Assessment of Chemicals*, John Wiley & Sons, New York, 1989.

Lamar, R. T. and Dietrich, D. M., *In situ* depletion of pentachlorophenol from contaminated soil by phanerochaete, *Appl. Environ. Microbiol.*, 56, 3039, 1990.

Lampe, M. A., Williams, M. L., and Elias, P. M., Human epidermal lipid: characterization and modulation during differentiation, *J. Lipid Res.*, 24, 131, 1983.

Langård, S., Gundersen, N., Tsalev, D. L., and Gylseth, B., Whole blood chromium level and chromium excretion in the rat after zinc chromate inhalation, *Acta. Pharmacol. Toxicol.*, 42, 142, 1978.

Langård, S., Role of chemical species and exposure characteristics in cancer among persons occupationally exposed to chromium compounds, *Scand. J. Work Environ. Health,* 19, Suppl. 1, 81, 1993.

Laug, E. P. and Kunze, F. M., The penetration of lead through the skin, *J. Ind. Hyg. Toxicol.*, 30, 256, 1948.

Lauwerys, R. R., Biological monitoring of exposure to inorganic and organometallic substances, in *Industrial Chemical Exposure: Guidelines for Biological Monitoring,* Davis, B. A., Ed., Biomedical Publications, New York, 1983, 9.

Le, X. C., Cullen, W. R., and Reimer, K. J., Determination of urinary arsenic and impact of dietary arsenic intake, *Talanta,* 40, 185, 1993.

Le, X. C., Cullen, W. R., and Reimer, K. J., Human urinary arsenic excretion after one-time ingestion of seaweed, crab, and shrimp, *Clin. Chem.*, 40, 617, 1994.

Lehman, L. D. and Klaassen, C. D., Dosage-dependent disposition of cadmium administered orally to rats, *Toxicol. Appl. Pharmacol.*, 84, 159, 1986.

Lewis, G. P., Jusko, W. J., Coughlin, L. L., and Hartz, S., Contribution of cigarette smoking to cadmium accumulation in man, *Lancet,* 1, 291, 1972.

Liao, T. F. and Oehme, F. W., Tissue distribution and plasma protein binding of ^{14}Cphenol in rats, *Toxicol. Appl. Pharmacol.*, 57, 220, 1981.

Lide, D. R., Ed., *CRC Handbook of Chemistry and Physics*, 73rd ed., CRC Press, Boca Raton, Florida, 1993.

Lieb, W. R. and Stein, W. D., Implications of two different type of diffusion for biological membranes, *Nature*, 234, 219, 1971.

Lindberg, E. and Vesterberg, O., Monitoring exposure to chromic acid in chrome plating by measuring chromium in urine, *Scand. J. Work Environ. Health,* 9, 333, 1983.

Lioy, P. J., Assessing total human exposure to contaminants. A multidisplinary approach, *Environ. Sci. Technol.,* 24, 938, 1990a.

Lioy, P. J., Airborne toxic elements and organic substances (ATEOS) project: overview, *Toxicol. Ind. Health,* 6(5), 53, 1990b.

Lioy, P. J., Freeman, N. C. G., Wainman, T., Stern, A. H., Boesch, R., Howell, T., and Shupack, S. I., Microenvironmental analysis of residential exposure to chromium-laden waste in and around New Jersey, *Risk Anal.,* 12, 287, 1992.

Lippmann, M., Yeates, D. B., and Albert, R. E., Deposition, retention, and clearance of inhaled particles, *Br. J. Ind. Med.,* 37, 337, 1980.

Loden, M., Detection methods, in *Methods for Skin Absorption,* Kemppainen, B. W. and Reifenrath, W. G., Eds., CRC Press, Boca Raton, Florida, 1990, 128.

Löf, A., Hansen, S. H., Näslund, P., Steiner, E., Wallén, M., and Hjelm, E. W., Relationship between uptake and elimination of toluene and debrisoquin hydroxylation polymorphism, *Clin. Pharmacol. Ther.,* 47, 412, 1990.

Löf, A., Hjelm, W., Colmsjö, A., Lundmark, B-O., Norström, Å., and Sato, A., Toxicokinetics of toluene and urinary excretion of hippuric acid after human exposure to $^2H_8$-toluene, *Br. J. Ind. Med.,* 50, 55, 1993.

Lovering, E. D., Black, D. B., and Rowe, M. L., Drug permeation through membranes IV: effect of excipients and various additives on permeation of chlordiazepoxide through polydimethyl-siloxane membranes, *J. Pharm. Sci.,* 63, 1224, 1974a.

Lovering, E. D. and Black, D. B., Diffusion layer effects on permeation of phenylbutazone through polydimethyl-siloxane, *J. Pharm. Sci.,* 63, 1399, 1974b.

Lu, H. H., Thomas, J. D., Tukker, J. J., and Fleisher, D., Intestinal water and solute absorption studies: comparison of *in situ* perfusion with chronic isolated loops in rats, *Pharm. Res.,* 9, 894, 1992.

Lucier, G. W., Rumbaugh, R. C., McCoy, Z., Hass, R., Harvan, D., and Albro, P., Ingestion of soil contaminated with 2,3,7,8-tetrachlorodibenzo-p-dioxin (TCDD) alters hepatic enzyme activities in rats, *Toxicol. Appl. Pharmacol.,* 6, 364, 1986.

Lucier, G. W., Trischer, A., Goldsworthy, T., Foley, J., Clark, G., Goldstein, J. A., and Maronpot, R., Ovarian hormones enhance 2,3,7,8-tetrachlorodibenzo-p-dioxin-mediated increases in cell proliferation and preneoplastic foci in a two-stage model for rat hepatocarcinogenesis, *Cancer Res.,* 51, 1391, 1991.

Mabuchi, K., Lillenfied, A. M., and Snell, L. M., Lung cancer among pesticide workers exposed to inorganic arsenicals, *Arch. Environ. Health,* 34, 312, 1979.

MacFarland, H. N., Designs and operational characteristics of inhalation exposure equipment, in *Inhalation Toxicology. Research Methods, Applications and Evaluation,* Salem, H., Ed., Marcel Dekker, Inc., New York, 1987, 93.

Macias, E. S. and Huster, R. B., A review of atmospheric particulate mass measurement via the beta-attenuation techniques, in *Fine Particles,* Liu, B. Y. H., Ed., Academic Press, New York, 1976, 536.

MacKenzie, R. D., Anwar, R. A., Byerrum, R. U., and Hoppert, C. A., Absorption and distribution of Cr^{51} in the Albino rat, *Arch. Biochem. Bioph.,* 79, 200, 1959.

Magos, L., Mercury-blood interaction and mercury uptake by the brain after vapor exposure, *Environ. Res.,* 1, 323, 1967.

Magos, L., Mercury, in *Handbook on Toxicity of Inorganic Compounds,* Seiler, H. G. and Sigel, H., Eds., Marcel Dekker, Inc., New York, 1988, 419.

Magos, L., Epidemiological and experimental aspects of metal carcinogenesis: physicochemical properties, kinetics, and the active species, *Environ. Health Persp.*, 95, 157, 1991.

Magos, L., Clarkson, T. W., and Greenwood, M. R., The depression of pulmonary retention of mercury vapor by ethanol: identification of the site of action, *Toxicol. Appl. Pharmacol.*, 26, 180, 1973.

Magos, L., Halbach, S., and Clarkson, T. W., Role of catalase in the oxidation of mercury vapor, *Biochem. Pharmacol.*, 27, 1373, 1978.

Maibach, H. I., Percutaneous Penetration of Benzene: Man, Report to American Petroleum Institute, April 18, 1981a. (Cited by Susten *et al.* 1985).

Maibach, H. I., Percutaneous Penetration of Benzene: Man II. Palm, Report to American Petroleum Institute, May 22, 1981b. (Cited by Susten *et al.* 1985).

Maibach, H. I. and Anjo, D. M., Percutaneous penetration of benzene and benzene contained in solvents used in the rubber industry, *Arch. Environ. Health*, 36, 256, 1981.

Malachowski, M. E., An update on arsenic, *Clin. Lab. Med.*, 10, 459, 1990.

Maloof, C. C., Use of edathmil calcium in the treatment of chrome ulcers of the skin, *AMA Arch. Ind. Health*, 11, 123, 1955.

Mannix, R. C., Phalen, R. F., Walters, R. B., and Kurosaki, T. T., Effects of sulfur dioxide and formaldehyde on particle clearance in the rats, *J. Toxicol. Environ. Health*, 12, 429, 1983.

Mappes, R., Experiments on excretion of arsenic in urine, *Int. Arch. Occup. Environ. Health*, 40, 267, 1977.

Marafante, E. and Vahter, M., Solubility, retention, and metabolism of intratracheally and orally administered inorganic arsenic compounds in the hamster, *Environ. Res.*, 42, 72, 1987.

Marceau, N., Aspin, N., and Kortsak, A., Absorption of copper 64 from gastrointestinal tract of the rat, *Am. J. Physiol.*, 218, 377, 1970.

Marty, J., Bucks, D. W., and Maibach, H. I., Noninvasive radioisotope counting on skin: surface or external counting? in *Percutaneous Absorption. Mechanisms-Methodology-Drug Delivery*, 2nd ed., Bronaugh, R. L. and Maibach, H. I., Eds., Marcel Dekker, Inc., New York, 1989, 435.

Marzulli, F. N., Callahan, J. F., and Brown, D. W. C., Chemical structure and skin penetrating capacity of a short series of organic phosphates and phosphoric acid, *J. Invest. Dermatol.*, 44, 339, 1965.

Mason, K. E., A conspectus of research on copper metabolism and requirements of man, *J. Nutr.*, 109, 1979, 1979.

Mass, M. J., Human carcinogenesis by arsenic, *Environ. Geochem. Health*, 14, 49, 1992.

McConnell, E. E., Lucier, G. W., Rumbaugh, R. C., Albro, P. W., Harvan, D. J., Hass, J. R., and Harris, M. W., Dioxin in soil: bioavailability after ingestion by rats and guinea pigs, *Science*, 223, 1077, 1984.

McConnell, E. E., Toxicology and Carcinogenesis Studies of Two Pentachlorophenol Technical-grade Mixtures in B6C3F1 Mice, No 89-2804, U.S. National Toxicology Program, Department of Health and Human Service, NIH Publication, 1989.

McDougal, J. N., Jepson, G. W., Clewell, H. J., III, and Andersen, M. E., Dermal absorption of dihalomethane vapor, *Toxicol. Appl. Pharmacol.*, 79, 150, 1985.

McLellan, J. S., Flanagan, P. R., Chamberlain, M. J., and Valberg, L. S., Measurement of dietary cadmium absorption in humans, *J. Toxicol. Environ. Health*, 4, 131, 1978.

Meerman, J. H., Sterenborg, H. M. J., and Mulder, G. J., Use of pentachlorophenol as long-term inhibitor of sulfation of phenols and hydroxanic acids in the rat *in vivo*, *Biochem. Pharmacol.*, 32, 1587, 1983.

Mehendale, H. M., Angevine, L. S., and Ohmiya, Y., The isolated perfused lung -- a critical evaluation, *Toxicology*, 2, 1, 1981.

Menzie, C. A., Potocki, B. B., and Santodonato, J., Exposure to carcinogenic PAHs in the environment, *Environ. Sci. Technol.*, 26, 1278, 1992.

Mercer, T. T., On the role of particle size in the dissolution of lung burdens, *Health Phys.*, 13, 1211, 1967.

Mertz, W., Roginski, E. E., and Reba, R. C., Biological activity and fate of trace quantities of intravenous chromium (III) in the rat, *Am. J. Physiol.*, 209, 489, 1965.

Mertz, W., Chromium occurrence and function in biological system, *Physiol. Rev.*, 49, 165, 1969.

Mertz, W., Effects and metabolism of glucose tolerance factor, *Nutr. Rev.*, 33, 126, 1975.

Mewhinney, J. A. and Griffith, W. C., Model of metabolism in beagles and humans, *Health Phys.*, 42, 629, 1982.

Meyer, M. C., Bioavailability of transdermal and topical dosage forms, in *Pharmaceutical Bioequivalence*, Welling, P. G., Tse, F. L. S., and Dighe, S. V., Eds., Marcel Dekker, Inc., New York, 1991, 169.

Miettinen, J. K., Absorption and elimination of dietary mercury (Hg^{2+}) and methylmercury in man, in *Mercury, Mercurials and Mercaptains*, Charles C Thomas Publisher, Miller, M. W. and Clarkson, T. W., Eds., Springfield, Illlinois, 1973, 233.

Miller, F. J., Menzel, D. B., and Coffin, D. L., Similarity between man and laboratory animals in regional pulmonary deposition of ozone, *Environ. Res.*, 17, 84, 1978.

Miller, F. J., Overton, J. H., Jr, Jaskot, R. H., and Menzel, D. B., A model of the regional uptake of gaseous pollutants in the lung. I. The sensitivity of the uptake of ozone in the human lung to lower respiratory tract secretions and exercise, *Toxic. Appl. Pharmacol.*, 79, 11, 1985.

Miller, F. J., Overton, J. H., Gerrity, T. R., and Graham, R. C., Interspecies dosimetry of reactive gases, in *Inhalation Toxicology. The Design and Interpretation of Inhalation Studies and Their Use in Risk Assessment*, Mohr, U. (Editor-in-chief), Springer-Verlag, Berlin, 1988, 139.

MOE (Ministry of the Environment), Review and Recommendations on a Lead in Soil Guideline, Report to the Ministry of the Environment by the Lead in Soil Committee, prepared by the Hazardous Contaminants Branch, Toronto, Ontario, 1987.

Moore, W., Stara, J. F., Crocker, W. C., Malanchuk, M., and Iltis, R., Comparison of ^{115m}Cadmium retention in rats following different routes of administration, *Environ. Res.*, 6, 473, 1973.

Moore, M. R., Hughes, M. A., and Goldberg, A., Lead absorption in man from dietary sources, *Int. Arch. Occup. Health*, 44, 81, 1979.

Moore, M. R., Meredith, P. A., Watson, W. S., Sumner, D. J., Taylor, M. K., and Goldberg, A., The percutaneous absorption of lead-203 in humans from cosmetic preparations containing lead acetate, as assessed by whole-body counting and other techniques, *Food. Cosmet. Toxicol.*, 18, 399, 1980.

Morrow, P. E., Clearance kinetics of inhaled particles, in *Respiratory Defense Mechanisms*, Part II. Brain, J. D., Proctor, D. F., and Reid, L. M., Eds., Marcel Dekker, Inc., New York, 1977, 491.

Morrow, P. E., Gibb, F. R., and Johnson, L., Clearance of insoluble dust from the respiratory tract, *Health Phys.,* 10, 543, 1964.

Morrow, P. E., Bates, D. V., Fish, B. R., Hatch, T. F., and Mercer, T. T., Deposition and retention models for internal dosimetry of the human respiratory tract, *Health Phys.,* 12, 173, 1966.

Morrow, P. E., Gibb, F. R., Davies, H., and Fisher, M., Dust removal from the lung parenchyma: an investigation of clearance stimulants, *Toxicol. Appl. Pharmacol.,* 12, 372, 1968.

Morrow, P. E., Gibb, R. F., Beiter, H., and Kilpper, R. W., Pulmonary retention of neutron-activated coal dust, *Arch. Environ. Health,* 34, 178, 1979.

Morrow, P. E., Beiter, H., Amato, F., and Gibb, F. R., Pulmonary retention of lead: an experimental study in man, *Environ. Res,* 21, 373, 1980.

Most, C. F., Jr., Filter effects on diffusion in silicone rubber, *J. Appl. Pharm. Sci.,* 14, 1019, 1970.

MSSC (Minister of Supply and Services Canada), Guidelines for Canadian Drinking Water Quality, 4th ed., Prepared by the Federal-Provincial Subcommittee on Drinking Water of the Federal-Provincial Advisory Committee on Environmental and Occupational Health, Published by authority of the Minister of National Health and Welfare, Ottawa, Canada, 1989.

Müller, L., Abel, J., and Ohnesorge, F. K., Absorption and distribution of cadmium (Cd), copper and zinc following oral subchronic low level administration to rats of different binding forms of cadmium (Cd-acetate, Cd-metallothionein, Cd-glutathione), *Toxicology,* 39, 187, 1986.

Murphy, B. L., Toole, A. P., and Bergstrom, P. D., Health assessment for arsenic contaminated soil, *Environ. Geochem. Health,* 11, 163, 1989.

Murphy, R. S., Kutz, F. W., and Strassman, S. C., Selected pesticide residues or metabolites in blood and urine specimens from a general population, *Environ. Health Persp.,* 48, 81, 1983.

Mussalo-Rauhamaa, H., Pyysalo, H., and Antervo, K., The presence of chlorophenols and their conjugates in Finnish human adipose and liver tissues, *Sci. Total. Environ.,* 83, 161, 1989.

Nacht, S. and Yeung, D., Artificial membranes and skin permeability, in *Percutaneous Absorption,* Bronaogh, R. L. and Reifenrath, W. G., Eds., Marcel Dekker, Inc., New York, 1985.

Nageswara Rao, C. and Nageswara Rao, B. S., Absorption and retention of magnesium and some trace elements by man from typical Indian diets, *Nutr. Metab.,* 24, 244, 1980.

Nakano, M., Kuchiki, A., and Arita, T., The effect of charges on permeabilities of drugs through collagen membranes, *Chem. Pharm. Bull.,* 24, 2345, 1976.

NAS (National Academy of Sciences), Chromium, National Academy of Sciences, Washington, D.C., 1974.

NAS (National Academy of Sciences), Arsenic, National Academy of Sciences, Washington, D.C., 1977a.

NAS (National Academy of Sciences), Copper, National Academy of Sciences, Washington, D.C., 1977b.

NAS (National Academy of Sciences), Risk Assessment in the Federal Government: Managing the Process, National Academy Press, Washington, D.C., 1983

NAS (National Academy of Sciences), Recommended Dietary Allowances, National Academy of Sciences, Washington, D.C., 1989.

Newton, D., Johnson, P., Lally, A. E., Pentreath, R. J., and Swift, D. J., The uptake by man of cadmium ingested in crab meat, *Human Toxicol.,* 3, 23, 1984.

Newton, D., Pickford, C. J., Chamberlain, A. C., Sherlock, J. C., and Hislop, J. S., Elevation of lead in human blood from its controlled ingestion in beer, *Hum. Exp. Toxicol.*, 11, 3, 1992.

Ng, K. M. E., Chu, I., Bronaugh, R. L., Franklin, C. A., and Somers, D. A., Percutaneous absorption/metabolism of phenanthrene in the hairless guinea pig: comparison of *in vitro* and *in vivo* results, *Fund. Appl. Toxicol.*, 16, 517, 1991.

Ng, K. M. E., Chu, I., Bronaugh, R. L., Franklin, C. A., and Somers, D. A., Percutaneous absorption/metabolism of pyrene, benzo[a]pyrene, and di[2-ethylhexyl)phthalate: comparison of *in vitro* and *in vivo* results in the hairless guinea pig, *Toxicol. Appl. Pharmacol.*, 115, 517, 1992.

NHW (National Health and Welfare, Canada), Chlorophenols and Their Impurities: a Health Hazard Evaluation, No. 84-EHD-110, Department of National Health and Welfare, Environmental Health Directorate, Health Protection Branch, Ottawa, 1988.

Nielsen-Kudsk, F., Absorption of mercury vapor from the respiratory tract in man, *Acta Pharmacol. Toxicol.*, 23, 250, 1965a.

Nielsen-Kudsk, F., The influence of ethyl alcohol on the absorption of mercury vapor from the lung in man, *Acta Pharmacol. Toxicol.*, 23, 263, 1965b.

Nielsen-Kudsk, F., Uptake of mercury in blood *in vivo* and *in vitro* from Hg-containing air, *Acta Pharmacol. Toxicol.*, 27, 149, 1969.

Nielsen, J. B., Toxicokinetics of mercuric chloride and methylmercuric chloride in mice, *J. Toxicol. Environ. Health*, 37, 85, 1992.

Nielsen, J. B. and Andersen, O., Oral mercuric chloride exposure in mice -- effects of dose on intestinal absorption and relative organ distribution, *Toxicology*, 59, 1, 1989.

Nielsen, J. B. and Andersen, O., Disposition and retention of mercuric chloride in mice after oral and parenteral administration, *J. Toxicol. Environ. Health*, 30, 167, 1990.

Nielsen, J. B. and Andersen, O., Methyl mercuric chloride toxicokinetics in mice: effects of strain, sex, route of administration and dose, *Pharmacol. Toxicol.*, 68, 201, 1991a.

Nielsen, J. B. and Andersen, O., Methyl mercuric chloride toxicokinetics in mice: sexual differences in whole-body retention and deposition in blood, hair, skin, muscles and fat, *Pharmacol. Toxicol.*, 68, 208, 1991b.

Niemeier, R. W. and Bingham, E., An isolated perfused lung preparation for metabolic studies, *Life Sci.*, 11, 807, 1972.

Niven, R. W. and Byron, P. R., Solute absorption from the airways of the isolated rat lung. I. The use of absorption data to quantify drug dissolution or release in the respiratory tract, *Pharmaceu. Res.*, 5, 574, 1988.

Nolan, R. J., Smith, F. A., and Hefner, J. G., Elimination and tissue distribution of 2,3,7,8-tetrachlorodibenzo-p-dioxin (TCDD) in female guinea pigs following a single oral dose, *Toxicol. Appl. Phamacol.*, 48, A162, 1979.

Nomiyama, K. and Nomiyama, H., Respiratory retention, uptake and excretion of organic solvents in man, *Int. Arch. Arbeitsmed.*, 32, 75, 1974.

Noonan, P. K. and Wester, R. C., Cutaneous biotransformations and some pharmacological and toxicological implications, in *Dermatotoxicology*, Marzulli, F. and Maibach, H., Eds., Hemisphere Publishing, New York, 1987, 71.

Noonan, P. K. and Wester, R. C., Cutaneous metabolism of xenobiotics, in *Percutaneous Absorption*, Bronaugh, R. and Maibach, H., Eds., Marcel Dekker, Inc., New York, 1989, 53.

Nordberg, G. F., Cadmium metabolism and toxicity, *Environ. Physiol. Biochem.*, 2, 7, 1972.

Nordberg, G. F., Kjellström, T., and Nordberg, M., Kinetics and metabolism, in *Cadmium and Health: a Toxicological and Epidemiological Appraisal. Exposure, Dose, and Metabolism,* Vol. I, Friberg, L., Elinder, C-G., Kjellström, T., and Nordberg, G. F., Eds., CRC Press, Boca Raton, Florida, 1985.

North, D. W., Risk assessment for ingested inorganic arsenic: a review and status report, *Environ. Geochem. Health,* 14, 59, 1992.

Notari, R. E., *Biopharmaceutics and Clinical Pharmacokinetics. An Introduction*, Fourth Edition, Marcel Dekker, Inc., New York, 1987.

NRC (U.S. National Research Council), Biologic Markers in Pulmonary Toxicology, National Academy Press, Washington, D.C., 1989.

NRC (U.S. National Research Council), Human Exposure Assessment for Airborne Pollutants. Advances and Opportunities, National Academy Press, Washington, D.C., 1991.

NRCC (National Research Council Canada), Effects of Arsenic in the Canadian Environment, Associate Committee on Scientific Criteria for Environmental Quality, National Research Council of Canada, Ottawa, 1978.

NRCC (National Research Council Canada), Effects of Cadmium in the Canadian Environment, NRCC 16743, Associate Committee on Scientific Criteria for Environmental Quality, Subcommittee on Heavy Metals and Certain Other Elements, 1979.

NRCC (National Research Council Canada), Chlorinated Phenols: Criteria for Environmental Quality, Associate Committee on Scientific Criteria for Environmental Quality, National Research Council of Canada, Ottawa, 1982.

Nugent, F. J. and Wood, J. A., Methods for the study of percutaneous absorption, *Can. J. Pharm. Sci.,* 13, 1, 1980.

Oberdörster, G., Airborne cadmium and carcinogenesis of the respiratory tract, *Scand. J. Work Environ. Health,* 12, 523, 1986.

Oberdörster, G., Deposition and retention modeling of inhaled cadmium in rat and human lung: an example for extrapolation of effects and risk estimation, in *Extrapolation of Dosimetric Relationships for Inhaled Particles and Gases*, Crapo, J. D., Smolko, E. D., Miller, F. J., Graham, J. A., and Hayes, A. W., Eds., Academic Press, Inc., San Diego, 1989a, 345.

Oberdörster, G., Equivalent oral and inhalation exposure to cadmium compounds: risk estimation based on route-to-route extrapolation, in *Principles of Route-to-Route Extrapolation for Risk Assessment*, Gerrity, T. R., Henry, C. J., Birnbaum, L., deRosa, C., Dutton, R. J., Miller, F., and Rhomberg, L., Eds., Elsevier, New York, 1989b, 217.

Oberdörster, G., Pulmonary deposition, clearance and effects of inhaled soluble and insoluble cadmium compounds, in *Cadmium in the Human Environment,* Nordberg, G. F., Herber, R. F. M. and Alessio, L., Eds., International Agency for Research on Cancer, Lyon, 1992, 189.

Oberdörster, G., Baumert, H. P., Hochrainer, D., and Stöber, W., The clearance of cadmium aerosols after inhalation exposure, *Am. Ind. Hyg. Assoc. J.,* 40, 443, 1979.

Offenbacher, E. G., Spencer, H., Dowling, H. J., and Pi-Sunyer, F. X., Metabolism chromium balances in men, *Am. J. Clin. Nutr.,* 44, 77, 1986.

Ogawa, E., Tsuzuki, H., and Yamazaki, T., Experimental study on absorption, distribution and excretion of trivalent and hexavalent chromes, *Japan. J. Pharmacol.,* 26, suppl., 92p, 1976.

Ogiso, T., Ito, Y., Iwaki, M., Atago, H., Tanaka, C., Maniwa, N., and Ishida, S., Percutaneous absorption of dexamethason acetate and palmitate, and the plasma concentration, *Chem. Pharm. Bull.(Tokyo),* 35, 4263, 1987.

Ohtsuji, H. and Ikeda, M., Quantitative relationship between atmospheric phenol vapor and phenol in the urine of workers in bakelite factories, *Br. J. Ind. Med.*, 29, 70, 1972.

Olson, J. R., Gasiewicz, T. A., and Neal, R. A., Tissue distribution, excretion, and metabolism of 2,3,7,8-tetrachlorodibenzo-p-dioxin (TCDD) in the Golden Syrian hamster, *Toxicol. Appl. Pharmacol.*, 56, 78, 1980.

Olson, J. R., Gasiewicz, T. A., Geiger, L. E., and Neal, R. A., The metabolism of 2,3,7,8-tetrachlorodibenzo-p-dioxin in mammalian systems, in *Accidental Exposure to Dioxins: Human Health Aspects,* Coulston, R. and Pocciari, F., Eds., Academic Press, New York, 1983, 81.

O'Neil, J. J. and Tierney, D. F., Rat lung metabolism: glucose utilization by isolated perfused lungs and tissue slices, *Am. J. Physiol.*, 226, 867, 1974.

O'Neill, W. P., Membrane system, in *Controlled Release Technologies: Methods, Theory, Application,* Vol. 1, Kydonieus, A. F., Ed., CRC Press, Boca Raton, FL, 1980.

OSHA (Occupational Safety and Health Administration), Occupational exposure to benzene, Final rule, Department of Labor, Occupational Safety and Health Administration, *Federal Register,* 52, 34460, 1987.

OSHA (Occupational Safety and Health Administration), Air contaminants, Final Rule, Occupational Safety and Health Administration, *Federal Register,* 54, 2332, 1989.

Östlund, K., Studies on the metabolism of methyl mercury in mice, *Acta Pharmacol. Toxicol.*, (Suppl 1) 27, 5, 1969.

Ott, M. G., Holder, B . B., and Gordon, H. L., Respiratory cancer and occupational exposure to arsenicals, *Arch. Environ. Health,* 29, 250, 1974.

Overton, J. H., Physicochemical processes and the formulation of dosimetry models, *J. Toxic. Environ. Health*, 13, 273, 1984.

Overton, J. H., Respiratory tract physiological processes and the uptake of gases, in *Principle of Route-to-Route Extrapolation for Risk Assessment,* Gerrity, T. R. and Henry, C. J., Eds., Elsevier, New York, 1990, 71.

Overton, J. H., Graham, R. C., and Miller, F. J., A model of the regional uptake of gaseous pollutants in the lung. II. The sensitivity of ozone uptake in laboratory animal lungs to anatomical and ventilatory parameters, *Toxic. Appl. Pharmacol.*, 88, 418, 1987.

Overton, J. H. and Miller, F. J., Absorption of inhaled reactive gases, in *Toxicology of the Lung*, Gradner, D. E., Crapo, J. M., and Massaro, E. J., Eds., Raven Press, New York, 1988, 477.

Palmes, E., Wang, C., Goldring, R., and Altshuler, B., Effect of depth of inhalation on aerosol persistence during breath holding, *J. Appl. Physiol.*, 34, 356, 1973.

Parke, D. V. and Williams, R. T., Studies in detoxification, the metabolism of benzene containing [^{14}C]benzene, *Biochem. J.*, 54, 231, 1953.

Pascoe, G. A., Blanchet, A. J., and Linder, G., Bioavailability of metals and arsenic to small mammals at a mining waste-contaminated wetland, *Arch. Environ. Contam. Toxicol.*, 27, 44, 1994.

Paterson, S. C. and Sarvesvaran, R., Plastic bag death: a toluene fatality, *Med. Sci. Law.*, 23, 64, 1983.

Pattle, R. E., The retention of gases and particles in the human nose, in *Inhaled Particles and Vapors,* Vol. 1, Davies, C. N., Ed., Pergamon Press, New York, 1961, 302.

Paustenbach, D. J., Shu, H. P., and Murray, F. J., A critical examination of assumption used in risk assessments of dioxin contaminated soil, *Regul. Toxicol. Pharmacol.*, 6, 248, 1986.

Paustenbach, D. J., Assessing the potential environmental and human health risk of contaminated soil, *Comments Toxicol.,* 1, 185, 1987.

Paustenbach, D. J., Rinehart, W. E., and Sheehan, P. J., The health hazards posed by chromium-contaminated soils in residential and industrial areas: conclusion of an expert panel, *Regul. Toxicol. Pharmacol.,* 13, 195, 1991a.

Paustenbach, D. J., Meyer, D. M., Sheehan, P. J., and Lau, V., An assessment and quantitative uncertainty analysis of the health risks to workers exposed to chromium contaminated soils, *Toxicol. Ind. Health,* 7, 159, 1991b.

Paxman, D. and Rappaport, S. M., Analysis of OSHA's short-term-exposure limit for benzene, *Regul Toxicol. Pharmacol.,* 11, 275, 1990.

Pegg, D. G., Zempel, J. A., Braun, W. H., and Watanabe, P. G., Disposition of tetrachloro[^{14}C]ethylene following oral and inhalation exposure in rats, *Toxicol. Appl. Pharmacol.,* 51, 465, 1979.

Pekari, K., Luotamo, M., Jarvisalo, J., and Aitio, A., Urinary excretion of chlorinated phenols in sawmill workers, *Int. Arch. Occup. Environ. Health,* 63, 57, 1991.

Pershagen, G., Line, B., and Björklund, N., Lung retention and toxicity of some inorganic arsenic compounds, *Environ. Res.,* 29, 425, 1982.

Petrilli, F. L. and De Flora, S., Metabolic reduction of chromium as a threshold mechanism limiting its *in vivo* activity, *Sci. Total Environ.,* 71, 357, 1988.

Phalen, R. F., *Inhalation Studies: Foundations and Techniques,* CRC Press, Boca Raton, Florida, 1984, 1.

Pinto, S. S., Varner, M. O., Nelson, K. W., Labbe, A. L., Che, B. S., and White, L. D., Arsenic trioxide absorption and excretion in industry, *J. Occup. Med.,* 18, 677, 1977.

Piotrowski, J. K., Quantitative estimation of aniline absorption through the skin in man, *J. Hyg. Epid. Micro. Immunol.,* 1, 23, 1957.

Piotrowski, J., Further investigations on the evaluation of exposure to nitrobenzene, *Br. J. Ind. Med.,* 24, 60, 1967.

Piotrowski, J. K., Evaluation of exposure to phenol: absorption of phenol vapor in the lungs through the skin and excretion of phenol in urine, *Br. J. Ind. Med.,* 28, 172, 1971.

Piper, W. N., Rose, J. Q., and Gehring, P. J., Excretion and tissue distribution of 2,3,7,8-tetrachlorodibenzo-p-dioxin in the rat, *Environ. Health Persp.,* 5, 241, 1973.

Poelma, F. G. J. and Tukker, J. J., Evaluation of a chronically isolated loop in the rat for the study of drug absorption kinetics, *J. Pharm. Sci.,* 76, 433, 1987.

Poiger, H., Weber, H., and Schlatter, C., Special aspects of metabolism and kinetics of TCDD in dogs and rats. Assessment of toxicity of TCDD-metabolite(s) in guinea pigs, in *Chlorinated Dioxins and Related Compounds. Impact on the Environment*, Hutzinger, O., Frei, R. W., Merian, E. and Pocchiari, F., Eds., Pergamon Press, New York, 1982, 317.

Poiger, H. and Schlatter, C., Influence of solvents and adsorbents on dermal and intestinal absorption of TCDD, *Food Cosmet. Toxicol.,* 18, 477, 1980.

Poiger, H. and Schlatter, C., Pharmacokinetics of 2,3,7,8-TCDD in man, *Chemosphere,* 15, 1489, 1986.

Poland, A., Palen, D., and Glover, E., Tumor promotion by TCDD in skin of HRS/J mice, *Nature,* 300, 271, 1982.

Potts, R. O. and Guy, R. H., Predicting skin permeability, *Pharma. Res.,* 9, 663, 1992.

Pounds, J. G., Marlar, R. J., and Allen, J. R., Metabolism of lead-210 in juvenile and adult rhesus monkeys (*Macaca Mulatta*), *Bull. Environ. Contam. Toxicol.*, 19, 684, 1978.

Pounds, J. G., Percutaneous Absorption of Lead, NCTR Technical Report, No. 199, National Center for Toxicological Research, Jefferson, 1979. (Cited by Hostynek *et al.*, 1993).

Pyykkö, K., Tähti, H., and Vapaatalo, H., Toluene concentrations in various tissues of rats after inhalation and oral administration, *Arch. Toxicol.*, 38, 169, 1977.

Quarterman, J. and Morrison, E., The effect of age on the absorption and excretion of lead, *Environ. Res.*, 17, 78, 1978.

Quarterman, J., Morrison, E., and Humphries, W. R., The influence of high dietary calcium and phosphate on lead uptake and release, *Environ. Res.*, 17, 60, 1978.

Raabe, O. G., Particle size analysis utilizing grouped data and the log-normal distribution, *J. Aerosol. Sci.*, 2, 289, 1971.

Raabe, O. G., Teague, S. V., Richardson, N. L., and Nelson, L. S., Aerodynamic and dissolution behavior of fume aerosols produced during the combustion of laser-ignited plutonium droplets in air, *Health Phys.*, 35, 663, 1978.

Raabe, O. G., Deposition and Clearance of Inhaled Aerosols, UCD-472-503, U.S. Department of Energy, National Technical Information Service, Springfield, Virginia, 1979.

Raabe, O. G., Deposition and clearance of inhaled aerosols, in *Mechanisms in Respiratory Toxicology*, Vol. 1, Witschi, H. and Nettesheim, P., Eds., CRC Press, Boca Raton, Florida, 1982, 28.

Rabinowitz, M. B., Wetherill, G. W., and Kopple, J. D., Kinetic analysis of lead metabolism in healthy humans, *J. Clin. Invest.*, 58, 269, 1976.

Rabinowitz, M. B., Kopple, J. D., and Wetherill, G. W., Effect of food intake and fasting on gastrointestinal lead absorption in humans, *Am. J. Clin. Nut.*, 33, 1784, 1980.

Rabinowitz, M. B., Toxicokinetics of bone lead. *Environ. Health Persp.*, 91, 33, 1991.

Rahman, A. and Barrowman, J. A., The influence of bile on the bioavailability of polynuclear aromatic hydrocarbons from the rat intestine, *Can. J. Physio. Pharmacol.*, 64, 1214, 1986.

Rahola, T., Aaran, R-K., and Miettinen, J. K., Retention and elimination of ^{115m}Cd in man, in *Health Physics Problems of Internal Contamination*, Bujdosó, E., Ed., Akadémiai Kiadó, Budapest, 1973a, 213.

Rahola, T., Korolainen, H. A., and Miettinen, J. K., Elimination of free and protein-bound ionic mercury ($^{203}Hg^{2+}$) in man, *Ann. Clin. Res.*, 5, 214, 1973b.

Rai, D., Eary, L. E., and Zachara, J. M., Environmental chemistry of chromium, *Sci. Total Environ.*, 86, 15, 1989.

Randall, J. A. and Gibson, R. S., Serum and urine chromium as indices of chromium status in tannery workers, *Proc. Soc. Exp. Bio. Med.*, 185, 16, 1987.

Rao, S. B. and Mehendale, H. M., Uptake and disposition of putrescine, spermidine, and spermine by rabbit lungs, *Drug Metabol. Dispos.*, 15, 189, 1987.

Rasetti, L., Cappellaro, F., and Gaido, P., Contribution to the study of saturnism by inhibited oils, *Rass. Med. Ind. Ig. Lavoro*, 30, 1961, 71. (Cited by Hostynek *et al.*, 1993).

Ravis, W. R., Data interpretation and analysis in percutaneous absorption studies, in *Methods for Skin Absorption*, Kemppainen, B. W. and Reifenrath, W. G, Eds., CRC Press, Boca Raton, Florida, 1990, 147.

Reifenrath, W. G., Chellquist, E. M., Shipwash, E. A., Jederberg, W. W., and Krueger, G. G., Percutaneous absorption in the hairless dog, weanling pig, and grafted athymic nude mouse:

evaluation of models for predicting skin penetration in man, *Br. J. Dermatol.*, 111, (Suppl.) 27, 123, 1984a.

Reifenrath, W. G., Chellquist, E. M., Shipwash, E. A., and Jederberg, W. W., Evaluation of animal models for predicting skin penetration in man, *Fund. Appl. Toxicol.*, 4, S224, 1984b.

Reigner, B. G., Gungon, R. A., Hoag, M. K., and Tozer, T. N., Pentachlorophenol toxicokinetics after intravenous and oral administration to rat, *Xenobiotica,* 21, 1547, 1991.

Reigner, B. G., Bois, F. Y., and Tozer, T. N., Assessment of pentachlorophenol exposure in humans using the clearance concept, *Hum. Exp. Toxicol.,* 11, 17, 1992a.

Reigner, B. G., Rigod, J. F., and Tozer, T. N., Disposition, bioavailability, and serum protein binding pentachlorophenol in the B6C3F1 mouse, *Pharma. Res.*, 9, 1053, 1992b.

Ridout, G. and Guy, R. H., Structure-penetration Relationship in Percutaneous Absorption, ACS Symposium Series, 371, 112, 1988.

Rigg, P. C. and Barry, B. W., Shed snake skin and hairless mouse skin as model membranes for human skin during permeation studies, *J. Invest. Dermatol.*, 94, 235, 1990.

Riihimäki, V. and Pfäffli, P., Percutaneous absorption of solvent vapor in man, *Scand. J. Work Environ. Health,* 4, 73, 1978.

Ritschel, W. A., Bioavailability in the clinical evaluation of drugs, *Drug Intell. Clin. Pharm.*, 6, 246, 1972.

Riviere, J. E. and Monteiro-Riviere, N. A., The isolated perfused porcine skin flap as an *in vitro* model for percutaneous absorption and cutaneous toxicology, *Crit. Rev. Toxicol.*, 21, 329, 1991.

Roberts, M. S. and Anderson, R. A., The percutaneous absorption of phenolic compounds: the effect of vehicles on the penetration of phenol, *J. Pharm. Pharmacol.*, 27, 599, 1975.

Roberts, M. S., Anderson, R. A., and Swarbrick, J., Permeability of human epidermis to phenolic compounds, *J. Pharm. Pharmacol.*, 29, 677, 1977.

Robertson, F. N., Hexavalent chromium in the ground water in Paradise Valley, Arizona, *Ground Water,* 13, 516, 1975.

Rohades, R. A., Isolated perfused lung preparation for studying altered gaseous environments, *Environ. Health Persp.*, 56, 43, 1984.

Rose, J. Q., Ramsey, J. C., Wentzler, T. H., Hummel, R. A., and Gehring, P. J., The fate of 2,3,7,8-tetrachlorodibenzo-p-dioxin following single and repeated oral doses to the rat, *Toxicol. Appl. Pharmacol.*, 36, 209, 1976.

Roskos, K. V., Maibach, H. I., and Guy, R. H., The effect of aging on percutaneous absorption in man, *J. Pharm. Biopharm.*, 17, 617, 1989.

Roth, R. and Bassett, D. J. P., The isolated perfused lung in toxicological studies, in *In Vitro Toxicology Model Systems and Methods,* McQueen, C. A., Ed., Telford Press, Caldwell, New Jersey, 1989, 3.

Ruby, M. V., Davis, A., Kempton, H., Drexler, J. H., and Bergstrom, P. D., Lead bioavailability: dissolution kinetics under simulated gastric conditions, *Environ. Sci. Technol.*, 26, 1242, 1992.

Rusch, G. M., O'Grodnick, J. S., and Rinehart, W. E., Acute inhalation study in the rat of comparative uptake, distribution and excretion for different cadmium containing materials, *Am. Ind. Hyg. Assoc. J.*, 47, 754, 1986.

Ryan, J. J., Schecter, A., Lizzote, R., Sun, W. F., and Miller, L., Tissue distribution of dioxins, and furans in humans from the general population, *Chemosphere,* 14, 929, 1985.

Ryan, J. J. and Norstrom, R. J., Occurrence of polychlorinated dibenzodibenzo-p-dioxins and dibenzofurans in humans and major exposure routes, in *Environmental Carcinogens Methods of Analysis and Exposure Measurement. Polychlorinated Dioxins and Dibenzofurans,* Vol. 11, IARC Scientific Publications No. 108, Rappe, C., Buser, H. R., Dodet, B., and O'Neill, I. K., Eds., International Agency for Research on Cancer, Lyon, 1991, 51.

SAB (The Science Advisory Board), Letter Report to the Administrator, EPA-SAB-89-038, US Environmental Protection Agency, Washington, D.C., 1989.

Sabourin, P. J., Chen, B. T., and Lucier, G., Effect of dose on the absorption and excretion of [14C]benzene administered orally or by inhalation in rats and mice, *Toxicol. Appl. Pharmacol.,* 87, 325, 1987.

Sabourin, P. J., Bechtold, W. E., Birnbaum, L. S., Lucier, G., and Henderson, R. F., Differences in the metabolism and disposition of inhaled [^{3}H]benzene by T344/N rats and B6C3F mice, *Toxicol. Appl. Pharmacol.,* 94, 128, 1988.

Sabourin, P. J., Bechtold, W. E., Griffith, W., Birnbaum, L. S., Lucier, G., and Henderson, R. F., Effect of exposure concentration, exposure rate, route of administration on metabolism of benzene by F344 rats and B6C3F1 mice, *Toxicol. Appl. Pharmacol.,* 99, 421, 1989.

Saidman, L. J., Anesthesia at a constant alveolar concentration, in *Modeling of Inhalation Exposure to Vapors: Uptake, Distribution, and Elimination,* Fiserova-Bergerova, V., Ed., CRC Press, Boca Raton, Florida, 1983, 132.

Salem, H., Principles of inhalation toxicology, in *Inhalation Toxicology. Research Methods, Applications and Evaluation,* Salem, H., Ed., Marcel Dekker, Inc., New York, 1987, 1.

Salem, H. and Katz, S. A., Speciation, bioavailability, and systemic distribution of chromium from Whetlerite dust, *Sci. Total Toxicol.,* 86, 59, 1989.

Samitz, M. H. and Shrager, J., Patch test reaction to hexavalent and trivalent chromium compounds, *Arch. Dermatol.,* 94, 304, 1966.

Sanders, C. L., Jackson, T. A., Adee, R. R., Powers, G. J., and Wehner, A. P., Distribution of inhaled metal oxide particles in pulmonary alveoli, *Arch. Intern. Med.,* 127, 1085, 1971.

Sanders, C. L., Skinner, C., and Gelman, R. A., Percutaneous absorption of 7,10-14C-benzp[a]pyrene and 7,12-dimethylbenzo[a]anthracene in mice, *Environ. Res.,* 33, 353, 1984.

Sanders, C. L., Skinner, C., and Gelman, R. A., Percutaneous absorption of 7,10-14C-benzp[a]pyrene and 7,12-dimethylbenzo[a]anthracene in mice, *JEPTO,* 7, 1-2, 25-34, 1986.

Sasser, L. B. and Jarboe, G. E., Intestinal absorption and retention of cadmium in neonatal rat, *Toxicol. Appl. Pharmacol.,* 41, 423, 1977.

Sato, A. and Nakajima, T., Differences following skin or inhalation exposure in the absorption and excretion kinetics of trichloroethylene and toluene, *Br. J. Ind. Med.,* 35, 43, 1978.

Sato, A. and Nakajima, T., Partition coefficients of some aromatic hydrocarbons and ketones in water, blood, and oil, *Br. J. Ind. Med.,* 36, 231, 1979.

Schäfer, S. G., Schwegler, U., and Schümann, K., Retention of cadmium in cadmium-naive, normal and iron-deficient rats as well as in cadmium-induced iron deficient animals, *Ecotoxicol. Environ. Safety,* 20, 71, 1990.

Scheuplein, R. J., Mechanism of percutaneous absorption. I. Routes of penetration and the influence of solubility, *J. Invest. Dermatol.,* 45, 334, 1965.

Scheuplein, R. J. and Blank, I. H., Permeability absorption, *Physiol. Rev.,* 51, 702, 1971.

Scheuplein, R. J. and Blank, I. H., Mechanism of percutaneous absorption. IV. Penetration of non-electrolytes (alcohols) from aqueous solutions and from pure liquids, *J. Invest. Dermatol.*, 60, 286, 1973.

Scheuplein, R. J. and Bronaugh, R. L., Percutaneous absorption, in *Biochemistry and Physiology of the Skin,* Goldsmith, L. A., Ed., Oxford University Press, Oxford, 1983, 1255.

Schlesinger, R. B., Comparative deposition of inhaled aerosols in experimental animals and humans: a review, *J. Toxicol. Health Environ.*, 15, 197, 1985.

Schlesinger, R. B., Biological disposition of airborne particles: basic principles and application to vehicular emission, in *Air Pollution, the Automobile, and Public Health,* Watson, A. Y., Bates, R. R. and Kennedy, D., Eds., National Academy Press, Washington, D.C., 1988, 239.

Schlesinger, R. B., Deposition and clearance of inhaled particles, in *Concepts in Inhalation Toxicology,* McClellan, R. O. and Henderson, R. F., Eds., Hemisphere Publishing, Co., New York, 1989, 163.

Schlesinger, R. B., The interaction of inhaled toxicants with respiratory tract clearance mechanisms, *Crit. Rev. Toxicol.*, 20(4), 257, 1990.

Schrenk, H. H., Yant, W. P., Pearce, J., Patty, F. A., and Sayers, R. R., Absorption, distribution and elimination of benzene by body tissues and fluids of dogs exposed to benzene vapor, *J. Ind. Hyg. Toxicol.*, 23, 20, 1941.

Schroeder, D. C. and Lee, G. F., Potential transformations of chromium in natural water, *Water Air Soil Pollut.*, 4, 355, 1975.

Schroy, J. M., Hileman, F. D., and Cheng, S. S., Physical/chemical properties of 2,3,7,8-TCDD, *Chemosphere,* 14, 877, 1985.

Schumann, A. M., Quast, J. F., and Wanabe, P. G., The pharmacokinetics and macromolecular interactions of perchloroethylene in mice and rats as related to oncogenicity, *Toxicol. Appl. Pharmacol.*, 55, 207, 1980.

Seech, A. G., Trevors, J. T., and Bulman, T. L., Biodegradation of pentachlorophenol in soil: the response to physical, chemical, and biological treatments, *Can. J. Microbiol.*, 37, 440, 1990.

Seiler, H. G. and Sigel, H., *Handbook on Toxicity of Inorganic Compounds*, Marcel Dekker, Inc., New York, 1988.

Selen, A., Factors influencing bioavailability and bioequivalence, in *Pharmaceutical Bioequivalence*, Welling, P. G., Tse, F. L. S. and Dighe, S. V., Eds., Marcel Dekker, Inc., New York, 1991, 117.

Shah, P. V. and Guthrie, F. E., Percutaneous absorption of three insecticides in rats: a comparison of two methods for *in vivo* determination, *J. Invest. Dermatol.*, 80, 291, 1983.

Shaikh, Z. A. and Smith, J. C., Metabolism of orally ingested cadmium in humans, in *Mechanisms of Toxicity and Hazard Evaluation,* Holmstedt, B., Lauwerys, R., Mercier, M., and Roberfroid, M., Eds., Elsevier/North-Holland, Amsterdam, 1980, 569.

Sheehan, P. J., Meyer, D. M., Sauer, M. M., and Paustenbach, D. J., Assessment of the human health risks posed by exposure to chromium-contaminated soils, *J. Toxicol. Environ. Health,* 32, 161, 1991.

Sherwood, R. J., Pharmacokinetics of benzene in a human after exposure at about the permissible limit, *Ann. N Y Acad. Sci.*, 534, 635, 1988.

Shinohara, K. and Cerutti, P. A., Formation of benzo[a]pyrene-DNA adducts in peripheral human lung tissue, *Cancer Lett.*, 3, 303, 1977.

Shiu,W. Y., Doucette, W., Gobas, F. A. P. C., Andren, A., and Mackay, D., Physical-chemical properties of chlorinated dibenzo-p-dioxins, *Environ. Sci. Technol.*, 22, 651, 1988.

Shu, H., Paustenbach, D., Murray, F. J., Mapple, L., Brunck, B., Dei Rossi, D., and Teitelbaum, P., Bioavailability of soil-bound TCDD: oral bioavailability in the rat, *Fund. Appl. Toxicol.*, 10, 648, 1988a.

Shu, H., Teilelbaum, P., Webb, A. S., Marple, L., Brunck, B., Dei Rossi, D., Murray, F. J., and Paustenabach, D., Bioavailability of soil-bound TCDD: dermal bioavailability in the rat, *Fund. Appl. Toxixol.*, 10, 335, 1988b.

Silkworth, J., McMartin, D., DeCaprio, A., Rej, R., O'Keefe, P., and Kaminsky, L., Acute toxicity in guinea pigs and rabbits of soot from a polychlorinated biphenyl-containing transformer fire, *Toxicol. Appl. Pharmacol.*, 65, 425, 1982.

Sittig, M., *Handbook of Toxic and Hazardous Chemicals and Carcinogens*, 3rd ed., Noyes Publications, New Jersey, 1991, 432.

Skene, S. A., Dewhurst, I. C., and Greengurg, M., Polychlorinated dibenzo-p-dioxins and polychlorinated dibenzofurans: the risk to human health. A review, *Hum. Toxicol.*, 8, 173, 1989.

Skog, E. and Wahlberg, J. E., A comparative investigation of the percutaneous absorption of metal compounds in the guinea pig by means of the radioactive isotopes: ^{51}Cr, ^{58}Co, ^{65}Zn, ^{110m}Ag, ^{115m}Cd, ^{203}Hg, *J. Invest. Dermatol.*, 43, 187, 1964.

Skowronsk, G. A., Turkall, R. M., and Abdel-Rahman, M. S., Soil adsorption alters bioavailability of benzene in dermally exposed male rats, *Am. Ind. Hyg. Assoc. J.*, 49, 506, 1988.

Smith, A. H., Hopenhayn-Rich, C., Bates, M. N., Goeden, H. M., Hertz-Picciotto, I., Duggan, H., Wood, R., Kosnett, M., and Smith, M. T., Cancer risks from arsenic in drinking water, *Environ. Health Persp.*, 97, 259, 1992.

Smith, K. L., Penetrant characteristics influencing skin absorption, in *Methods for Skin Absorption,* Kemppainen, B. W. and Reifenrath, W. G., Eds., CRC Press, Boca Raton, Florida, 1990, 23.

Smith, P. L., Gastrointestinal physiology, in *Gastrointestinal Toxicology*, Rozman, K. and Hänninen, O., Eds., Elsevier, New York, 1986, 1.

Smith, T. J., Crecelius, E. A., and Reading, J. C., Airborne arsenic exposure and excretion of methylated arsenic compounds, *Environ. Health Persp.*, 19, 89, 1977.

Smolen, V. F., and Ball, L., *In-vitro* drug product dissolution testing: apparatus and methodologies, in *Controlled Drug Bioavailability,* Smolen, V. F. and Ball, L., Eds., John Wiley & Sons, New York, 1984, 135.

Snelling, W. M. and Dodd, D. E., Inhalation studies, in *Handbook of In Vivo Toxicity Testing,* Arnold, D. L., Grice, H. C., and Krewski, D. R., Eds., Academic Press, Inc., New York, 1990, 89.

Snipes, M. B., Long-term retention and clearance of particles inhaled by mammalian, *Crit. Rev. Toxicol.*, 20(3), 175, 1989.

Snow, E. T., A possible role for chromium in genetoxicity, *Environ. Health Persp.*, 92, 75, 1991.

Snyder, C. A., Erlichman, M. N., Laskin, S., Goldstein, B. D., and Albert, R. E., The pharmacokinetics of repetitive benzene exposures at 300 and 100 ppm in AKR mice and Sprague-Dawley rats, *Toxicol. Appl. Pharmacol.*, 57, 164, 1981.

Southwell, D., Barry, B. W., and Woodruff, R., Variations in the permeability of human skin within and between specimens, *Int. J. Pharm.*, 18, 299, 1984.

Speizer, F. E. and Frank, N. R., The uptake and release of SO_2 by the human nose, *Arch. Environ. Health,* 12, 725, 1966.

Srbova, J., Teisinger, J., and Skramovsky, S., Absorption and elimination of inhaled benzene in man, *Arch. Ind. Hyg. Occup. Med.,* 2, 1, 1950.

Srikrishna, V., Riviere, J. E., and Monteiro-Riviere, N. A., Cutaneous toxicity and absorption of paraquat in porcine skin, *Toxicol. Appl. Pharmacol.,* 115, 89, 1992.

Starich, G. H. and Blincoe, C., Dietary chromium - forms and availabilities, *Sci. Total Environ.,* 28, 443, 1983.

Stauber, J. L., Florence, T. M., Gulson, B. L., and Dale, L. S., Percutaneous absorption of inorganic lead compounds, *Sci. Total Environ.,* 145, 55, 1994.

Steele, M. J., Beck, B. D., Murphy, B. L., and Strauss, H. S., Assessing contribution from lead in mining wastes to blood lead, *Regul. Toxicol. Pharmacol.,* 11, 158, 1990.

Stein, W. D., The transport carrier principle, in *Pharmacology of Intestinal Permeation I,* Csáky, T. Z., Ed., Springer-Verlag, Berlin, 195, 1984.

Stephens, R. D., Harnly, M., Hayward, D. G., Chang, R. R., and Flattery, J., Bioaccumulation of dioxins in food animals II: controlled exposure studies, *Chemosphere,* 20, 1091, 1990.

Stern, A. H., Freeman, N. C. G., Pleban, P., Boesch, R. R., Wainman, T., Howell, T., and Shupack, S. I., Residential exposure to chromium waste-urine biological monitoring in conjunction with environmental exposure monitoring, *Environ. Res.,* 58, 147, 1992.

Sternlieb, I., Gastrointestinal copper absorption in man, *Gastroenterology,* 52, 1038, 1967.

Stevens, J. T., Hall, L. L., Farmer, J. D., DiPasquale, L. C., Chernoff, N., and Durham, W. F., Disposition of ^{14}CAs-cacodylic acid in rats after intravenous, intratracheal, or peroral administration, *Environ. Health Persp.,* 19, 151, 1977.

Stiff, M. J., The chemical states of copper in polluted fresh water and a scheme of analysis to differentiate them, *Water Res.,* 5, 585, 1977.

Storer, J. S., DeLeon, I., Millikan, L. E., Laseter, J. L., and Griffing, C., Human absorption of crude coal, *Arch. Dermatol.,* 120, 874, 1984.

Storm, J. E., Collier, S. W., Stewart, R. F., and Bronaugh, R. L., Metabolism of xenobiotics during percutaneous penetration: role of absorption rate and cutaneous enzyme activity, *Fund. Appl. Toxicol.,* 15, 132, 1990.

Stradling, G. N., Ham, G. J., Smith, H., Cooper, J., and Breadmore, S. E., Factors affecting the mobility of plutonium-238 dioxide in the rat, *Int. Radiat. Biol.,* 34, 37, 1978.

Strickland, G. T., Beckner, W. M., and Leu, M., Absorption of copper in homozygotes and heterozygotes for Wilson's disease and controls: isotope tracer studies with ^{67}Cu and ^{64}Cu, *Clin. Sci.,* 43, 617, 1972.

Stuart, B. O., Deposition and clearance of inhaled particles, *Environ. Health Persp.,* 55, 369, 1984.

Sugden, K. D., Burris, R. B., and Rogers, S. J., An oxygen dependence in chromium mutagenesis, *Mutat. Res.,* 244, 239, 1990.

Sullivan, M. F., Miller, B. M., and Goebel, J. C., Gastrointestinal absorption of metals (^{51}Cr, ^{65}Zn, ^{95m} Tc, ^{109}Cd, ^{113}Sn, ^{147}Pm, and ^{283}Pu) by rats and swine, *Environ. Res.,* 35, 439, 1984.

Sun, J. D., Wolff, R. K., and Kanapily, G. M., Deposition, retention, and biological fate of inhaled benzo[a]pyrene adsorbed onto ultrafine particles and as a pure aerosol, *Toxicol. Appl. Pharmacol.,* 65, 231, 1982.

Susten, A. S., Dames, B. L., Burg, J. R., and Niemeier, R. W., Percutaneous penetration of benzene in hairless mice: an estimate of dermal absorption during tire-building operations, *Am. J. Ind. Med.*, 7, 323, 1985.

Susten, A. S., Niemeier, R. W., and Simon, S. D., *In vivo* percutaneous absorption studies of volatile organic solvents in hairless mice II. Toluene, ethylbenzene and aniline, *J. Appl. Toxicol.*, 10, 217, 1990.

Suzuki, S., Taguchi, T., and Yokohashi, G., Dietary factors influencing upon the retention rate of orally administered ^{115m}CdCl2 in mice, *Ind. Health,* 7, 155, 1969.

Takahata, N., Hayashi, H., Watanabe, S., and Anso, T., Accumulation of mercury in the brains of two autopsy cases with chronic inorganic mercury poisoning, *Folia. Psychiatr. Neurol. Jpn.*, 24, 59, 1970.

Takeichi, S., Yamada, T., and Shikata, I., Acute toluene poisoning during painting, *Forensic Sci. Int.*, 32, 109, 1986.

Takenaka, S., Oldiges, H., König, H., Hochrainer, D., and Oberdörster, G., Carcinogenicity of cadmium chloride aerosols in W. rat, *J. Natl. Cancer Inst.*, 70, 367, 1983.

Tam, K. H., Charbonneau, S. M., Bryce, F., and Lacroix, G., Metabolism of inorganic arsenic (74As) in humans following oral ingestion, *Toxicol. Appl. Pharmacol.*, 50, 319, 1979.

Tam, K. H. and Lacroix, G., Determination of arsenic in urine and feces by dry ashing, atomic absorption spectrometry, *Int. J. Environ. Anal. Chem.*, 8, 283, 1980.

Tanaka, M., Fukuda, H., and Nagai, T., Permeation of drug through a model membrane consisting of millipore filter with oil, *Chem. Pharm. Bull. (Tokyo)*, 26, 9, 1978.

Tauber, J., Instant benzol death, *J. Occup. Med.*, 12, 91, 1970.

Taylor, E. J., *Dorland's Illustrated Medical Dictionary,* 27th ed., W.B. Saunders, Philadelphia, 1988, iii.

Teisinger, J. and Fiserova-Bergerova, V., Pulmonary retention and excretion of mercury vapor in man, *Ind. Med. Surg.*, 34 580, 1965.

Teraoka, H., Distribution of 24 elements in the internal organs of normal males and the metallic workers in Japan, *Arch. Environ. Health,* 36, 155, 1981.

Thomas, R. G., Ewing, W. G., Cartron, D. L., and McClellan, R. O., *In vivo* solubility of four forms of barium determined by scanning techniques, *Am. Ind. Hyg. Assoc. J.*, 34, 350, 1973.

Thomson, A. B. R. and Dietschy, J. M., The role of the unstirred water layer in intestinal permeation, in *Pharmacology of Intestinal Permeation II,* Csáky, T. Z., Ed., Springer-Verlag, Berlin, 1984, 165.

Thornton, I., Sources and pathways of cadmium in the environment, in *Cadmium in the Human Environment,* Nordberg, G. F., Herber, R. F. M., and Alessio, L., Eds., International Agency for Research on Cancer, Lyon, 1992, 149.

Thun, M. J., Schnorr, T. M., Smith, A. B., Halperin, W. E., and Lemen, R. A., Mortality among a cohort of U.S. cadmium production workers: an update, *J. Natl. Cancer Inst.*, 74, 325, 1985.

Tinwell, H., Stephens, S. C., and Ashby, J., Arsenite as the probable active species in the human carcinogenicity of arsenic: mouse micronucleus assays on Na and K arsenite, orpiment, and Fowler's solution, *Environ. Health Persp.*, 95, 205, 1991.

Törnquist, S., Nicklasson, M., Söderkvist, P., Nybom, L., and Toftgård, R., Bioavailability of benzo[a]pyrene deposited in the lung, *Drug Metab. Dispos.*, 16, 842, 1988.

Travis, C. and Bowers, J., Altered pharmacokinetics of soil-adsorbed benzene administered orally in the rat, *Bull. Environ. Contam. Toxicol.*, 45, 197, 1990.

Travis, C. C., Quillen, J. L., and Arms, A. D., Pharmacokinetics of benzene, *Toxicol. Appl. Pharmacol.*, 102, 400, 1990.

Travis, C. C. and Hattemer-Frey, H. A., Human exposure to dioxin, *Sci. Total Environ.*, 104, 97, 1991.

Triebig, G. and Schaller, K. H., Copper, in *Biological Indicators for the Assessment of Human Exposure to Industrial Chemicals,* Alessio, L., Berlin, A., and Roi, M. R., Eds., EUP 8903 EN, Directorate-General for Employment and Social Affairs, Health and Safety Directorate, Directorate-General for Science Research and Development, Commission of the European Communities, Brussels-Luxembourg, 1984.

Tsai, J. C., Cappel, M. J., Flynn, G. L., Weiner, N. D., Krenter, J., and Ferry, J. J., Drug and vehicle deposition from topical applications: use of *in vitro* mass balance technique with minoxidil solution, *J. Pharm. Sci.*, 81, 736, 1992.

Tseng, W. P., Effects and dose-response relationship of skin and blackfoot disease with arsenic, *Environ. Health Persp.*, 19, 109, 1977.

Tseng, W. P., Chu, H. M., How, S. W., Fong, J. M., Lin, C. S., and Yen, S., Prevalence of skin cancer in an endemic area of chronic arsenism in Taiwan, *J. Natl. Cancer Inst.*, 40, 453, 1968.

Tsuda, T., Babazono, A., Yamamoto, E., Kurumatani, N., Mino, Y., Ogawa, T., Kishi, Y., and Aoyama, H., Ingested arsenic and internal cancer: a historical cohort study followed for 33 years, *Am. J. Epidemiol.*, 141, 198, 1995.

Tsuruta, H., Skin absorption of organic solvent vapors in nude mice *in vivo*, *Ind. Health*, 27, 37, 1989.

Tucker, L. D. and Shertzer, H. G., Split lung perfusion: a new method to test the effects of chemical agents on pulmonary metabolism, *Toxicol. Appl. Pharmacol.*, 55, 353, 1980.

Turkall, R. M., Skowronski, G., Gerges, S., Von Hagen, S., and Abdel-Rahman, M. S., Soil adsorption alters kinetic and benzene in orally exposed male rats, *Arch. Environ. Contam. Toxicol.*, 17, 159, 1988.

Turkall, R. M., Skowronski, G., and Abdel-Rahman, M. S., The effect of soil type on absorption of toluene and its bioavailability, in *Petroleum Contaminated Soils,* Vol. III, Kostecki, P. T. and Calabrese, E. J., Eds., Lewis Publishers, Chelsea, Michigan, 1990, 283.

Turkall, R. M., Skowronski, G., and Abdel-Rahman, M. S., Differences in kinetics of pure and soil-adsorbed toluene in orally exposed male rats, *Arch. Environ. Contam. Toxicol.*, 20, 155, 1991.

Turkall, R. M., Skowronski, G., Kadry, A. M., and Abdel-Rahman, M. S., A comparative study of the kinetics and bioavailability of pure and soil-adsorbed naphthalene in dermally exposed male rats, *Arch. Environ. Contam. Toxicol.*, 26, 504, 1994.

Turnheim, K., Intestinal permeation of water, in *Pharmacology of Intestinal Permeation I*, Csáky, T. Z., Ed., Springer-Verlag, Berlin, 1984, 381.

Turnlund, J. R., Stable isotope studies of the effect of dietary copper on copper absorption and excretion, *Adv. Exp. Med. Biol.*, 258, 21, 1989.

Turnlund, J. R., Michel, M. C., Keyes, W. R., Schutz, Y., and Margen, S., Copper absorption in elderly men determined by using stable ^{65}Cu, *Am.J. Clin. Nutr.*, 36, 587, 1982.

Turnlund, J. R., Swanson, C. A., and King, J. C., Copper absorption and retention in pregnant women fed diets based on animal and plant proteins, *J. Nutr.*, 113, 2346, 1983.

Turnlund, J. R., King, C. J., Gong, B., Keyes, W. R., and Michel, M. C., A stable isotope study of copper absorption in young men: effect of phytate and alpha-cellulose, *Am. J. Clin. Nutr.*, 42, 18, 1985.

Turnlund, J. R., Wada, L., King, C. J., Keyes, W. R., and Acord, L. L., Copper absorption in young men fed adequate and low zinc diets, *Biol. Trace Element Res.*, 17, 31, 1988a.

Turnlund, J. R., Reager, R. D., and Costa, F., Iron and copper absorption in young and elderly men, *Nutr. Res.*, 8, 333, 1988b.

Turnlund, J. R., Keyes, W. R., Anderson, H. L., and Acord, L. L., Copper absorption and retention in young men at three levels of dietary copper using the stable isotope ^{65}Cu, *Am. J. Clin. Nutr.*, 49, 870, 1989.

Uhl, S., Schmid, P., and Schlatter, C., Pharmacokinetics of pentachlorophenol in man, *Arch. Toxicol.*, 38, 182, 1986.

Ultman, J. S., Transport and uptake of inhaled gases, in *Air Pollution, the Automobile, and Public Health*, Watson, A. Y., Bates, R. R., and Kennedy, D., Eds., National Academy Press, Washington, D.C., 1988, 323.

Umbreit, T. H., Patel, D., and Gallo, M. A., Acute toxicity of TCDD contaminated soil from an industrial site, *Chemosphere*, 14, 945, 1985.

Umbreit, T. H., Hesse, E. J., and Gallo, M. A., Bioavailability of dioxin in soil from a 2,4,5-T manufacturing site, *Science*, 232, 497, 1986.

Umbreit, T. H., Hesse, E. J., and Gallo, M. A., Reproductive toxicity in female mice of dioxin-contaminated soils from a 2,4,5-trichlorophenoxyacetic acid manufacturing site, *Arch. Environ. Contam. Toxicol.*, 16, 461, 1987.

Underwood, E. J., *Trace Elements in Human and Animal Nutrition*, 4th ed., Academic Press, New York, 1977, 258.

UNEP (United Nations Environment Program), Environmental Data Report, Report Prepared by the GEMS Monitoring and Assessment Research Center, London, UK, 1991.

USEPA (U.S. Environmental Protection Agency), Assessment of Human Exposure to Atmospheric Benzene, EPA 450/3-78-031, U.S. Environmental Protection Agency, Research Triangle Park, NC, 1978.

USEPA (U.S. Environmental Protection Agency), Water-related Environmental Fate of 129 Priority Pollutants, EPA 440/4-79-029A, Office of Water Planning and Standards, Washington, D.C., 1979a.

USEPA (U.S. Environmental Protection Agency), Reviews of the Environmental Effects of Pollutants XI. Chlorophenols, EPA 600-1-79-012, Health Effects Research Laboratory, 1979b.

USEPA (U.S. Environmental Protection Agency), Exposure and Risk Assessment for Copper, EPA 440/4-81-015, Cincinnati, OH, 1980a.

USEPA (U.S. Environmental Protection Agency), Ambient Water Quality Criteria for Benzene, EPA 440/5-80-018, Washington, D.C., 1980b.

USEPA (U.S. Environmental Protection Agency), Ambient Water Quality Criteria for Lead, PB 81-117681, Springfield, VA, National Technical Information Service, 1980c.

USEPA (U.S. Environmental Protection Agency), Health Assessment Document for Chromium, EPA 540/1-86-019, Environmental and Criteria Assessment Office, Research Triangle Park, NC, 1984a.

USEPA (U.S. Environmental Protection Agency), Health Assessment Document for Inorganic Arsenic, EPA 600/8-83-021F, Environmental and Criteria Assessment Office, Research Triangle Park, NC., 1984b.

USEPA (U.S. Environmental Protection Agency), Mercury Health Effects Update: Health Issue Assessment, EPA 600/8-84-019F, U.S. Environmental Protection Agency, Office of Health and Environmental Assessment, Research Triangle Park, NC, 1984c.

USEPA (U.S. Environmental Protection Agency), Health Advisories for 52 Chemicals which have been Detected in Drinking Water, NTIS PE86-118338, Office of Drinking Water, Washington, D.C., 1985a.

USEPA (U.S. Environmental Protection Agency), Drinking Water Criteria Document for Toluene, ECAO-CIN-408, Office of Drinking Water, Washington, D.C., 1985b.

USEPA (U.S. Environmental Protection Agency), Health Assessment Document (HAD) of Polychlorinated Dibenzo-p-dioxins, EPA 600/8-84-014F, Office of Health and Environmental Assessment, Environmental Criteria and Assessment Office, Cincinnati, OH, 1985c.

USEPA (U.S. Environmental Protection Agency), Air Quality Criteria for Lead, EPA 600/8-83-028F, U.S. Environmental Protection Agency, Office of Research and Department, Office of Health and Environmental Assessment, Environmental Criteria and Assessment Office, Research Triangle Park, NC, 1986.

USEPA (U.S. Environmental Protection Agency), Estimating Exposures to 2,3,7,8-TCDD (Draft), OH, EPA 600/8-88-005A, U.S. Environmental Protection Agency, Office of Health and Environmental Assessment, Cincinnati, 1988.

USEPA (U.S. Environmental Protection Agency), Risk Assessment Guidance for Superfund Volume I., Human Health Evaluation Manual (Part A), Interim Final, EPA 540/1-89 002, Office of Emergency and Remedial Response, Washington, D.C., 1989a.

USEPA (U.S. Environmental Protection Agency), Interim Methods for Development of Inhalation Reference Dose, EPA 600/8-88 066F, Office of Research and Development, Office of Health and Environmental Assessment, Environmental Criteria and Assessment Office, Research Triangle Park, NC, 1989b.

USEPA (U.S. Environmental Protection Agency), Exposure Factor Handbook, EPA 600/8-89 1043, Office of Health and Environmental Assessment, Washington, D.C., 1989c.

USEPA (U.S. Environmental Protection Agency), ROD Annual Report: FY 1989, EPA 540/8-90 006, Office of Emergency and Remedial Response, Washington, D.C., 1990a.

USEPA (U.S. Environmental Protection Agency), Drinking Water Criteria Document on Chromium, NTIS PE91-142844, Prepared by Life System, Inc., Cleveland, OH, 1990b.

USEPA (U.S. Environmental Protection Agency), Metals Bioavailability and Disposition Kinetics. Research Needs Workshop, Environmental Toxicology Division, Health Effects Research Laboratory, Office of Research and Development, Environmental Research Center, Research Triangle Park, NC, 1990c

USEPA (U.S. Environmental Protection Agency), EPA guideline on exposure assessment, *Federal Register*, May 29, 57, 22890, 1992a.

USEPA (U.S. Environmental Protection Agency), Dermal Exposure Assessment: Principles and Applications, Interim Report, EPA 600/8-91 011B, Office of Health and Environmental Assessment, Washington, D.C., 1992b.

USEPA (U.S. Environmental Protection Agency), A review and update of current knowledge and methods for performing exposure assessment for 2,3,7,8-TCDD, exposure assessment group, Office of Health and Environmental Assessment, in *Dioxin Perspectives. A Pilot Study on International Information Exchange on Dioxins and Related Compounds,* NATO Committee on the Challenges of Modern Society, Bretthauer, E. W., Kraus, H. W., and di Domenico, A., Eds., Plenum Press, New York, 1992c, 123.

Vahter, M., Metabolism of arsenic, in *Biological and Environmental Effects of Arsenic,* Fowler, B., Ed., Elsevier, New York, 1983, 171.

Vahter, M. E., Arsenic, in *Biological Monitoring of Toxic Metals,* Clarkson, T. W., Friberg, L., Nordberg, G. F., and Sager, P. R., Eds., Plenum Press, New York, 1988, 303.

Vahter, M. and Norin, H., Metabolism of 74As-labeled trivalent and pentavalent inorganic arsenic in mice, *Environ. Res.,* 21, 446, 1980.

Valberg, L. S., Sorbie, J., and Hamilton, D. L., Gastrointestinal metabolism of cadmium in experimental iron deficiency, *Am. J. Physiol.,* 231, 462, 1976.

van den Berg, M., Olie, K., and Hutzinger, O., Uptake and selective retention in rats of orally administered chlorinated dioxins and dibenzofurans from fly-ash and fly-ash extract, *Chemosphere,* 12, 535, 1983.

van den Berg, M., de Vroom, E., van Greevenbroek, M., Olie, K., and Hutzinger, O., Bioavailability of PCDDs and PCDFs adsorbed on fly-ash in rat, guinea pig and syrian golden hamster, *Chemosphere,* 14, 865, 1985.

van den Berg, M., Heeremans, C., Meerman, L., Veenhoven, E., Olie, K., and Hutzinger, O., Some pharmacokinetic aspects of PCDDs and PCDFs in mammals after administration of a fly-ash extract from a municipal incinerator, *Chemosphere,* 15, 1477, 1986.

van den Berg, M., Sinke, M., and Wever, H., Vehicle dependent bioavailability of polychlorinated dibenzo-p-dioxins (PCDDs) and dibenzofurans (PCDFs) in the rat, *Chemosphere,* 16, 1193, 1987.

Van den Heuvel, J. and Lucier, G., Environmental toxicology of polychlorinated dibenzo-p-dioxins and polychlorinated dibenzofurans, *Environ. Health Persp.,* 100, 189, 1993.

VanRooij, J. G. M., De Roos, J. H., and Bodelier-Bade, M. M., Absorption of polycylic aromatic hydrocarbons through human skin: difference between anatomical sites and individuals, *J. Toxicol. Environ. Health,* 38, 355, 1993a.

VanRooij, J. G. M., Bodelier-Bade, M. M., and Jongeneelen, F. J., Estimation of individual dermal and respiratory uptake of polycylic aromatic hydrocarbons in 12 coke oven workers, *Br. J. Ind. Med.,* 50, 623, 1993b.

Veulemans, H. and Masschelein, R., Experimental human exposure to toluene, *Int. Arch. Occup. Environ. Health,* 42, 91, 1978.

Vincent, J. H. and Armbruster, L., On the quantitative definition of the inhalability of airborne dust, *Ann. Occup. Hyg.,* 24, 245, 1981.

Visek, W. J., Whitney, I. B., Kuhn, U. S. G., and Comar, C. L., Metabolism of Cr^{51} by animals as influenced by chemical state, *Proc. Soc. Exp. Biol. Med.,* 84, 610, 1953.

Waalkes, M. P., Coogan, T. P., and Barter, R. A., Toxicological principles of metal carcinogenesis with special emphasis on cadmium, *Crit. Rev. Toxicol.,* 22(3), 175, 1992.

Wagner, J., Method of estimating relative absorption of a drug in a series of clinical studies in rich blood levels measured after single and/or multiple doses, *J. Pharm. Sci.,* 56, 652, 1967.

Wagner, J., An overview of the analysis and interpretation of bioavailability studies in man, *Pharmacology,* 8, 102, 1972.

Wagner, J., Bioavailability assessment: methods to estimate total area (AUC 0-∞) and total amount excreted (Ae∞) and importance of blood and urine sampling scheme with application to digoxin, *J. Pharm. Biophar.*, 5, 533, 1977.

Wahlberg, J. E., Percutaneous absorption of sodium chromate (51Cr), cobaltous (58Co), and mercuric (203Hg) chloride through excised human and guinea pig skin, *Acta Derm. Venereol.*, 45, 415, 1965.

Wahlberg, J. E. and Skog, E., Percutaneous absorption of trivalent and hexavalent chromium, *Arch. Dermatol.*, 92, 315, 1965.

Walker, W. R. and Reeves, R. R., Perfusion of intact skin by a saline solution of bis(glycinato) copper (II), *Bioinorganic Chem.*, 7, 271, 1977.

Wallace, L., Personal exposure, indoor and outdoor air concentrations, and exhaled breath concentrations of selected volatile organic compounds measured for 600 residents of New Jersey, North Dakota, North Carolina and California, *Toxicol. Environ. Chem.*, 12, 215, 1986.

Wallace, L. A., The exposure of the general population to benzene, *Cell Biol. Toxicol.*, 5, 297, 1989.

Walsh, L. M., Sumner, M. E., and Keeney, D. R., Occurrence and distribution of arsenic in soils and plants, *Environ. Health Persp.*, 19, 67, 1977.

Walsh, C. T., The influence of age on the gastrointestinal absorption of mercuric chloride and methyl mercury chloride in the rats, *Environ. Res.*, 27, 412, 1982.

Washitake, M., Takashima, T., Tanaka, S., Anmo, T., and Tanaka, S., Drug permeation through egg shell membranes, *Chem. Pharm. Bull. (Tokyo)*, 28, 2855, 1980.

Watanabe, K. H., Bois, F. Y., Daisey, J. M., Auslander, D. M., and Spear, R. C., Benzene toxicokinetics in humans: exposure of bone marrow to metabolites, *Occup. Environ. Med.*, 51, 414, 1994.

Watson, M. S. and Brain, S. D., Uptake of iron oxide aerosols by mouse airway epithelium, *Lab. Invest.*, 40, 450, 1979.

Watson, W. S., Hume, R., and Moore, M. R., Oral absorption of lead and iron, *Lancet*, II, 236, 1980.

Webber, C. E. and Harvey, J. W., Accidental human inhalation of ruthenium tetroxide, *Health Phys.*, 30, 357, 1976.

Weber, H., Long-term study of the distribution of soluble chromate-51 in the rat after a single intratracheal administration, *J. Toxicol. Environ. Health*, 11, 749, 1983.

Weber, P. M., O'Reilly, S., Pollycove, M., and Shipley, L., Gastrointestinal absorption of copper: studies with ^{64}Cu, ^{95}Zr, a whole-body counter and the scintillation camera, *J. Nutr. Med.*, 10, 591, 1969.

Weber, L. W. D., Zesch, A., and Rozman, K., Penetration, distribution and kinetics of 2,3,7,8-tetrachlorodibenzo-p-dioxin in human skin *in vitro*, *Arch Toxicol.*, 65, 421, 1991.

Weibel, E. R., Morphology estimation of pulmonary diffusion capacity, *Respir. Physiol.*, 11, 54, 1970.

Welling, P. G., Effects of gastrointestinal disease on drug absorption, in *Pharmacokinetic Basis for Drug Treatment*, Benet, L., Ed., Raven Press, New York, 1984.

Welling, P. G., Dosage route, bioavailability, and clinical efficacy, in *Pharmacokinetics: Regulatory, Industrial, Academic Perspectives*, Welling, P. G. and Tse, F. L. S., Eds., Marcel Dekker, Inc., New York, 1988, 97.

Wells, A. C., Venn, J. B., and Heard, M. J., Deposition in the lung and uptake to blood of motor exhaust labeled with ^{203}Pb, in *Inhaled Particles IV,* Walters, W. H., Ed., Pergamon Press, New York, 1977, 175.

Wester, R. C. and Noonan, P. K., Topical bioavailability of a potential antiacne agent (SC-23110) as determined by cumulative excretion and areas under plasma concentration-time curves, *J. Invest. Dermatol.*, 70, 92, 1978.

Wester, P. O., Brune, D., and Nordberg, G., Arsenic and selenium in lung, liver, and kidney tissue from dead smelter workers, *Br. J. Ind. Med.*, 38, 179, 1981.

Wester, R. C., Noonan, P. K., Smeach, S., and Kosobud, L., Pharmacokinetics and bioavailability of intravenous and topical nitroglycerin in the rhesus monkey. Estimate of percutaneous first-pass metabolism, *J. Pharm. Sci.*, 72, 745, 1983.

Wester, R. C. and Maibach, H. I., Dermatotoxicology, in *Environmental Pathology*, Oxford Univ. Press, Oxford, 1985, 181.

Wester, R. C. and Maibach, H. I., Regional variation in percutaneous absorption, in *Percutaneous Absorption. Mechanisms-Methodology-Drug Delivery,* 2nd ed., Bronaugh, R. L. and Maibach, H. I., Eds., Marcel Dekker, Inc., New York, 1989a, 111.

Wester, R. C. and Maibach, H. I., Dermal decontamination and percutaneous absorption, in *Percutaneous Absorption. Mechanisms-Methodology-Drug Delivery,* 2nd ed., Bronaugh, R. L. and Maibach, H. I., Eds., Marcel Dekker, Inc., New York, 1989b, 335.

Wester, R. C., Maibach, H. I., Bucks, D. A. W., Sedik, L., Melendres, J., Lioa, C., and Dizio, S., Percutaneous absorption of [14C]DDT and [14C]benzp[a]pyrene from soil, *Fund. Appl. Toxicol.*, 510, 1990.

Wester, R. C., Maibach, H. I., Sedik, L., Melendres, J., Dizio, S., and Wade, M., *In vitro* percutaneous absorption of cadmium from water and soil into human skin, *Fund. Appl. Toxicol.*, 19, 1, 1992.

Wester, R. C., Maibach, H. I., Sedik, L., Melendres, J., and Wade, M., *In vivo* and *in vitro* percutaneous absorption and skin decontamination of arsenic from water and soil, *Fund. Appl. Toxicol.*, 20, 336, 1993a.

Wester, R. C., Maibach, H. I., Sedik, L., Melendres, J., Wade, M., and Dizio, S., Percutaneous absorption of pentachlorophenol from soil, *Fund. Appl. Toxicol.*, 20, 68, 1993b.

Weyand, E. H. and Bevan, D. R., Benzo[a]pyrene disposition and metabolism in rats following intratracheal instillation, *Cancer Res.*, 46, 5655, 1986.

Weyand, E. H. and Bevan, D. R., Species differences in disposition of benzo[a]pyrene, *Drug Metab. Dispos.*, 15, 442, 1987.

WHO (World Health Organization), Mercury, World Health Organization, Geneva, 1976.

WHO (World Health Organization), Lead, World Health Organization, Geneva, 1977.

WHO (World Health Organization), Arsenic, World Health Organization, Geneva, 1981.

WHO (World Health Organization), Toluene, World Health Organization, Geneva, 1985.

WHO (World Health Organization), Pentachlorophenol, World Health Organization, Geneva, 1987.

WHO (World Health Organization), Chromium, World Health Organization, Geneva, 1988.

WHO (World Health Organization), Chlorophenols other than Pentachlorophenol, World Health Organization, Geneva, 1989a.

WHO (World Health Organization), Pentachlorophenol Health and Safety, World Health Organization, Geneva, 1989b.

WHO (World Health Organization), Lead - Environmental Aspects, World Health Organization, Geneva, 1989c.

WHO (World Health Organization), Mercury - Environmental Aspects, World Health Organization, Geneva, 1989d.

WHO (World Health Organization), Polychlorinated Dibenzo-*para*-dioxins and Dibenzofurans, World Health Organization, Geneva, 1989e.

WHO (World Health Organization), Methylmercury, World Health Organization, Geneva, 1990.

WHO (World Health Organization), Inorganic Mercury, World Health Organization, Geneva, 1991.

WHO (World Health Organization), Cadmium, World Health Organization, Geneva, 1992.

Wieczeorek, H., Evaluation of low exposure to styrene. II. Dermal absorption of styrene vapor in humans under experimental condition, *Int. Arch. Occup. Environ. Health,* 57, 71, 1985.

Wiegand, H. J., The reduction of chromium (VI) to chromium (III) by glutathione: an intracellular redox pathway in the metabolism of the carcinogen chromium, *Toxicol.,* 33, 341, 1984.

Wiegand, H. J., Ottenwälder, H., and Bolt, H. M., Disposition of intratracheally administered chromium (III) and chromium (VI) in rabbit, *Toxicol. Let.*, 22, 273, 1984.

Williams, P. L., Pentachlorophenol, an assessment of the occupational hazard, *Am. Ind. Hyg. Assoc. J.,* 43, 799, 1982.

Williams, G. M. and Weisburger, J. H., Chemical carcinogenesis, in *Casarett and Doull's Toxicology. The Basic Science of Poisons*, Fourth Edition, Amdur, M. O., Doul, J., and Klaassen, C. D., Eds., Pergamon Press, New York, 1990, 127.

Winek, C. L. and Collom, W. D., Benzene and toluene fatalities, *J. Occup. Med.*, 13, 259, 1971.

Withey, J. R., Shedden, J., Law, F. C. P., and Abedini, S., Distribution of benzo[a]pyrene in pregnant rats following inhalation exposure and a comparison with similar data obtained with pyrene, *J. Appl. Toxicol.,* 13, 193, 1993.

Woodruff, T. J., Bois, F. Y., Auslander, D., and Spear, R., Structure and parameterization of pharmacokinetic models: their impact on model predictions, *Risk Anal.,* 12, 189, 1992.

Woolson, E. A., Emissions, cycling and effects of arsenic in soil ecosytems, in *Biological and Environmental Effects of Arsenic,* Fowler, B., Ed., Elseier, New York, 1983, 51.

Xu, G. B., Effects of age on deposition of inhaled aerosols in the human lung, *Aerosol. Sci. Technol.,* 5, 349, 1986.

Yamagata, N., Iwashima, K., and Nagai, T., Gastrointestinal absorption of ingested ^{115m}cadmium in man, *Bull. Inst. Publ. Health,* 24, 1, 1975.

Yang, J. J., Roy, T. A., and Mackerer, C. R., Percutaneous absorption of anthracene in the rat: comparison of *in vivo* and *in vitro* results, *Toxicol. Ind. Health*, 2, 79, 1986a.

Yang, J. J., Roy, T. A., and Mackerer, C. R., Percutaneous absorption of benzo[a]pyrene in the rat: comparison of *in vivo* and *in vitro* results, *Toxicol. Ind. Health,* 2, 409, 1986b.

Yang, J. J., Roy, T. A., Krueger, A. J., Neil, W., and Mackerer, C. R., *In vivo* and *in vitro* percutaneous absorption of benzo[a]pyrene from petroleum crude-fortified soil in the rat, *Bull. Environ. Contam. Toxicol.,* 43, 207, 1989a.

Yang, J. J., Roy, T. A., Krueger, A. J., Neil, W., and Mackerer, C. R., Percutaneous absorption of benzo[a]pyrene from soils, in *Petroleum Contaminated Soils*, Vol. 2, Calabrese, E. J. and Kostecki, P. T., Eds., Lewis Publishers, Chelsea, Michigan, 1989b.

Yannai, S. and Sachs, K. M., Absorption and accumulation of cadmium, lead and mercury from foods by rats, *Food Chem. Toxicol.*, 31, 351, 1993.

Ziegler, E. E., Edwards, B. B., Jensen, R. L., Mahaffey, K. R., and Fomon, S. J., Absorption and retention of lead by infants, *Pediat. Res.*, 12, 29, 1978.

GLOSSARY OF TERMS

Absolute bioavailability - the fraction or percentage of external dose that is present in the systemic circulation, that is, the ratio of internal dose to external dose. This is also described as the *bioavailability factor*.

Absorbed dose - see internal dose.

Absorption - a process whereby xenobiotics cross body boundaries and reach the systemic circulation.

Absorption barrier - any biological barrier of the body where various substances can cross to reach the bloodstream. The most common absorption barriers include: alveolar wall (*air-blood barrier*), gut wall (*gut-blood barrier*) and skin.

Active transport - a process whereby a xenobiotic that forms a complex with a carrier on one side of a cell membrane crosses absorption barriers against a concentration or electrochemical gradient, requiring expenditure of energy.

Adherence fraction for soil - the proportion of soil contacted that will adhere to the skin and become available for absorption.

Administered dose - the amount of a xenobiotic that is actually inhaled, ingested or applied on the skin. This dose, sometimes, is called *potential dose* in the literature.

Aerosol - a relatively time-stable suspension of solid and liquid particles in the atmosphere, with a size range of 0.001-100 μm in diameter. Aerosols often refer to various aerodisperse systems which reflect their origin, physical state, and range of particle size, as well as the performance characteristics of the analyzer. Aerosols include *dust, smoke, fume, mist, fog, cloud, smog, fine particles, coarse particles, total suspended particles, primary aerosols, secondary aerosols, and inhalable particulate matter.*

Air-blood barrier - see absorption barrier.

Alveolar ventilation - the respiratory tidal volume (see ventilation volume) minus the volume of the dead space (such as nasopharyngeal and tracheobrochial regions) in which gas exchange does not occur.

Applied dose - the amount of a xenobiotic contact with the surface of primary absorption sites (e.g. alveoli, gut wall and skin).

Area under blood concentration-time curve (AUC) - the concentration of a xenobiotic in blood (plasma or serum) integrated over time.

Average daily dose - the dose per unit body weight over an average time, including the average daily internal dose and the average daily external dose.

Bioavailability - the extent and rate of absorption for a xenobiotic (a foreign substance) which enters the systemic circulation in the unaltered (parent) form from the applied exposure site.

Bioavailability factor - see absolute bioavailability.

Bioavailable dose - see internal dose.

Biologically effective dose - the delivered mass of active toxic species at the sites of toxic action.

Blood-air partition coefficient - a constant representing the solubility ratio of a xenobiotic between the blood and air.

Cloud - any free aerosol system with a definite size and form, without regard to particle size.

Coarse particles - aerodynamic diameters of aerosols between 2.5-15 μm, which are produced by a mechanical process.

Diffusion coefficient - a constant governing mobility of a gas by diffusion at equal pressure differentials determined by the ratio of water solubility (S) of a xenobiotic and its molecular weight (MW), as: $\frac{S}{\sqrt{MW}}$.

Deposition coefficient - a fraction of total inhaled gas or vapor that deposits on the airway wall of respiratory tract, also called *total intake.*

Dissolution - a process by which water soluble constituents are eluted from the matrix of particles or solid material.

Distribution - a process whereby the absorbed xenobiotics are transported by the bloodstream to various organs and tissues.

Dose - the total mass of a contaminant which actually crosses a body boundary, is absorbed or deposited into the body, and ultimately reaches the target sites over a given period.

Dose-response assessment - a process of assessment of the quantitative relationship between any dose (external dose, internal dose, or biologically effective dose) and the likelihood of an adverse health effect.

Dose-response relationship - a correlative relationship between the magnitude of various doses and intensity of toxic response following exposure.

Dust - dry solid particles dispersed in the air formed by grinding, milling, crushing, or blasting without regard to particle size.

Enterohepatic circulation - a process whereby xenobiotics and their metabolites entering the intestine with biliary excretion may be reabsorbed into the portal blood and returned to the liver.

Excretion - a process whereby xenobiotics are eliminated from the body by several routes, such as urine, feces and lungs.

Exposure - the environmental contaminants experienced by an individual or a population.

Exposure assessment - a process of qualitative and/or quantitative evaluation of the magnitude of human exposure to xenobiotics, exposure duration and frequency, sources of exposure, routes of exposure, characteristics (size, nature and class) of the potentially exposed population.

Exposure dose - the total mass of a xenobiotic in the environmental media that an individual or population is exposed to over time.

External dose - the total mass of a xenobiotic at the applied site available for absorption. External dose in the literature generally implies a variety of doses, such as *applied dose*, *administered dose*, and *exposure dose*, depending on the designs and analytical methods involved in experiments.

Facilitated diffusion - a process whereby a xenobiotic that forms a complex with a carrier on one side of the cell membrane crosses absorption barriers but is not against a concentration or electrochemical gradient and does not require expenditure of energy.

Filtration - a process whereby a xenobiotic penetrates the absorption barriers following bulk water flux without expenditure of energy.

Fine particles - aerodynamic diameters of aerosols between 0-2.5 μm, which are formed by condensation and coagulation.

First-pass effect - a process whereby some xenobiotics will be metabolized at the applied sites (the airway wall, the gut wall and the skin) or during passage through the liver from the portal vein prior to the systemic circulation. It is also called presystemic metabolism.

Fogs - finely dispersed liquid droplets which appear as a cloud.

Fumes - condensation aerosols consisting of small liquid and solid particles formed by condensation of vapors, combustion, and sublimation, with a range of 0.01-1.0 μm in diameter.

Gut-blood barrier - see absorption barrier.

Hazard identification - a process of identification of xenobiotics that are harmful to human health, and a qualitative evaluation of their character and importance.

Health risk assessment - a process that seeks to estimate the adverse health consequences of a specified set of conditions. The assessment process includes: (1) hazard identification; (2) dose-response assessment; (3) exposure assessment; and (4) risk characterization.

Hepatic clearance - the rate of total body clearance via the liver.

Hygroscopicity - some water-soluble aerosols will grow in moist air until they reach equilibrium, e.g. sea salt, tobacco smoke and sulfuric acid droplets.

Inertial impaction - this process means that particulates (ranging from 5 to 30 μm) continue to travel along their original direction because of their inertia despite changes in the airstream direction.

Inhalability - see inhalable fraction.

Inhalable fraction - the portion of the total mass of airborne aerosols that actually cross the nose and mouth. It is also called *intake efficiency* or *inhalability*.

Inhalable particulate matter (IPM) - aerosols less than 10-15 μm in diameter (IPM_{10} or IPM_{15}) that may be inhaled into the respiratory tract by humans.

Intake efficiency - see inhalable fraction.

Interception - a process which occurs at bifurcations of bronchial trees primarily for long fibrous particulates (e.g. asbestos).

Internal dose - the total mass of a xenobiotic that is actually absorbed into the systemic circulation. Internal dose, sometimes, is called the "absorbed dose" or "bioavailable dose".

Lifetime average daily dose -the dose per unit body weight over an average lifetime, including either the lifetime average daily internal dose or the lifetime average daily external dose.

Mass median aerodynamic diameter (MMAD) - this measure reflects the reality that sampling methods used to characterize particle size distribution depend upon the aerodynamic properties rather than physical size of particles. The aerosol size parameters are aerodynamic diameters for aerosols ≥ 0.5 μm.

Mass median diffusive diameter - a diffusion diameter for aerosols < 0.5 μm that reflects the diffusion properties of aerosols.

Mechanical translocation - a process whereby a xenobiotic is removed from the respiratory tract by various physical processes, such as sneezing, wiping, blowing, coughing, mucociliary transport and phagocytosis.

Metabolism - a process whereby the absorbed xenobiotics are converted from one form to other forms by enzymatic systems in living organisms.

Mists - invisible liquid droplets formed by condensation of vapor, having diameter greater than 10 μm.

Passive transport - a process whereby a xenobiotic crosses absorption barriers without expenditure of energy, including *simple diffusion* and *filtration.*

Peak height (C_{max}) - the measured values of peak concentration of a xenobiotic in the blood. This is one of the parameters used for estimating the rate of absorption of a xenobiotic.

Peak time (t_{max}) - time until the peak concentration is reached. This is one of the parameters used for estimating the rate of absorption of a xenobiotic.

Permeability - the rate of diffusion of a xenobiotic through each unit area of the barrier (usually the skin) for a unit concentration difference.

Persorption - a process whereby large molecules, e.g. starch granules and diatomaceous earth particles, may enter the intercellular space from a gap at the tip of the villi between two cells which are sloughed off.

Phagocytosis - a process whereby a solid particle is engulfed by a cell.

Pinocytosis - a process whereby a liquid droplet is engulfed by a cell.

Portal circulation - portal venous blood between the gut and liver.

Potential dose - see administered dose.

Presystemic metabolism - see first-pass effect.

Primary aerosols - dispersion aerosols, directly emitted into the air without change of shape in the air.

Rate-limiting step (absorption) - the slowest stage in the absorption process which will control the overall rate.

Relative bioavailability - When intravenous administration is not feasible (because of a solution problem), the bioavailability factor may be estimated from relative bioavailability. It is determined by a comparison of the extent of absorption among several doses of the same xenobiotic, or two or more forms of the same xenobiotic, or different vehicles (food, soil and water) of the same xenobiotic, with the same subject. The extent of absorption for the test material (f_2) is compared to that for the standard material (f_1), as: relative bioavailability $= \frac{f_2}{f_1}$.

Risk characterization - a description of the nature and extent of human risk by comparison of all the results from hazard identification, dose-response assessment and exposure assessment with a discussion of attendant uncertainties and analytical assumptions.

Secondary aerosols - dispersion aerosols formed by chemical reactions.

Sedimentation - a process whereby small particles (1 to 5 μm) settle in the small bronchi, bronchioles and the alveolar space by gravitational force.

Simple diffusion - a process whereby a xenobiotic penetrates the absorption barriers following Fick's law without an expenditure of energy.

Smog - highly visible mixtures of particles and gases associated with urban aerosol formation (e.g. automobile exhaust).

Smokes - visible, dispersion aerosols containing a variety of solids, liquids, and gases formed by condensation of supersaturated vapors, with a range of 0.1-10 μm in diameter.

Systemic circulation - both the arterial and venous blood except the portal blood.

Total intake - see deposition coefficient.

Total suspended particles (TSP) - aerosols collected by the high-volume sampler.

Toxicokinetics - the study of kinetics of xenobiotics, including quantitation and determination of the time course of disposition (absorption, distribution, metabolism and excretion) of xenobiotics in the organism.

Ventilation/perfusion ratio (V_A/Q) - the ratio of ventilation (V_A) to blood flow (Q) which remains constant although the ventilation or blood flow change with time.

Ventilation volume - the volume of air inspired or expired during different inhalation stages, including the tidal volume, inspiratory reserve volume, expiratory reserve volume, residual volume, inspiratory capacity, functional residual capacity, vital capacity, and total lung capacity. *Tidal volume*: the volume of air inspired or expired during restful breathing, about 500 ml for a healthy male adult. Minute volume can be calculated by the product of the tidal volume and the breathing frequency (15 times/min). *Inspiratory reserve volume*: the extra volume of air that can be inspired by forceful inspiration, about 3000 ml. *Expiratory reserve volume*: the extra volume of air that can be expired by forceful expiration, about 1100 ml. *Residual volume*: the volume of air remaining in the respiratory tract following forceful expiration, about 1200 ml. *Inspiratory capacity*: the sum of the tidal volume and inspiratory reserve volume, about 3500 ml. *Functional residual capacity*: the sum of the expiratory reserve volume and residual volume, about 2300 ml. *Vital capacity*: the sum of the inspiratory reserve volume, expiratory reserve volume, and tidal volume, about 4600 ml. *Total lung capacity*: the sum of the vital capacity and residual volume, about 5800 ml.

INDEX

Absolute bioavailability, *see* bioavailability
Absorbed dose, *see* dose
Administered dose, *see* dose
Air-blood barrier, *see* barrier
Alveolar-macrophage associated particle clearance, *see* clearance
Aniline, 210-213
Applied dose, *see* dose
Area under blood concentration-time curve (AUC), *see* bioavailability
Arsenic, 75-90
Artificial membrane models, *see in vitro* method
Average daily external dose, *see* dose
Average daily internal dose, *see* dose
Barrier
  air-blood, 19, 21-22
  biological, 19
  gut/blood, 38
Benzene, 174-183
Bioavailability
  absolute, 10-11
  AUC, 8, 52
  concept, 7-12
  extent, 9-11
  factor, 10, 52-53, 55, 64
  peak height, 8-9
  rate, 8-9
  relative, 11-12, 52-53
  time of peak concentration, 8-9
Bioavailability factor, *see* bioavailability
Bioavailable dose, *see* dose
Biological barriers, *see* barrier
Biological membrane models, 62
Biologically effective dose, *see* dose
Blood/air partition coefficient, 22-24
Blood level measurement, *see in vivo* methods
Cadmium, 91-100
Chelation, 231
Chemical reactivity, 26
Chromium and chromium compounds, 101-116
Clearance (lung)
  alveolar-macrophage associated particle, 33
  dissolution-absorption process, 33, 34
  lymphatic system, 34
  mechanical translocation (non-absorption process), 33
Closed loop preparation, *see in situ* methods
Conducting Airways, 22
Congenitally athymic (nude) animal models, *see in vitro* methods, biological membrane models
Copper, 117-124
Deposition (lung)
  diffusion, 29
  inertial impaction, 29
  interception, 29
  nasopharyngeal, 30
  pulmonary, 30
  sedimentation, 29
  total, 30
  tracheobronchial, 30
Dermis, 45
Dialysis membrane, *see in vitro* methods, artificial membrane models
Diffusion, *see* deposition
Diffusion cells, *see in vitro* methods
Diffusion coefficient, 26
Diffusivity, 47
Dispersion, 22
Dissociation constant, 42

Dissolution rate, 41-42, 60
Dissolution testing, *see in vitro* methods
Dissolution-absorption clearance, *see* clearance
Dose
  absorbed, 10
  administered, 10-11
  applied, 10-11
  average daily external, 16
  average daily internal, 14-15
  bioavailable, 10
  biologically effective, 13
  concept, 12-13
  exposure, 10-11
  external, 10-11,16-17
  intravenous, 11
  labeled, 10
  potential, 10
  reference, 2
  risk specific, 2
Dose-response assessment, *see* risk
Enterohepatic circulation, 44
Epidermis, 45
Equilibrium lung burden, 34
Everted intestinal sac, *see in vitro* methods
Exposure, 12
Exposure assessment, *see* risk
Exposure dose, *see* dose
External counting technique, *see in vivo* methods, surface disappearance
External dose, *see* dose
Fecal analysis, *see in vivo* methods, lung clearance and retention
Fecal excretion measurement, *see in vivo* methods
First pass effect (presystemic metabolism)
  hepatic, 41
  respiratory, 36
  skin, 48
Full-thickness skin resection, *see in vitro* methods, biological membrane
Gut/blood barrier, *see* barrier
Hazard identification, *see* risk
Head-only exposure, *see in vivo methods*, inhalation exposure methods
Hepatic first-pass effect, *see* first pass effect
Hepatic presystemic metabolism, *see* first pass effect
Histological method, *see in vivo* methods, surface disappearance
Hygroscopicity, 36
In situ methods
  closed loop preparation, 62, 63
  isolated loop with complete collection of mesenteric venous, 62, 63
  perfused loop preparation, 62, 63
In vitro methods
  artificial membrane models, 69-70
    dialysis membrane, 69
    partitioning membrane, 70
  biological membrane models, 67-69
    congenitally athymic (nude) animal models, 68
    full-thickness skin resection, 68
    isolated perfused porcine skin flap, 68
    split resection, 68
  diffusion cells, 68-69
  dissolution testing, 65
  inhalation exposure methods, 56
    head-only, 56
    nose-only, 56
    intratracheal instillation, 56,
    whole-body, 56

isolated perfused intestinal segment, 62
isolated perfused lung, 60-61
pH partition, 66

In vitro-in vivo correlation, 65

In vivo methods
- blood level measurement, 51-53
- everted intestinal sac, 62, 63
- liver ratio, 65
- lung clearance and retention, 59-60
  - fecal analysis, 60
  - macrophage function, 60
  - mucociliary transport velocity, 59
  - whole lung clearance, 60
- mass-balance technique, 54-56
- surface disappearance, 66-67
  - external counting technique, 67
  - histological method, 67
  - tape stripping, 67
- toxic response, 56
- urinary excretion, 53-54

Inertial impaction, *see* deposition
Inhalable fraction, 27
Inhalable particulate matter, 26
Intercellular route, 41
Interception, *see* deposition
Internal dose, *see* dose
Intratracheal instillation, *see in vivo* methods, inhalation exposure methods
Intravenous dose, *see* dose
Ionization, 48
Isolated loop with complete collection of mesenteric venous, *see in situ* methods
Isolated perfused intestinal segment, *see in vitro* methods
Isolated perfused lung, *see in vitro* methods
Isolated perfused porcine skin flap, *see in vitro* methods, biological membrane models
Labeled dose, *see* dose
Lead, 134-144
Lung overloading, 35
Lymphatic system clearance, *see* clearance
Macrophage function, *see in vivo* methods, lung clearance and retention
Mass median aerodynamic diameter, 27
Mass-balance technique, *see in vivo* methods
Mayonnaise sandwich, 19
Measurement of blood levels, *see in vivo* methods
Measurement of fecal excretion, *see in vivo* methods
Measurement of liver ratio, *see in vivo* methods
Measurement of the cumulative urinary excretion, *see in vivo* methods
Measurement of toxic response, *see in vivo* methods
Mechanical translocation, *see* clearance
Mercury, 125-133
Mucociliary transport velocity, *see in vivo* methods, lung clearance and retention
Mucos, 36
Muscularis externa, 38
Nasopharyngeal deposition, *see* deposition
Nasopharyngeal region, 19
Non-absorption processes, *see* clearance
Nose-only exposure, *see in vivo* methods, inhalation exposure methods
Octanol/water partition coefficient, 47
Partitioning membrane, *see in vitro* methods, artificial membrane models

Peak height, *see* bioavailability
Perfused loop preparation, *see in situ* methods
Permeability coefficient, 48
pH partition method, *see in vitro* methods
Pharmacodynamics, 7
Pharmacokinetics, 7
Phenol, 203-209
Physiologically based-pharmacokinetic (PBPK) modeling, 7
Polychlorinated phenols, 193-202
Polynuclear aromatic hydrocarbons (PAHs), 155-173
Potential dose, *see* dose
Pulmonary deposition, *see* deposition
Pulmonary region, 19
Reference dose, *see* dose
Relative bioavailability, *see* bioavailability
Respiratory presystemic metabolism, *see* first pass effect
Retention, 34
Risk
  dose-response assessment, 2, 12
  exposure assessment, 2-4, 12
  hazard identification, 2, 12
  health risk assessment, 1, 12
  risk characterization, 2, 4, 12
Risk characterization, *see* risk
Risk specific dose, *see* dose
Sedimentation, *see* deposition
Serosa, 38
Skin presystemic metabolism, *see* first pass effect
Split Resection, *see in vitro* methods, biological membrane
Stratum corneum, 45
Submucosa, 38
Surface disappearance, *see in vivo* methods
Systemic availability, 8
Systemic circulation, 8
Tape Stripping, *see in vivo* methods, surface disappearance
2,3,7,8-Tetrachlorodibenzo-p-dioxin (TCDD), 145-154
Time of peak concentration, *see* bioavailability
Toluene, 184-192
Total deposition, *see* deposition
Toxic response measurement, *see in vivo* methods
Toxicity, 7
Toxicokinetics, 7
Tracheobronchial deposition, *see* deposition
Tracheobronchial region, 19
Transappendageal route, 45
Transcellular route, 41
Transepidermal route, 46
Transport of the xenobiotics, 19
Unaltered form, 7
Unstirred water layer, 40
Urinary excretion measurement, *see in vivo* methods
Ventilation, 26
Ventilation-perfusion ratio, 26
Whole body exposure, *see in vivo* methods, inhalation exposure methods
Whole lung clearance, *see in vivo* methods, lung clearance and retention